T0136978

# Lecture Notes in Networks and Systems

Volume 81

The series "Lecture Notes in Networks and Systems" publishes the latest developments in Networks and Systems—quickly, informally and with high quality. Original research reported in proceedings and post-proceedings represents the core of LNNS.

Volumes published in LNNS embrace all aspects and subfields of, as well as new challenges in, Networks and Systems.

The series contains proceedings and edited volumes in systems and networks, spanning the areas of Cyber-Physical Systems, Autonomous Systems, Sensor Networks, Control Systems, Energy Systems, Automotive Systems, Biological Systems, Vehicular Networking and Connected Vehicles, Aerospace Systems, Automation, Manufacturing, Smart Grids, Nonlinear Systems, Power Systems, Robotics, Social Systems, Economic Systems and other. Of particular value to both the contributors and the readership are the short publication timeframe and the world-wide distribution and exposure which enable both a wide and rapid dissemination of research output.

The series covers the theory, applications, and perspectives on the state of the art and future developments relevant to systems and networks, decision making, control, complex processes and related areas, as embedded in the fields of interdisciplinary and applied sciences, engineering, computer science, physics, economics, social, and life sciences, as well as the paradigms and methodologies behind them.

**** Indexing: The books of this series are submitted to ISI Proceedings, SCOPUS, Google Scholar and Springerlink ****

More information about this series at http://www.springer.com/series/15179

Yousef Farhaoui
Editor

# Big Data and Networks Technologies

*Editor*
Yousef Farhaoui
Department of Computer Science,
Faculty of Sciences and Techniques
Moulay Ismail University
Errachidia, Morocco

ISSN 2367-3370 ISSN 2367-3389 (electronic)
Lecture Notes in Networks and Systems
ISBN 978-3-030-23671-7 ISBN 978-3-030-23672-4 (eBook)
https://doi.org/10.1007/978-3-030-23672-4



This Springer imprint is published by the registered company Springer Nature Switzerland AG
The registered company address is: Gewerbestrasse 11, 6330 Cham, Switzerland

# Contents

# Experimental Validation of New SIP Authentication Protocol

Mourade Azrour[1(✉)], Yousef Farhaoui[1], and Azidine Guezzaz[2]

[1] Faculty of Sciences and Techniques, Department of Computer Science, IDMS Team, Moulay Ismail University, Errachidia, Morocco
azrour.mourade@gmail.com

[2] Technology High School, Essaouira, Cadi Ayyad University, Marrakech, Morocco

**Abstract.** In the last decade, Session Initiation Protocol (SIP) is the most popular application layer protocol created in order to manage multimedia sessions over IP protocol. SIP is not used only by telephony over IP (ToIP), but it can be used also by other in line application such as instant message, video conferences, and others. Since SIP inherits the security threats of IP which are added to SIP owner problems, the security of SIP services must be enhanced. Recently, we have designed a new SIP authentication protocol. Then, we have proved theoretically that our protocol is secured against various attacks. In this paper, we use the API JAIN SIP to implement our protocol. Therefore, we have tested developed applications in local area network. The obtained results confirm that our proposed protocol is efficient when it is compared with some implemented protocols.

**Keywords:** Authentication protocol · Security · JAIN SIP · Session initiation protocol · Telephony over IP

## 1 Introduction

Nowadays, internet is a continuously developing tool. It is not only used in order to look for information, but it provides multiple services that have created new ways of communicating and interacting with other people, such as e-mail, instance messaging, video-conferencing, and telephony over IP. To establish session of communication a signal protocol is necessary. For this reason, divers signaling protocol have been developed, such as: SIP, H323, MGCP, SKYPE, and others. However, Session Initiation Protocol (SIP) [1] is the most popular signaling protocol. SIP is developed by IETF with objective to initiate, change, and stop a multimedia session. It requires fewer resources and it is considered less complex if it is compared with the ITU's protocol H323 [2].

Due to its deliverance in unsecured public networks, SIP security is coming more in the more important. Nevertheless, authentication is the most security service recommended for SIP. The original SIP authentication protocol suggested is HTTP Digest authentication [3] but it was discovered vulnerable. Then, a multiple researchers have proposed many protocols [4–17] which are based on diverse mechanisms. The other


Y. Farhaoui (Ed.): BDNT 2019, LNNS 81, pp. 1–11, 2020.
https://doi.org/10.1007/978-3-030-23672-4_1

works are focusing on securing SIP against malformed message attack [18, 19], flooding attack [20, 21], and SPIT [22, 23].

Recently, we have introduced a new SIP authentication protocol [24] that is proved theoretically secured against various attacks. In this work, we have tested our protocol in real local area network after developing two SIP applications (Sever and Client) under programming language JAVA and API JAIN SIP. The experimental results confirm that our protocol is efficient when it is compared to various implemented protocols.

The rest of this paper is organized as following. In the section two, we introduce general information about protocol SIP. In the section three, we review briefly our proposed protocol. The section four presents the developed prototype as it delivers the result of various tests done in local area network. Then, we conclude our work in the last section.

## 2 SIP Overview

SIP architecture is illustrated in Fig. 1. It composes on user agent and four servers: proxy server, redirect server, registrar server and location serve. User Agent can be User Agent Client which generates the requests messages or User Agent Server that generates responses messages. Registrar Server is a server that is responsible to register users' locations. Proxy Server is a server which is used to achieve call set-up functions. Redirect Server translates the SIP address of an endpoint to IP address then returns it to the client. Location server is used by the tree first servers in order to allow them to look for, register, or update the user's agent location.

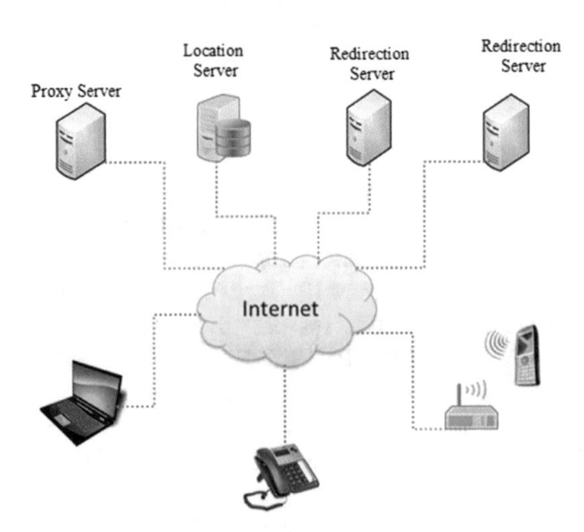


**Fig. 1.** SIP architecture

SIP message is a request sent by the client to a server, but it can be also a response from a server to a client. SIP message is composed on three essential parties, which are:

start-line, header fields, and message-body. SIP requests are characterized by having Request-Line in the first-line. Request-Line includes three types of information: method name, Request-URI, and the protocol version separated by a single space character. The six SIP original methods are defined in RFC3261; these methods are: REGISTER, INVITE, ACK, CANCEL, BYE and OPTIONS. The other methods SUBSCRIBE, REFER, MESSAGE, NOTIFY, UPDATE, INFO and PRACK are described as optional in other RFC's.

# 3 Review of Our Proposed Protocol

Our proposed protocol consists of four phases: system setup phase, registration phase, login and authentication phase, and password change phase.

## 3.1 System Setup Phase

In this section, the server selects an elliptic curve equation $\mathrm{E_p(a, b)}$, over a finite field $\mathrm{F_p}$, an additive group G of order p and P a base point generator with order n over equation $\mathrm{E_p(a, b)}$, n is a large prime of height entropy. Then, the server selects an integer $\mathrm{K_s} \in (1, \mathrm{q})$ as the long-live secret key, and computes its public key $Q_s = K_s P$. Next, the server chooses three one-way hash functions $h(.)$, $h_1(.)$and $h_2(.)$. Finally, S saves secretly its private key $K_s$ and publishes the following parameters $\{(E(GF_q), P, q, h(.), h_1(.), h_2(.), Q_s\}$.

## 3.2 Registration Phase

When user $U$ wants to register in the server $S$. *user* and *Server* perform the following steps via a secured channel.

***Reg 1***: *User U* chooses freely its username and password $PW$, selects randomly a number $a \in_R Z_p^*$, computes $HPW = h(PW||a)P$. Then, $U$ sends username and HPW to $S$.

***Reg 2***: *Server S* computes $VPW = h(username||K_s)P \oplus HPW$. Then, $S$ inserts $VPW$ in a smart card. Finally, $S$ sends Smart Card to $U$.

## 3.3 Login and Authentication Phase

This phase is illustrated in Fig. 2. If a legitimate user $U$ wants to login to remote server S, he/she inserts its Smart Card in card reader, then inputs username and password PW. After that the server and Smart Card perform the following steps:

- ***Auth 1:*** $User \rightarrow Server : REQUEST(V, bP, I, T_1)$
  User's Smart Card chooses randomly a number $b \in_R Z_p^*$, computes $bP$, $L = b\,Q_s$, $M = b(VPW \oplus h(PW||a)P)$, $I = username + L$, and $V = h(username||bP||L||M)$. Then, it sends to server message $REQUEST(V, bP, I, T_1)$,$T_1$ refers to the current timestamp.

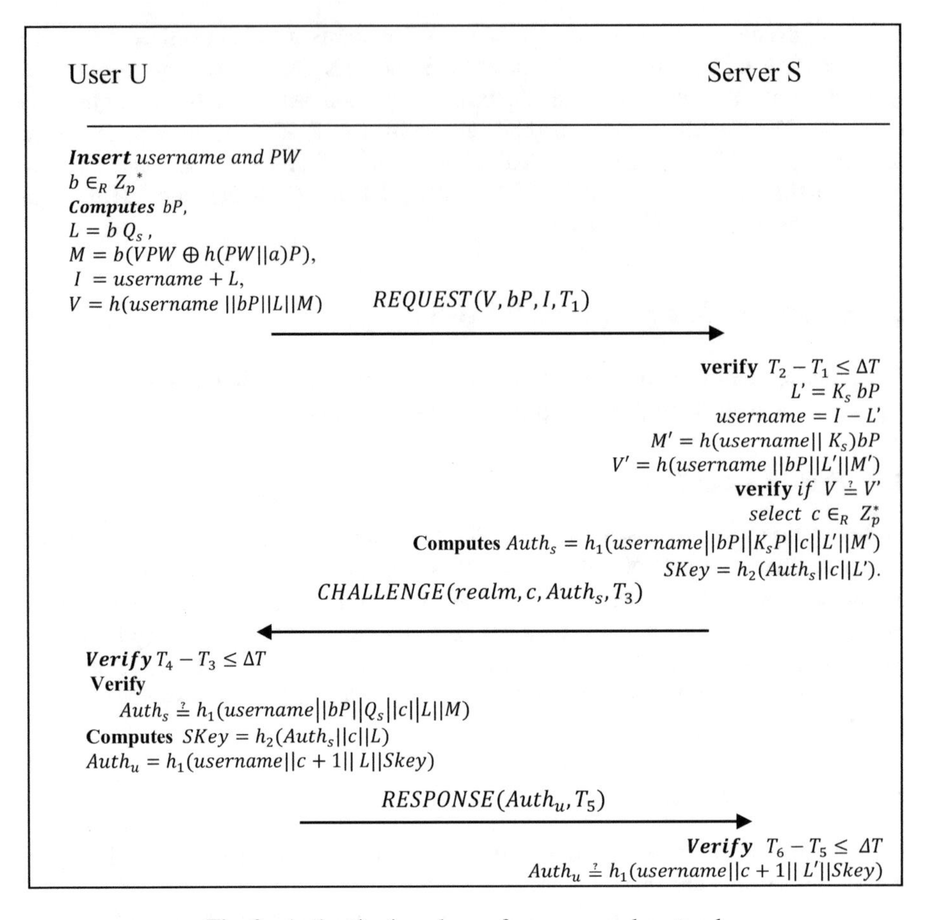


**Fig. 2.** Authentication phase of our proposed protocol

- ***Auth 2:*** $Server \rightarrow User : CHALLENGE(realm, c, Auth_s, T_3)$: After the reception of message $REQUEST$at $T_2$, the server S checks the validity of $T_2 - T_1 \leq \Delta T$. If is not fresh, S stops the process. If is valid, S computes $L' = K_s\,bQ_s$, $username = I - L'$,$M' = h(username||K_s)bP$ and $V' = h(username||bP||L'||M')$, then S verifies the validity of $V \stackrel{?}{=} V'$. If it is true, user $U$ is authenticated by the server S. Next, server S select randomly a number $c \in_R Z_p^*$, and computes $Auth_s = h_1(username||bP||K_sP||c||L'||M')$ and $SKey = h_2(Auth_s||c||K_sbP)$. Finally, S sends $CHALLENGE(realm, c, Auth_s, T_3)$ back to U.
- ***Auth 3:*** $User \rightarrow Server : RESPONSE(Auth_u, T_5)$: Upon receiving message $CHALLENGE$ at time $T_4$ from server S, user's Smart Card verifies the validity of $T_4 - T_3 \leq \Delta T$. If is not fresh, user $U$ stops the process. If is valid, user's Smart Card verifies $Auth_s \stackrel{?}{=} h_1(username||bP||Q_s||c||L||M)$. If it is true, $S$ is authenticated

by user, then $U$ computes $SKey = h_2(Auth_s||c||bQ_s)$ and $Auth_u = h_1(username ||c+1||L||Skey)$. Finally user's Smart Card sends to $S$ response message $RESPONSE(Auth_u, T_5)$.

- ***Auth 4:*** Upon receiving message $RESPONSE$ at time $T_6$ from user $U$, server $S$ verifies the validity of $T_6 - T_5 \leq \Delta T$. If is not fresh, $S$ stops the process. If is valid, $S$ verifies the validity of $Auth_u \stackrel{?}{=} h_1(username||c+1||L'||Skey)$.

### 3.4 Password Change Phase

In this phase, if user U wants to change his password he/she must firstly execute the login and authentication phase. Then execute the following steps:

- ***Pass*** **1**: $US : \{a*, Rep_u\}$
  $U$ Inserts her/his *username* and old $PW$, chooses freely his/her new password $PW*$, and selects randomly $a*, b \in_R Z_p^*$. Then computes, $HPW = h(PW*||a*)P$, $tag_u = h(username||b||HPW)$, and it then uses the session key $SKey$ to encrypt the new parameters $Rep_u = E_{SKey}(username||tag_u||b||HPW)$.
- ***Pass*** **2**: $S \rightarrow U : \{Rep_s\}$
  Upon receiving the information, the server decrypts the message and then checks the validity of the authentication $\mathrm{tag_u} \stackrel{?}{=} \mathrm{h(username||b||HPW)}$. If it is valid, the server computes the new secret information $\mathrm{VPW^* = h(username\ ||K_s)P \oplus HPW}$ and $\mathrm{tag_s = h(username||b+1||VPW^*)}$. Then, it sends encryption information $Rep_s = E_{KSey}(VPW^*||tag_s)$ back to the user $U$.
- ***Pass 3***:
  The user $U$ decrypts received message and verifies the validity of $tag_s \stackrel{?}{=} h(username||b+1||VPW^*)$. If it is valid, the user $U$ stores $VPW^*$ *and* $a^*$ in its smart card.

## 4 Implementation of Our Proposed Protocol

### 4.1 Developed Prototype

In order to test the efficiency of our proposed protocol, we have developed a client-server application in java based on API JAIN SIP. The prototype of our developed application is illustrated in Fig. 3.

Our developed prototype consists of two distinct applications: server (Fig. 4) and client which can communication via IP network. The developed client can exchange the SIP message not only with developed server but also with other SIP servers (example: miniSIPServer, Star Trinity SIP Tester…) as well as our developed server can response to various SIP client application (X-lite, SIPScan, Sipdroid…).

Each application is composed on different modules. The client has two modules: the first one is the request module that is responsible to create, treat, and send the SIP requests (REGISTER, INVITE…). The second one is the security module which implements the functions of our proposed protocol. This module is also implemented in

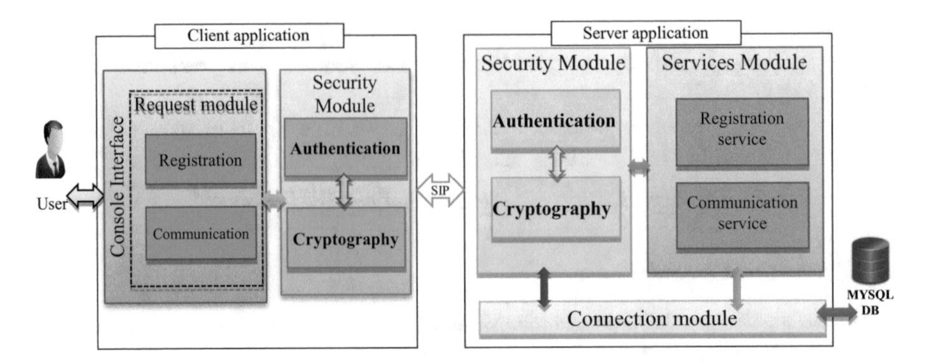


**Fig. 3.** Architecture of developed prototype

the server application. In addition to the security module the server application has two other modules the services module and the connection module. The services module provides the services requested by the clients, while the connection module plays the role of intermediate between the two first modules and the MYSQL database. The details of each module are giving in the following.

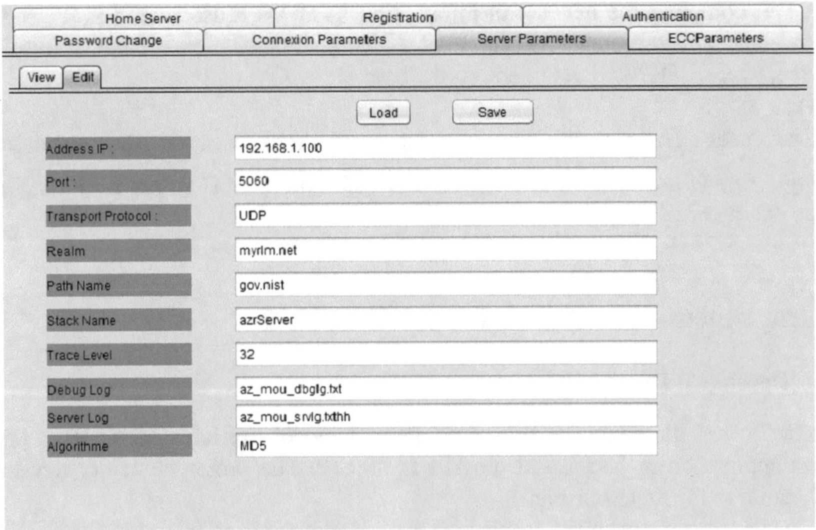

**Fig. 4.** Server application interface

**Security Module:** It is the most important element in our application because it contains the implementation of our authentication protocol. Both the client and server integrate this module, since the client and server must authenticate each other (mutual authentication). The security module is composed on two sub-elements: authentication and cryptography. The authentication sub-element includes the implementation of authentication phase of our protocol. The cryptography sub-element contains the principal function of cryptography such as one-way functions (MD5, SHA1…) and the implementation of Elliptic Curve Cryptography (ECC).

**Request Module:** It is a part of a console interface that allows to the user to send a SIP request messages (INVITE, REGISTER…) after inputting its parameters in specific fields.

**Service Module:** It is developed with the aim to give to the server application the ability to treat and response to the client SIP requests. In our developed application, we have implemented two services: the registration and the communication. The first one is responsible to register and update the user information. It receives the SIP request REGISTER, treat the request, then send the SQL request to the connection module. The second one receive, treat, and response to the SIP requests (INVITE, CANCEL, BYE, ACK…) that are used to control (open, update, terminate) a multimedia session.

**Connection Module:** Based on the security requirement, we have separated between the two first server's modules and the access to the database. So, to connect our application to the MYSQL database, we have created a connection module which receives the SQL request from the service module and security module, verifies its content, then request directly the database. The received response is translated to the module that sends the SQL request.

### 4.2 Tested Scenarios

With a hope to test our developed application in local network, we have experiment the developed application as shown in Fig. 5. The experimental architecture consists of two SIP servers and four SIP clients. The Clients and servers software are installed in one smart phone, two laptops and two computers.

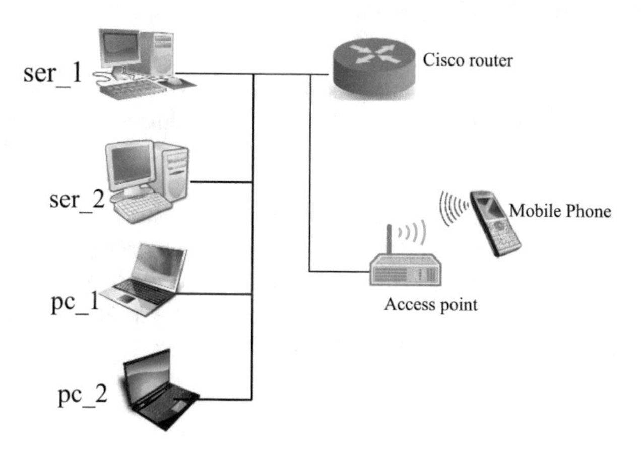


**Fig. 5.** Architecture of experimental test

We have installed the developed sever in computer ser_1, and installed the SIP server miniSIPServer V24.0.4 in computer ser_2. In the smart phone, we have installed the SIP Client SIPDroid. The developed SIP client application is installed in laptop PC_1. In the laptop PC_2 we have installed the SIP Client X-lite and Star Trinity SIP Tester.

After installing the software and connecting all the materials, we have make all possible tests as it is clear in Table 1.

**Table 1.** Tested scenarios.

| Scenario N° | SIP client | SIP server |
|---|---|---|
| 1 | Our developed client | Our developed server |
| 2 | SIPDroid | Our developed server |
| 3 | X-lite | Our developed server |
| 4 | Star trinity SIP tester | Our developed server |
| 5 | Our developed client | miniSIPServer V24.0.4 |
| 6 | SIPDroid | miniSIPServer V24.0.4 |
| 7 | X-lite | miniSIPServer V24.0.4 |
| 8 | Star trinity SIP tester | miniSIPServer V24.0.4 |

### 4.3 Results and Discussions

As shown in Table 1, we have tested our developed server with our developed client and with some existed clients. The tests done confirm that our server can interact with the existing client; this is possible because our server have three options for authenticate the client message:

- No authentication
- Authentication with HTTP Digest
- Authentication with our protocol

To show the performance of our proposed protocol, we have also tested miniSIPServer V24.0.4 with our developed client and with some existed clients. Then, we have compared the generated time between the request and response. The obtained results are illustrated in Figs. 6 and 7.

As we can see the first test (between our developed client and our developed server) generates the minimum time (0,20 s to 0,45 s) so the average of 0,26 s. We can see also that our server can response to the client requests in minimum time compared to miniSIPServer V24.0.4 (test 1, 2, 3 and 4 are best that the tests 5, 6, 7 and 8). These results confirm our theatrical demonstration in [23].

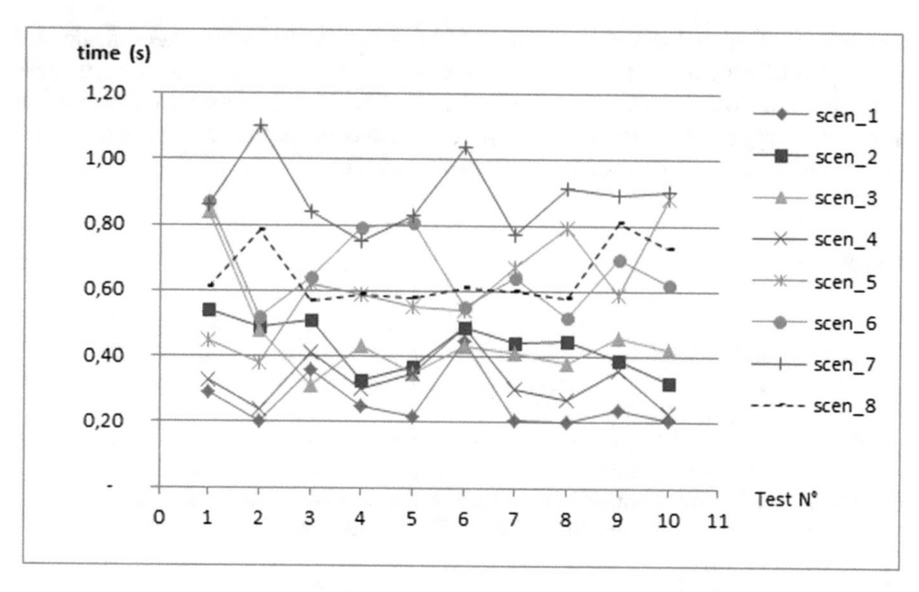


**Fig. 6.** Consumption time between the request and response

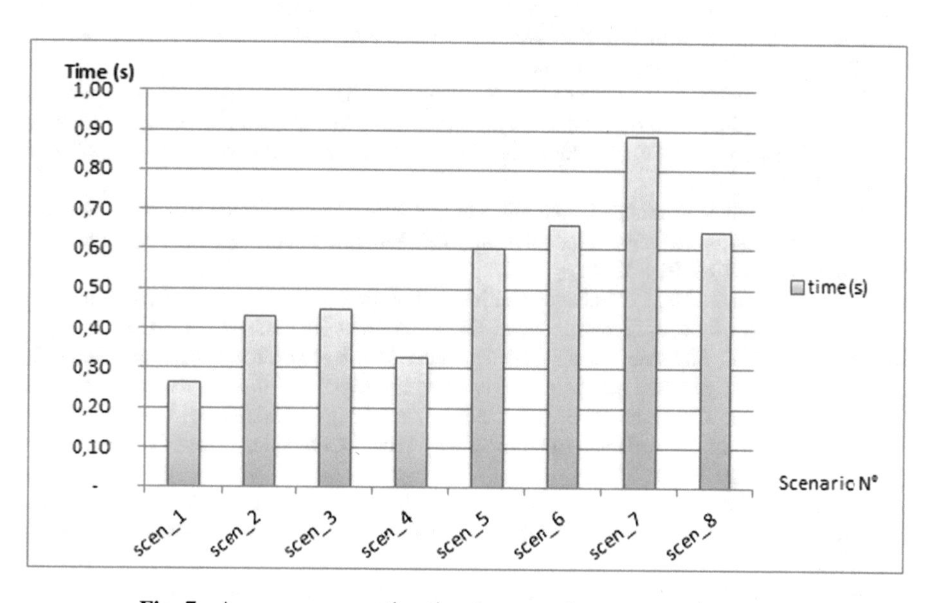


**Fig. 7.** Average consumption time between the request and response

## 5 Conclusion

In this paper, we present a practical implementation of our proposed SIP authentication protocol using Elliptic Curve Cryptography. The proposed protocol has been effectively developed based on JAVA programming language and API Jain SIP. Developed applications client and server are tested in Local Area Network with other downloaded Server and Client SIP. The experimental results confirm the theoretical security analysis. The comparison between our protocol and HTTP Digest protocol implemented in downloaded application shows that our protocol can enhance the authentication in SIP environment, as it can be used to share the session key.

## References


1. Rosenberg, J., Schulzrinne, H., Camarillo, G., Johnston, A., Peterson, J., Sparks, R., Handley, M., Schooler, E: SIP: session initiation protocol. No. RFC 3261 (2002)
2. Thom, G.A.: H. 323: the multimedia communications standard for local area networks. IEEE Commun. Mag. **34**(12), 52–56 (1996)
3. Franks, J., Hallam-Baker, P., Hostetler, J., Lawrence, S., Leach, P., Luotonen, A., Stewart, L.: HTTP authentication: basic and digest access Authentication, June 1999
4. Arshad, H., Nikooghadam, M.: An efficient and secure authentication and key agreement scheme for session initiation protocol using ECC. Multimedia Tools Appl. **75**(1), 181–197 (2016)
5. Lin, H., Wen, F., Du, C.: An anonymous and secure authentication and key agreement scheme for session initiation protocol. Multimedia Tools and Appl. **76**(2), 2315–2329 (2017)
6. Farash, M.: Security analysis and enhancements of an improved authentication for session initiation protocol with provable security. Peer-to-Peer Networking and Appl. **9**, 1–10 (2014). https://doi.org/10.1007/s12083-014-0315-x
7. Azrour, M., Farhaoui, Y., Ouanan, M.: Weakness in Zhang et al.'s authentication protocol for session initiation protocol. In: Ezziyyani, M., Bahaj, M., Khoukhi, F. (eds.) AIT2S 2017. LNNS, vol. 25, pp. 239–251. Springer, Cham (2018). https://doi.org/10.1007/978-3-319-69137-4_22
8. Arshad, H., Nikooghadam, M.: An efficient and secure authentication and key agreement scheme for session initiation protocol using ECC. Multimed Tools Appl. (2014). https://doi.org/10.1007/s11042-014-2282-x
9. Zhu, W., Chen, J., He, D.: Enhanced authentication protocol for session initiation protocol using smart card. Int. J. Electron. Secur. Digit. Forensics **7**(4), 330–342 (2015)
10. Chaudhry, S.A., Mahmood, K., Naqvi, H., Sher, M.: A secure authentication scheme for session initiation protocol based on elliptic curve cryptography. In: 2015 IEEE International Conference on Computer and Information Technology, Ubiquitous Computing and Communications. Dependable, Autonomic and Secure Computing, Pervasive Intelligence and Computing (2015)
11. Kumari, S., Chaudhry, A., Wu, F., Li., X., Farash, M.S., Khan, M.K.: An improved smart card based authentication scheme for session initiation protocol. Peer-to-Peer Netw. Appl. (2015)
12. Wu, K., Gong, P., Wang, J., Yan, X., Li, P.: An improved authentication protocol for session initiation protocol using smart card and elliptic curve cryptography. Rom. J. Inf. Sci. Technol. **16**(4), 324–335 (2013)
13. Azrour, M., Ouanan, M., Farhaoui, Y.: SIP authentication protocols based on elliptic curve cryptography: survey and comparison. Indonesian J. Electric. Eng. Comput. Sci. **4**(1), 231–239 (2016)
14. Azrour, M., Farhaoui, Y., Ouanan, M.: A server spoofing attack on Zhang et al. SIP authentication protocol. Int. J. Tomogr. Simul. **30**(3), 47–58 (2017)

15. Azrour, M., Farhaoui, Y., Ouanan, M.: Cryptanalysis of Farash et al.'s SIP authentication protocol. Int. J. Dyn. Syst. Differ. Equ. **8**(1–2), 77–94 (2018)
16. Azrour, M., Ouanan, M., Farhaoui, Y., Guezzaz, A.: Security analysis of Ye et al. authentication protocol for internet of things. In: Farhaoui, Y., Moussaid, L. (eds.) ICBDSDE 2018. SBD, vol. 53, pp. 67–74. Springer, Cham (2019). https://doi.org/10.1007/978-3-030-12048-1_9
17. Azrour, M., Ouanan, M., Farhaoui, Y.: A new efficient SIP authentication and key agreement protocol based on chaotic maps and using smart card. In: Proceedings of the 2nd International Conference on Computing and Wireless Communication Systems, p. 70. ACM, November 2017
18. Azrour, M., Ouanan, M., Farhaoui, Y.: Survey of SIP malformed messages detection. Indonesian J. Electric. Eng. Comput. Sci. **7**(2), 457–465 (2017)
19. Tsiatsikas, Z., Kambourakis, G., Geneiatakis, D., Wang, H.: The devil is in the detail: sdp-driven malformed message attacks and mitigation in sip ecosystems. IEEE Access **7**, 2401–2417 (2019)
20. Azrour, M., Ouanan, M., Farhaoui, Y.: A new architecture to protect sip server against flooding attack. Int. J. Tomogr. Simul. **31**(3), 15–28 (2018)
21. Dassouki, K., Safa, H., Hijazi, A., El-Hajj, W.: A SIP delayed based mechanism for detecting VOIP flooding attacks. In: 2016 International Wireless Communications and Mobile Computing Conference (IWCMC), pp. 588–593. IEEE. (2016)
22. Azrour, M., Farhaoui, Y., Ouanan, M., Guezzaz, A.: SPIT detection in telephony over ip using k-means algorithm. Procedia Comput. Sci. **148**, 542–551 (2019)
23. Chikha, R.J.B., Abbes, T., Chikha, W.B., Bouhoula, A.: Behavior-based approach to detect spam over IP telephony attacks. Int. J. Inf. Secur. **2**(15), 131–143 (2016)
24. Azrour, M., Farhaoui, Y., Ouanan, M.: A new secure authentication and key exchange protocol for session initiation protocol using smart card. Int. J. Network Secur. **195**(6), 870–879 (2017). https://doi.org/10.6633/IJNS.201711.19(6).02

# Promoting Health and Well-Being Using Wearable and Smartphone Technologies for Ambient Assisted Living Through Internet of Things

Gonçalo Marques[1,2(✉)] and Rui Pitarma[1]

[1] Polytechnic Institute of Guarda – Unit for Inland Development, Av. Dr. Francisco Sá Carneiro, nº 50, 6300–559 Guarda, Portugal
goncalosantosmarques@gmail.com, rpitarma@ipg.pt
[2] Instituto de Telecomunicações, Universidade da Beira Interior, Covilhã, Portugal

**Abstract.** Ambient Assisted Living is an emerging multi-disciplinary field which addresses several technologies to increase the quality of life for enhanced living environments and occupational health. It is well documented that poor indoor environmental quality has a significant impact on people's health and productivity. This is particularly relevant because most people spend more than 90% of their time inside buildings. In many buildings, indoor environmental quality can be highly poor. Thus, it is crucial to monitoring the indoor living environments in order to detect problems and act accordingly in useful time. This paper aims to present the ***iAmb***, a solution for indoor environmental quality monitoring based on Internet of Things (IoT). This solution is composed of a hardware prototype for ambient data acquisition denominated by ***iAmb*** and software applications for data consulting. The monitored data are stored in a ThingSpeak platform and are available for data consulting trough Web, smartphone and smartwatch. The results obtained are very promising, representing a significant contribution to indoor environmental quality supervision systems based on IoT. Compared to other systems the ***iAmb*** is based on open-source technologies, is a wireless solution that provides modularity, scalability, easy installation and has smartwatch compatibility.

**Keywords:** Ambient Assisted Living · Enhanced living environments · Indoor environment quality · Internet of Things · Occupational health · Smart cities

## 1 Introduction

Ambient Assisted Living (AAL) is an emerging multi-disciplinary field aiming at providing an ecosystem of different types of sensors, computers, mobile devices, wireless networks and software applications for personal healthcare monitoring and telehealth systems [1]. AAL aims to addresses the technologies that can be used to increase the quality of life of elders or disabled by providing a secure and protected living environment [2]. Ambient intelligence (AmI) represents a generation of


Y. Farhaoui (Ed.): BDNT 2019, LNNS 81, pp. 12–22, 2020.
https://doi.org/10.1007/978-3-030-23672-4_2

intelligent computing where a great diversity of sensors and computers stand everywhere, and for everyone, AAL is a variety of AmI systems [3]. AAL solutions aim to assist the older adults independently and actively living should leverage the efforts from both the technical side and social side [3].

The basic idea of the Internet of Things (IoT) is the pervasive presence of a variety of objects with interaction and cooperation capabilities among them to reach a common objective [4]. IoT and AAL technologies provide many benefits to the healthcare domain in activities such as tracking of objects, patients and staff, identification and authentication of people, automatic data collection and sensing [5, 6].

Indoor environmental quality (IEQ) in buildings includes indoor air quality (IAQ), acoustics, thermal comfort, and lighting [7]. It is well known that poor IEQ has a negative impact on occupational health, particularly on children and old people. IAQ is a significant determinant of personal exposure to pollutants because people spend about 90% of their time in indoor environments. In the case of older people and new-borns who are most likely affected by this pollutant may spend all their time in indoor environments [8]. The assessment that IAQ indicators must thereby determine how well indoor air (a) satisfies thermal and respiratory requirements, (b) prevents the unhealthy accumulation of pollutants, and (c) allows for a sense of well-being is proposed by [9]. In the USA, air quality is regulated by the Environmental Protection Agency (EPA). This organization consider that indoor levels of pollutants may be up to 100 times higher than outdoor pollutant level and ranked poor air quality as one of the top 5 environmental risks to the public health [10]. The problem of poor IAQ is of utmost importance affecting especially severe form the poorest people in the world who are most vulnerable presenting itself as a serious problem for world health such as tobacco use or the problem of sexually transmitted diseases [11]. IEQ real-time monitoring is assumed as an essential tool of extreme importance to study and analyse the IEQ in order to plan interventions for enhanced occupational health. The continuous technological advancements turn possible to develop smart devices with great capabilities for sensing, to connect and, consecutively, several advancements in AAL systems architectures particularly in the area of IEQ.

This paper aims to present the ***iAmb***, a solution for IEQ monitoring based on IoT architecture. This solution is composed by a hardware prototype for ambient data acquisition and Web/smartphone/smartwatch compatibility for data consulting. This system uses an open-source Arduino UNO as processing unit, an ESP8266 for Wi-Fi 2.4GHZ as communication unit and incorporates a temperature, humidity, $CO_2$, dust and light sensors as sensing unit. The ***iAmb*** incorporates wearable (***iAmbWear***) and smartphone (***iAmbMobile***) technology to provide enhanced visualization and analytics features to the end user for enhanced living environments.

## 2 Related Work

Wearable sensors are used on a large scale for AAL. The use of wearable sensors along with data treatment algorithms and visualization tools allow to measure better and describe real-life environments, mobility, physical activity, and physiological responses. Wearable motion sensors can help to detect behavioral anomaly situation in smart AAL by collect motion data combined with locational context [12].

Wearable technologies can be used to reduce the expenditure on health care by enabling people to be monitored in their own homes, rather than in hospitals, for a fraction of the cost [13]. Non-invasive wearable sensors for health monitoring are expected to provide low-cost solutions for remote monitoring of older adults at home or in nursing homes and lead to major improvements in patient monitoring and care [14]. Therefore, mobile computing has great importance in the monitoring of indoor living environments and can support the creation of important systems for IEQ supervision.

A dedicated, miniaturized, low-cost electronic nose based on metal oxide sensors and signal processing techniques that are capable of measuring carbon monoxide and nitrogen dioxide in mixtures with relative humidity and volatile organic compounds by using an optimized gas sensor array and highly effective pattern recognition techniques is presented by [15]. Another wireless solution for IEQ monitoring that is capable of measuring the environmental parameters such as temperature, humidity, gaseous pollutants, aerosol/particulate matter is proposed by [16]. A monitoring system that uses a low-cost wireless sensor network, to collect IAQ information developed using Arduino, XBee modules and microsensors, for storage and availability of monitoring data in real-time is presented by [17]. A real-time system for environmental monitoring based on wireless sensor network (WSN) capable of monitoring ambient temperature, relative humidity, acoustic levels and concentration of dust particles in the air for smart cities is proposed by [18]. A WSN approach to monitor the temperature distribution in a large-scale indoor space, which aims to improve the quality of the measurements transmitted by wireless signals, identify the temperature distribution pattern of the large-scale space, and optimize the allocation of supply air rated flow rate to multiple supply air terminals that serve the space according to the identified temperature distribution pattern is proposed by [19]. Several IoT architectures for indoor quality monitoring that incorporates open-source technologies for processing and data transmission and microsensors for data acquisition but also allows access to data collected from different places simultaneously through Web access and mobile applications in real-time are proposed by [20–29].

The literature review shows intelligent systems and projects that confirm these wearable sensors as having an extremely important role in the future and present of the AAL systems. Taking into account the importance of IEQ supervision for AAL, the advantages of wearable technologies, and to allow ubiquitous and pervasive access to IEQ data, the ***iAmbWear*** and ***iAmbMobile*** applications have been developed by the authors to provide fast and easy access to monitored data.

## 3 Materials and Methods

In general, the indoor quality covers visual and thermal comfort but also air quality. In this context, environmental parameters such as temperature, humidity, air speed, lighting level and pollutants concentrations are crucial for a proper evaluation of indoor environments.

The ***iAmb*** is a real-time IEQ monitoring solution that is capable of measuring temperature, humidity, $PM_{10}$, $CO_2$ and luminosity in real-time. It is a completely wireless solution. This solution is based on open-source technologies, it uses an

Arduino UNO [30] as a microcontroller and an ESP8266 module for data communication. The wireless communication implements using the IEEE 802.11 b/g/n networking protocol. The IEEE 802.11 standard supports radio transmission within the 2.4 GHz band [31].

The data collected by the system is stored in a ThingSpeak platform. ThingSpeak is an open source IoT application that provides developers with APIs to store and retrieve data from sensors and devices using HTTP over the Internet [32]. The Arduino UNO is responsible for data collection and is connected to the communication unit (ESP8266) by serial communication. The ESP8266 is responsible for uploading this data to the ThingSpeak platform. The end user can access the data from the Web page provided by ThingSpeak platform or via smartphone and smartwatch applications. The ***iAmbWear*** and ***iAmbMobile*** have been developed in Swift, an open-source programming language, using Xcode integrated development environment (IDE), which is compatible with iOS 12 and above [33].

The smartwatch pairs with the smartphone using Bluetooth 4.0 Low Energy (BLE) standard or via Wi-Fi as long as both devices are on the same Wi-Fi network. BLE is a low-power wireless technology for single-hop communication which can be assumed as a significant approach for IoT [34]. Zigbee and BLE were developed for battery-powered systems. However, BLE outperforms ZigBee in terms of power consumption [35]. BLE uses the 2.4 GHz ISM frequency band with adaptive frequency hopping to reduce interference and includes 24 bit CRC and AES 128 bit encryption technique on all packets to guarantee robustness and authentication [36].

The system provides a history of changes in the indoor conditions in order to help the building manager to make a precise analysis of the indoor environment. Further, the monitored data can also be used to support decision-making on possible interventions for enhanced living environments and occupational health.

The ***iAmb*** system architecture is based on IoT; Fig. 1 represents the system architecture.

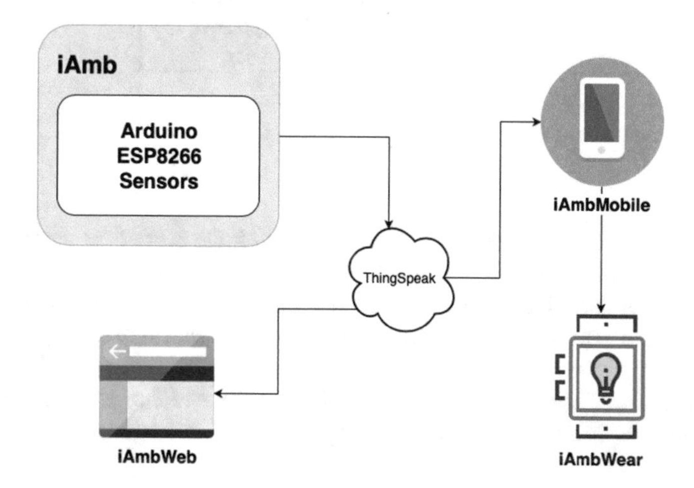


**Fig. 1.** ***iAmb*** system architecture.

The ***iAmb*** can be divided into three major parts: a processing unit, a sensing unit and a communication unit. This system is built using the embedded Arduino UNO microcontroller and an open-source platform that incorporates an Atmel AVR microcontroller [30] as a processing unit. The sensing unit incorporates a temperature and humidity sensor, a dust concentration sensor, a light sensor and a $CO_2$ sensor. It also incorporates an ESP8266 as a communication unit. Figure 2 represents ***iAmb*** prototype; a brief description of the used sensors is presented in Table 1.

**Fig. 2.** ***iAmb*** prototype

**Table 1.** ***iAmb*** sensors data.

| Sensor | Description | Range/accuracy |
|---|---|---|
| TH2 | Temperature and relative humidity sensor | 0–100%/±4.5% RH<br>0–70 °C/ ± 0.5 °C |
| Shinyei Model PPD42NS | Dust sensor that can measure the dust concentration in the air. It has stable and sensitive detection as it is responsive to a PM of diameter 1 μm and major. | 0–8000 pcs/L |
| TSL2561 | Digital light sensor | 0.1–40,000 lx |
| MG-811 | Analogue $CO_2$ sensor | 350–10000 ppm |

The selection of the sensors was made focusing on the cost of the system since the main objective was to test the functional architecture of the solution. Considering that the system is projected to be used in indoor environments where energy is easily accessible, there was no great concern regarding the selection of energy efficient sensors.

The firmware of the ***iAmb*** is implemented using the Arduino platform language in the Arduino IDE. It belongs to the C-family programming languages.

The ESP8266 has an important functionality that provides to the end user an easy configuration of the Wi-Fi network to which it will be connect. The ESP8266 is by default a Wi-Fi client but if it is unable to connect to the Wi-Fi network or if there are no wireless networks available the ESP8266 will turn to hotspot mode and will create a Wi-Fi network with a SSID "iAmb". At this point, the end user can connect to this Wi-Fi network which permits to configure the Wi-Fi network to which the ***iAmb*** is going to connect through the introduction of the network SSID and password.

## 4 Results and Discussion

The ***iAmb*** allows viewing the data as graphical and numerical values by using a Web browser or a smartphone and smartwatch applications. A sample of the data collected by ***iAmb*** is showed in Figs. 3, 4 and 5. Figure 3 represents luminosity data measured in lux; the Fig. 4 represents temperature data measured in Celsius and Fig. 5 represents dust sensor data measured in μg/m$^3$. It should be noted that the graphs displayed the results obtained in the real environment with induced simulations.

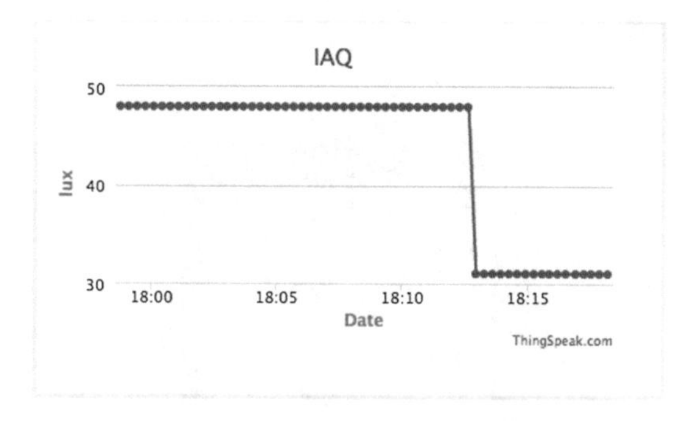


**Fig. 3.** Luminosity (lux)

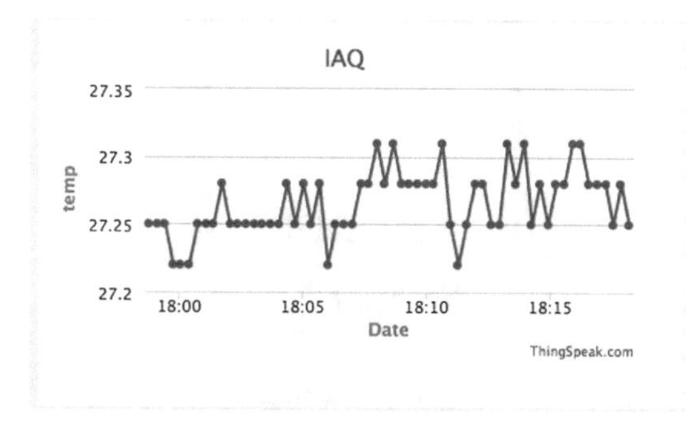


**Fig. 4.** Temperature (°C)

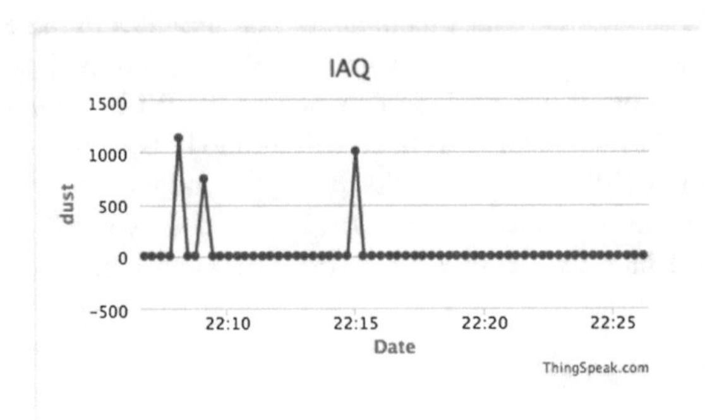


**Fig. 5.** Dust Sensor $PM_{10}$ (μg/m3)

The smartphone application allows quick, simple access, intuitive and real-time access to the monitored data (Fig. 6). Actually, smartphones have excellent processing and storage capabilities, and people carry them in their daily lives. For all these reasons, a mobile application has been created to allow quick, easy and intuitive access to the monitoring data. In this way, the user can carry the IEQ data of their home with him for everyday use.

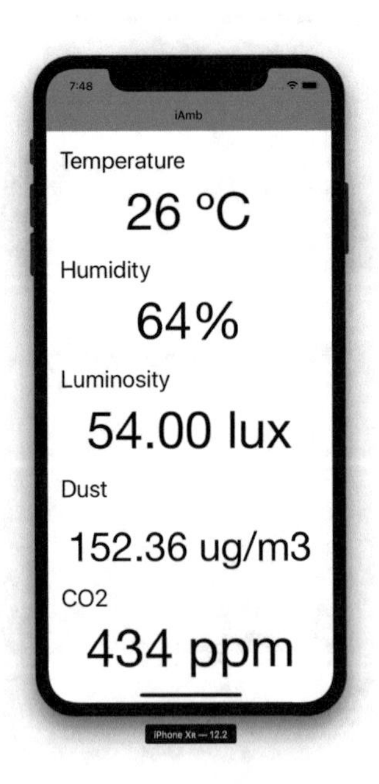


**Fig. 6.** ***iAmbMobile***

The Web/smartphone/smartwatch software also allows the user to access the data, which enables a more precise analysis of the detailed temporal evolution. Thus, the system is a powerful tool for analyse IEQ and to support decision making on possible interventions to improve a healthy and more productive indoor environment.

Smartwatch enables to receive unobtrusive notifications in a great diversity of situations and also could reduce smartphone dependency [37]. In that context, the ***iAmbWear*** application has been developed. The user can check the type of data, the data and an emotion that gives him simple and intuitive feedback to parameter quality (Fig. 7).

**Fig. 7.** ***iAmbWear***

As referred by [38] emoticons in instant messenger applications are examples of emotionally expressive semantic interaction in the domain of on-screen interaction. The ***iAmbWear*** uses emoticons to provide ubiquitously the current information about the indoor health living environment. The use of the emoticons in the smartwatch application, this functionality aims to provide an understanding information to the end user which does not require technical knowledge about IEQ. Assumed the ubiquity of smartwatches, this enhanced user interface is an ideal solution for wireless health applications with low user burden [39].

The smartwatches are a ubiquitous and pervasive method for data consulting. The wrist is an amazing location to notice changes in the air quality to the user in efficient and fast modes [40]. The emergence and importance of smartwatch usability and application are also shown by a diversity of US patents, such as a patent for remote monitoring of a medical patient using a smartwatch described by [41] or a patent for receiving and displaying unassociated information elements in useful ways on a wearable computing device [42].

The ***iAmbWear*** application aims to improve the IEQ for enhanced living environments and occupational health. Compared to other systems, the ***iAmbWear*** provides a ubiquitous and pervasive solution for IEQ visualization in order to act in a useful time to increase for enhanced living environments and occupational health [21]. Improvements to the system hardware and software are planned to make it much more appropriate for specific purposes such as hospitals, schools and offices. As future work, the main goal is to make technical improvements, including the development of important alerts and notifications to notify the user when the living environment has serious deficiencies.

## 5 Conclusion

A complete wireless solution for IEQ monitoring (***iAmb***) based in IoT has presented. This solution is composed by a hardware prototype for ambient data collection and a Web, smartphone and smartwatch compatibility for data consulting. The smartwatch application provides ubiquitous access to IEQ data.

The results obtained are very promising, representing a significant contribution to IEQ monitoring systems based on IoT. Compared to existing systems, it has great importance due to the use of low-cost and open-source technologies. Notice that the system has advantages both for easy installation and configuration. The system not only uses wireless communication technology but also supports smartphone and smartwatch compatibility. Furthermore, the proposed system has been developed to be compatible with traditional houses and not only for smart homes.

Despite all the advantages in the use of IoT architecture, still exist many open issues like scalability, quality of service problems and security and privacy issues. The proposed system should find ways to respond to these problems.

As future work is expected to introduce new sensors to this system for monitoring other IEQ parameters as well as the development of a platform that allows to share in a secure way the collected data to health professionals. In order to adapt the system to specific cases, physical system and related software improvements have been planned.

We believe that in the future, systems as the ***iAmb*** will be an important part of the living environments, particularly, wearable technology must be used in the near future for enhanced occupational health and well-being. On the one hand, the monitored data can be extremely useful to provide support to a clinical analysis by health professionals. On the other hand, through monitoring, it is possible to perceive correctly the indoor conditions that influence the occupant's health and plan the interventions for enhanced living environments and occupational health.

## References

1. Marques, G., Pitarma, R.: An indoor monitoring system for ambient assisted living based on internet of things architecture. Int. J. Environ. Res. Public. Health **13**(11), 1152 (2016)
2. Bacciu, D., Barsocchi, P., Chessa, C., Gallicchio, A.: Micheli: an experimental characterization of reservoir computing in ambient assisted living applications". Neural Comput. Appl. **24**(6), 1451–1464 (2014)
3. Cubo, J., Nieto, A., Pimentel, E.: A cloud-based internet of things platform for ambient assisted living. Sensors **14**(8), 14070–14105 (2014)
4. Giusto, D. (ed.): The Internet Of Things: 20th Tyrrhenian Workshop On Digital Communications. Springer, New York (2010). https://doi.org/10.1007/978-1-4419-1674-7
5. Atzori, L., Iera, A., Morabito, G.: The Internet of Things: a survey. Comput. Netw. **54**(15), 2787–2805 (2010)

6. Marques, G.: Ambient assisted living and Internet of Things. In: Cardoso, P.J.S., Monteiro, J., Semião, J., Rodrigues, J.M.F. (eds.) Harnessing the Internet of Everything (IoE) for Accelerated Innovation Opportunities, pp. 100–115. IGI Global, Hershey (2019)
7. Vilcekova, S., Meciarova, L., Burdova, E.K., Katunska, J., Kosicanova, D., Doroudiani, S.: Indoor environmental quality of classrooms and occupants' comfort in a special education school in Slovak Republic. Build. Environ. **120**, 29–40 (2017)
8. Walsh, P.J., Dudney, C.S., Copenhaver, E.D.: Indoor Air Quality. CRC Press, Boca Raton (1983)
9. Gold, D.R.: Indoor air pollution. Clin. Chest Med. **13**(2), 215–229 (1992)
10. Seguel, J.M., Merrill, R., Seguel, D., Campagna, A.C.: Indoor air quality. Am. J. Lifestyle Med. **11**(4), 284–295 (2016). 1559827616653343
11. Bruce, N., Perez-Padilla, R., Albalak, R.: Indoor air pollution in developing countries: a major environmental and public health challenge. Bull. World Health Organ. **78**(9), 1078–1092 (2000)
12. Zhu, C., Sheng, W., Liu, M.: Wearable sensor-based behavioral anomaly detection in smart assisted living systems. Autom. Sci. Eng. IEEE Trans. On **12**(4), 1225–1234 (2015)
13. Aced López, S., Corno, F., De Russis, L.: Supporting caregivers in assisted living facilities for persons with disabilities: a user study. Univers. Access Inf. Soc. **14**(1), 133–144 (2015)
14. Bandodkar, A.J., Wang, J.: Non-invasive wearable electrochemical sensors: a review. Trends Biotechnol. **32**(7), 363–371 (2014)
15. Zampolli, S., et al.: An electronic nose based on solid state sensor arrays for low-cost indoor air quality monitoring applications. Sens. Actuators B Chem. **101**(1–2), 39–46 (2004)
16. Bhattacharya, S., Sridevi, S., Pitchiah, R.: Indoor air quality monitoring using wireless sensor network, pp. 422–427 (2012)
17. Marques, G., Pitarma, R.: Health informatics for indoor air quality monitoring, pp. 1–6 (2016)
18. Sanchez-Rosario, F., et al.: A low consumption real time environmental monitoring system for smart cities based on ZigBee wireless sensor network. In: 2015 International Wireless Communications and Mobile Computing Conference (IWCMC), Dubrovnik, Croatia, pp. 702–707 (2015)
19. Zhou, P., Huang, G., Zhang, L., Tsang, K.-F.: Wireless sensor network based monitoring system for a large-scale indoor space: data process and supply air allocation optimization. Energy Build. **103**, 365–374 (2015)
20. Marques, G., Pitarma, R.: IAQ evaluation using an IoT CO2 monitoring system for enhanced living environments. In: Rocha, Á., Adeli, H., Reis, L.P., Costanzo, S. (eds.) Trends and Advances in Information Systems and Technologies, vol. 746, pp. 1169–1177. Springer International Publishing, Cham (2018)
21. Pitarma, R., Marques, G., Ferreira, B.R.: Monitoring indoor air quality for enhanced occupational health. J. Med. Syst. **41**(2), 23 (2017)
22. Marques, G., Pitarma, R.: Smartwatch-based application for enhanced healthy lifestyle in indoor environments. In: Omar, S., Haji Suhaili, W.S., Phon-Amnuaisuk, S. (eds.) CIIS 2018. AISC, vol. 888, pp. 168–177. Springer, Cham (2019). https://doi.org/10.1007/978-3-030-03302-6_15
23. Marques, G., Pitarma, R.: Using IoT and social networks for enhanced healthy practices in buildings. In: Rocha, Á., Serrhini, M. (eds.) EMENA-ISTL 2018. SIST, vol. 111, pp. 424–432. Springer, Cham (2019). https://doi.org/10.1007/978-3-030-03577-8_47
24. Marques, G., Pitarma, R.: Monitoring health factors in indoor living environments using Internet of Things. In: Rocha, Á., Correia, A.M., Adeli, H., Reis, L.P., Costanzo, S. (eds.) WorldCIST 2017. AISC, vol. 570, pp. 785–794. Springer, Cham (2017). https://doi.org/10.1007/978-3-319-56538-5_79

25. Marques, G., Roque Ferreira, C., Pitarma, R.: A system based on the internet of things for real-time particle monitoring in buildings. Int. J. Environ. Res. Public. Health **15**(4), 821 (2018)
26. Salamone, F., Belussi, L., Danza, L., Galanos, T., Ghellere, M., Meroni, I.: Design and development of a nearable wireless system to control indoor air quality and indoor lighting quality. Sensors **17**(5), 1021 (2017)
27. Akkaya, K., Guvenc, I., Aygun, R., Pala, N., Kadri, A.: IoT-based occupancy monitoring techniques for energy-efficient smart buildings. In: 2015 IEEE Wireless Communications and Networking Conference Workshops (WCNCW), New Orleans, LA, USA, 2015, pp. 58–63 (2015)
28. Marques, G.M.S., Pitarma, R.: Smartphone application for enhanced indoor health environments. J. Inf. Syst. Eng. Manag. **1**, 4 (2016)
29. Marques, G., Pitarma, R.: Monitoring and control of the indoor environment. In: 2017 12th Iberian Conference on Information Systems and Technologies (CISTI), Lisbon, Portugal, pp. 1–6 (2017)
30. D'Ausilio, A.: Arduino: a low-cost multipurpose lab equipment. Behav. Res. Methods **44**(2), 305–313 (2012)
31. Bhoyar, R., Ghonge, M., Gupta, S.: Comparative study on IEEE standard of wireless LAN/Wi-Fi 802.11 a/b/g/n. Int. J. Adv. Res. Electron. Commun. Eng. IJARECE **2**(7), 687–691 (2013)
32. Doukas, C., Maglogiannis, I.: Bringing IoT and cloud computing towards pervasive healthcare, pp. 922–926 (2012)
33. Neuburg, M.: iOS 7 Programming Fundamentals: Objective-c, xcode, and cocoa basics. O'Reilly Media Inc, Sebastopol (2013)
34. Gomez, C., Oller, J., Paradells, J.: Overview and evaluation of bluetooth low energy: an emerging low-power wireless technology. Sensors **12**(9), 11734–11753 (2012)
35. Jawad, H., Nordin, R., Gharghan, S., Jawad, A., Ismail, M.: Energy-efficient wireless sensor networks for precision agriculture: a review. Sensors **17**(8), 1781 (2017)
36. Ojha, T., Misra, S., Raghuwanshi, N.S.: Wireless sensor networks for agriculture: the state-of-the-art in practice and future challenges. Comput. Electron. Agric. **118**, 66–84 (2015)
37. Cecchinato, M.E., Cox, A.L., Bird, J.: Smartwatches: the good, the bad and the ugly? pp. 2133–2138 (2015)
38. Ross, P.R., Overbeeke, C.J., Wensveen, S.A.G., Hummels, C.M.: A designerly critique on enchantment. Pers. Ubiquitous Comput. **12**(5), 359–371 (2008)
39. Kalantarian, H., Alshurafa, N., Nemati, E., Le, T., Sarrafzadeh, M.: A smartwatch-based medication adherence system, pp. 1–6 (2015)
40. Kerber, F., Hirtz, C., Gehring, S., Löchtefeld, M., Krüger, A.: Managing smartwatch notifications through filtering and ambient illumination, pp. 918–923 (2016)
41. Subramaniam, S.: Smartwatch with a multi-purpose sensor for remote monitoring of a patent. Google Patents (2015)
42. Vonshak, I., Damir, L.M., Knight, M.E., Levak, H.: Smartwatch or other wearable device configured to intelligently interact with a user. Google Patents (2016)

# Implicit JSON Schema Versioning Driven by Big Data Evolution in the τJSchema Framework

Zouhaier Brahmia[1(✉)], Safa Brahmia[1], Fabio Grandi[2], and Rafik Bouaziz[1]

[1] University of Sfax, Sfax, Tunisia
{zouhaier.brahmia, safa.brahmia}@fsegs.rnu.tn, rafik.bouaziz@usf.tn

[2] University of Bologna, Bologna, Italy
fabio.grandi@unibo.it

**Abstract.** In JSON-based NoSQL data stores, Big Data instance documents and their JSON schemas must evolve over time to reflect changes in the real world. When a JSON instance document, valid with respect to a JSON schema, is updated giving rise to a new document no longer valid with respect to the schema, the update is usually rejected also resulting in user frustration. In such a case, the JSON schema has to be explicitly changed by an administrator in order to become compliant with the new Big Data format before the update can be effected by the user. The different approach we propose in this work is to privilege the user actions and accept in a transparent way any update he/she wants to apply to the instance documents: violation of the validity of an updated instance document with respect to its JSON schema is automatically detected and schema changes necessary to produce a new schema version compliant with the new Big Data format are automatically applied by the system, producing a new JSON schema version. Hence, in this work, we deal with *implicit JSON schema versioning* driven by updates to JSON-based Big Data instance documents. Our proposed solution consists in an extension of the τJSchema (Temporal JSON Schema) framework we previously introduced to create and validate temporal JSON documents and to allow classical temporal JSON schema versioning, to also support implicit JSON schema versioning.

**Keywords:** JSON · JSON schema · τJSchema · Instance update operation · Instance versioning · JSON instance version · Schema change operation · Schema versioning · JSON schema version

## 1 Introduction

Nowadays, Big Data [13, 14] are being exploited by several computer applications in various areas. NoSQL database management systems [6, 9, 15, 16] are used to define, store, and manipulate such data. JSON [11] is one of the most popular semistructured data formats that are used to store Big Data, due to its user-friendliness and independence from programming languages. In JSON-based Big Data NoSQL data stores,


Y. Farhaoui (Ed.): BDNT 2019, LNNS 81, pp. 23–35, 2020.
https://doi.org/10.1007/978-3-030-23672-4_3

Big Data are stored in JSON instance documents whose structure is specified in a JSON Schema [12] document. Since the information contents of data stores model some application reality, both JSON-based Big Data instance documents and their JSON schemas evolve over time to reflect changes in such a reality.

For several reasons (e.g., to recover any desired JSON schema/instance document versions, to track JSON schema/instance document changes over time, to execute temporal and multi-schema queries), advanced applications exploiting JSON-based Big Data NoSQL data stores require maintaining a complete history of changes performed both at instance and schema levels [8]. Such a requirement can be fulfilled with the adoption of the temporal schema versioning technique [1, 4], as proposed in our previous work [3, 5]. As a result, a set of JSON resource (i.e., a JSON instance document or a JSON schema file) versions is kept in the data store, where each JSON instance document version is valid with respect to a JSON schema version.

Usually, in existing NoSQL environments supporting temporal JSON schema versioning [3, 5], changes to a JSON schema version, when necessary, are performed by the NoSQL DataBase Administrator (NSDBA) in an explicit manner; these changes consist of a valid sequence of JSON schema change operations that are applied by the system to generate a new JSON schema version. Furthermore, the update of a JSON instance document, which is valid with respect to the current JSON schema version, gives rise to a new one which must be valid with respect to the same schema version; otherwise, the update is rejected, which may also result in user frustration. However, in some cases, performing updates of a JSON instance document has priority over a continued compliance with its JSON schema version (e.g., when it is required to store data even if they are incomplete, or to provide some flexibility with respect to stored schemas of JSON data in a heterogeneous/collaborative environment) and, thus, the new instance document resulting from the updates must be accommodated anyway in the Big Data store. In this paper, we deal with this issue and propose an approach that consists, in such a case, in automatically and implicitly (i.e., transparently to the user and without requiring the NSDBA intervention) generating a new JSON schema version for the new instance document. Contrarily to existing explicit (semistructured) database schema versioning approaches [3, 5, 7, 10, 17], our proposal deals with implicit JSON schema versioning driven by updates to JSON-based Big Data instance documents. It focuses on how to automatically produce a new JSON schema version as a result of some JSON instance changes that do not respect the compliance with the current JSON schema version. Moreover, we integrate our approach in an extension of our τJSchema (Temporal JSON Schema) framework [2], previously introduced to create and validate temporal JSON documents and to allow explicit temporal JSON schema versioning, to also support implicit temporal JSON schema versioning driven by JSON instance evolution.

By developing this approach, we have tried to fill a gap in both the state-of-the-art and the state-of-the-practice in JSON-based NoSQL databases, which consists in the lack of theoretical proposals and practical tools, respectively, for managing in a consistent and user-friendly manner instance updates that must be performed but that would produce a new JSON instance document no longer valid with respect to the current JSON schema version. Precisely, our approach enables the system to generate, in an implicit and automatic manner, the necessary sequence of JSON schema change operations that must be executed to produce a new JSON schema version compliant with the structure of the new JSON instance document resulting from the updates.

The rest of the paper is organized as follows. Section 2 illustrates our approach for implicitly changing JSON schemas in response to some specific updates to JSON-based Big Data instance documents, in a JSON schema versioning setting. Section 3 extends our τJSchema framework, initially designed to only allow explicit temporal JSON schema versioning, to also support our new approach for implicit temporal JSON schema versioning. Section 4 presents an example illustrative of the approach functioning. Conclusions can be found in Sect. 5.

## 2 JSON Schema Changes Generated by Updates to JSON-Based Big Data

In this section, we present and exemplify our approach for automatically managing JSON schema changes that are required by updates to Big Data stored as JSON instance documents in order to maintain compliance with their JSON schema.

Our contribution concerns an aspect that has not been previously considered in JSON schema management, and deals with JSON schema evolution that is not caused by explicit schema changes but is implicitly driven by instance changes. Indeed, we ensure that JSON instance update operations specified by the user, even leading to a new JSON instance document version that is no longer valid to the current JSON schema version, are always accepted. The proposed solution is based on the automatic generation of a new JSON schema version compliant to the new JSON instance document structure, through the application of a suitable sequence of schema changes in a transparent way. The JSON instance document update operations that trigger implicit schema changes are as follows:

- insertion of components (properties, objects, arrays) that do not belong to the current JSON schema;
- rename operations involving components having a name (e.g., properties);
- update, delete or insert operations involving components (properties, objects, arrays) present in the current JSON schema, which violate some JSON schema constraints. For example:
  - update operations changing the value of a property or of an array item, to a value that is not compatible with the type of this property or this array item, respectively, as defined in the current JSON schema;
  - delete operations acting on components which are declared as required in the current JSON schema;
  - insert operations adding components which are not defined in the current JSON schema or having values that are not compatible with the types of these components or violate some constraints defined in the current JSON schema.

To illustrate our approach, let us assume that the user has to insert, in the current JSON instance document version JID_V1, a new instance of an object O having a property P1, whereas the object O does not have a property P1 defined in the current JSON schema version JS_V1 with respect to which JID_V1 is valid. Thus, in order to store the new object instance in JID_V1, a JSON schema change on JS_V1 is required to specify that the object O also has the property P1. This change generates both a new JSON schema version JS_V2 and a new JSON instance document version JID_V2 to which the instance of O can be added. Notice here that the old instances of O, stored in JID_V1, will still be valid with respect to JS_V2 if the property P1 has been defined without specifying that it is required. However, such old instances will be no longer valid with respect to JS_V2 if P1 has been declared as required. In this case, the instances of O for which a value for the property P1 cannot be provided by the user are removed from JID_V2, in order to guarantee the validity of the new JSON instance document version JID_V2 with respect to its JSON schema version JS_V2. However, if a value for the property P1 of an old instance of O is provided by the user, this instance is updated and kept in JID_V2.

We propose that the system allows the user to choose the most suitable option (i.e., whether P1 is required or not), according to the requirements of the application.

Besides, if the object O has a required property P2 in JS_V1 and the user has to insert in JID_V1 a new instance of O that does not include the property P2, a JSON schema change on JS_V1 is also required to drop the property P2 from O or to specify that the property P2 of O is not required, by removing P2 from the array of required properties of O. This schema change gives rise to both a new JSON schema version JS_V2 and a new JSON instance document version JID_V2, in which the new instance of O can be inserted. Similarly to the above situation, the user can choose between keeping or not the old instances of O (coming from JID_V1), in JID_V2. In fact, if the user chooses to drop the property P from the object O, the old instances of O will not be kept in JID_V2. But, if the user chooses to declare that the property P of the object O is not required, the old instances of O will also be kept in JID_V2.

## 3 τJSchema Extension to Support Implicit Schema Versioning

In this section, we extend our τJSchema framework [2, 3, 5] to support our approach, allowing the user to perform JSON-based Big Data updates that may give rise to the execution of implicit JSON schema changes.

Since in τJSchema both conventional JSON schemas and their underlying Conventional JSON-based Big Data instance documents are temporally versioned along the transaction-time dimension, changes at schema or instance level are performed only on the current JSON schema (JSON instance document, respectively) version. Changing the current version of a JSON instance document consists in inserting some new Big Data, or modifying or deleting some already stored Big Data.

The user can specify a sequence of updates (JIUO) to be applied to a JSON instance document and ask the system to execute them; the system processes them via the procedure "ApplyUpdatesWithImplicitSchemaVersioning" whose algorithm is shown in Fig. 1.

This procedure uses the following variables and calls the following procedures and function (see [2, 3, 5] for details on the τJSchema components, as temporal JSON document, squashed JSON document and temporal JSON schema).

```
Procedure ApplyUpdatesWithImplicitSchemaVersioning
Inputs: ConvJDoc_v, ConvJSch_w, JIUO
Outputs: ConvJDoc_v+1, ConvJSch_w+1
Begin
01:  CopyJDoc(ConvJDoc_v, ConvJDoc_v+1);
02:  ExecJInstanceUpdateOps(JIUO, ConvJDoc_v+1, JSCO);
03:  If (CompareJInstanceDocs(ConvJDoc_v+1, ConvJDoc_v))
04:  Then DeleteJInstanceDoc(ConvJDoc_v+1);
05:       Display("No changes have been actually applied
                   to ConvJDoc_v");
06:  Else UpdateTemporalJDoc(TempJD, ConvJDoc_v+1);
07:    If (JSCO is empty)
08:    Then UpdateSquashedJDoc(SquashJD_w, ConvJDoc_v+1);
09:    Else CopyJSchema(ConvJSch_w, ConvJSch_w+1);
10:         ExecJSchemaChangeOps(JSCO, ConvJSch_w+1);
11:         PropagateToInstances(ConvJSch_w+1,
                                JSCO,ConvJDoc_v+1);
12:         UpdateTemporalJSchema(TempJS, ConvJSch_w+1);
13:         CreateSquashedJDoc(SquashJD_w+1,
                             ConvJSch_w+1, ConvJDoc_v+1);
14:    End If
15:  End If
End
```

**Fig. 1.** The procedure for applying updates to a conventional JSON-based Big Data instance document version with implicit schema versioning

- Variables:
  - ConvJDoc_v (input): the current conventional JSON instance document version (number v).
  - ConvJSch_w (input): the current conventional JSON schema version (number w) with respect to which ConvJDoc_v is valid.
  - JIUO (input): a valid sequence of JSON instance document update operations specified by the user.ConvJDoc_v+1 (output): the new conventional JSON instance document version (number v+1).
  - ConvJSch_w+1 (output): the new conventional JSON schema version (number w+1) with respect to which ConvJDoc_v+1 is valid.
  - JSCO: a valid sequence of JSON schema change operations automatically generated by the system during the application of the updates.
  - TempJD: the temporal JSON document that ties the conventional JSON document versions and the temporal JSON schema together.
  - SquashJD_w: the squashed JSON document number w.
  - SquashJD_w+1: the squashed JSON document number w+1.
  - TempJS: the temporal JSON schema that ties the conventional JSON schema and the temporal logical and physical characteristics together.

- Procedures:
  - CopyJDoc(ds, dt): it copies the contents of the JSON instance document source (ds) into another empty JSON instance document target (dt).
  - ExecJInstanceUpdateOps(iuo, d, sco): it applies the sequence of JSON instance document update operations (iuo) on the instance document passed as argument (d) and generates a sequence of implicit schema changes (sco) if necessary (may require interaction with the user).
  - DeleteJInstanceDoc(d): it deletes the conventional JSON instance document whose name is passed as argument (d).
  - Display(message): it displays the character string passed as argument (message).
  - UpdateTemporalJDoc(tjd, cd): it adds, to the temporal JSON document passed as argument (tjd), a new slice associated to a new conventional JSON instance document version (cd).
  - UpdateSquashedJDoc(sjd, cd): it adds, to the squashed JSON document passed as argument (sjd), a new slice associated to a new conventional JSON instance document version (cd).
  - CopyJSchema(ss, st): it copies the contents of the JSON schema source (ss) into another empty JSON schema target (st).
  - ExecJSchemaChangeOps(sco, js): it applies the sequence of JSON schema change operations (sco) on the JSON schema version passed as argument (js).
  - PropagateToInstances(cjs_nv, seq_sc, cjd_nv): it applies the sequence of JSON schema changes (seq_sc), which have been executed on the previous conventional JSON schema version to obtain the new one (cjs_nv), on the conventional JSON document version "cjd_nv" so that it becomes valid to the new JSON schema version.
  - UpdateTemporalJSchema(tjs, cjs): it adds, to the temporal JSON schema passed as argument (tjs), a new slice associated to a new conventional JSON schema version (cjs).
  - CreateSquashedJDoc(sjd, tjs, cjd): it creates a new squashed JSON document (sjd) based on a new conventional JSON instance document version (cjd) and its temporal JSON schema (tjs).
- Function:
  - CompareJInstanceDocs(d1, d2): a function that compares two JSON instance documents (d1 and d2) and returns true or false according to the fact that these documents are identical or not.

As shown above, our JSON instance document change algorithm is based on (i) the current conventional JSON instance document version, (ii) the current conventional JSON schema version, and (iii) the sequence of JSON instance document update operations (specified by the user). These inputs allow the system to detect instance update operations that require JSON schema changes. In the following, we discuss the JSON instance document update operations, defined in our previous work [3, Sect. 4.4], and show in which cases they trigger JSON schema change operations (also defined in the same work [3, Sect. 4.2]). For the sake of simplicity and due to space limitations, we focus in this paper only on JSON instance document update operations dealing with simple type properties. We will complete the study with the presentation of the operations dealing with the other JSON components (i.e., objects and arrays), in a future work that will be an extension of the current one.

**InsertSimpleTypeProperty(JID.json, targetPropertyPath, position, propertyName, propertyValue)**
This operation generates JSON schema change operations, in the following cases:

– *Case 1: There is no property with the same name and at the same path, in the corresponding JSON schema version.*

In this case, the system generates the following JSON schema change operation:

```
AddPropertyToConventionalJSONSchema(CJS.json,
                   targetComponentPath, position,
                   propertyName, propertyType)
```

such that “CJS.json” is the new JSON schema version, “targetComponentPath” is deduced from “targetPropertyPath”, the “position” and “propertyName” arguments are equal to the corresponding arguments of the instance update operation, and the “propertyType” argument is deduced from the value of the “propertyValue” argument.

– *Case 2: There is a property with the same name and at the same path, in the corresponding JSON schema version, but with a type that is different from the type of the provided value.*

In this case, the system generates the following sequence of JSON schema change operations to update the type of the concerned property

```
i) DropPropertyFromConventionalJSONSchema(CJS.json,
                   propertyPath);
ii) AddPropertyToConventionalJSONSchema(CJS.json,
                   targetComponentPath, position,
                   propertyName, propertyType);
```

such that “CJS.json” is the new JSON schema version, “propertyPath” is deduced from the “targetPropertyPath” and “position” arguments, “targetComponentPath” is deduced from “targetPropertyPath”, the “position” and “propertyName” arguments are equal to the corresponding arguments of the instance update operation, and the “propertyType” argument is deduced from the value of the “propertyValue” argument.

**UpdateSimpleTypeProperty(JID.json, propertyPath, newPropertyValue)**
This operation generates JSON schema change operations, in case *there is a property with the same name and at the same path, in the corresponding JSON schema version, but with a type that is different from the type of the provided value*. In this case, the system generates the same sequence of schema change operations that has been presented in case 2 of the previous instance update operation.

**RenameProperty(JID.json, propertyPath, newPropertyName)**
This operation generates always the following JSON schema change operation

```
RenamePropertyInConventionalJSONSchema(CJS.json,
                         propertyPath, newPropertyName)
```

such that "CJS.json" is the new JSON schema version, "propertyPath" is deduced from the "propertyPath" argument of the JSON instance update operation, and the "newPropertyName" argument is equal to the value of the "newPropertyName" argument of the instance update operation.

**DeleteProperty(JID.json, propertyPath)**
This operation generates a JSON schema change operation in case *there is a property at the same path, in the corresponding JSON schema version, but which is declared as required.* In this case, the system generates the following JSON schema change operation

```
DropItemFromArrayTypeKeywordInConventionalJSONSchema(
                  CJS.json, arrayPath, propertyName)
```

such that "CJS.json" is the new JSON schema version, "arrayPath" is the path of the corresponding "required" array, and "propertyName" is deduced from the "propertyPath" argument of the JSON instance update operation.

## 4 Illustrative Example

To illustrate our approach, we consider the example of a τJSchema-based NoSQL data store used for managing computer science journals. Suppose that this data store contains, in its initial state, a conventional JSON instance document version, as shown in Fig. 2, and its conventional JSON schema version, as shown in Fig. 3, both created on April 01, 2019. Notice that the conventional JSON instance document version stores the details of a journal named "Non-standard Databases" having "Ahmad Jamil" as editor, "Q2" as Impact Factor quartile, and 4 as periodicity. Notice also that due to space limitations we will not present here the temporal JSON schema for journals, named "JournalsTemporalJSONSchema.json", and the temporal JSON document for journals, named "JournalsTemporalJSONDocument.json".

```
{"journals":[
   {"journal":{"name":"Non-standard Databases",
               "editor":"Ahmad Jamil",
               "quartile":"Q2",
               "periodicity":4,
} } ] }
```

**Fig. 2.** The first conventional JSON instance document version for journals ("Journals_V1.json"), on April 01, 2019

```
{"$schema":"http://json-schema.org/draft-04/schema#",
 "id":"http://jsonschema.net",
 "type":"object",
 "properties":
    {"journals":
        {"id":"http://jsonschema.net/journals",
         "type":"array",
         "items":
             {"type":"object",
              "properties":
                  {"journal":{"type":"object",
                              "properties":
                                 {"name":{"type":"string"},
                                  "editor":{"type":"string"},
                                  "quartile":{"type":"string"},
                                  "periodicity":{"type":"number"}
                                  },
                                  "required":["name","editor",
                                              "periodicity"]
                  } },
                  "required":["journal"]
    } } },
    "required":["journals"]}
```

**Fig. 3.** The first conventional JSON schema version of journals (“JournalsSchema_V1.json”), defining the structure of “Journals_V1.json”, on April 01, 2019

Hence we assume that, on April 15, 2019, the user has performed the following updates on the first conventional JSON instance document version (shown in Fig. 2):

- changing the name of the property “editor” (of the object “journal”) to “editor-in-chief”;
- changing the name of the property “quartile” (of the object “journal”) to “currentQuartile”;
- changing the value of the property “periodicity” (of the object “journal”) from 4 (with number type) to “quarterly” (with string type);
- adding a new property to the object “journal”, named “ISSN”, with string type and value “1234-567X”;
- adding a new property to the object “journal”, named “publisher”, with string type and value “IZI Global”;
- adding a new property to the object “journal”, named “scopus-indexed”, with boolean type and value true.

The execution of these updates gives rise not only to a new conventional JSON instance document version, as shown in Fig. 4, but also to a new conventional JSON schema version, as shown in Fig. 5, since the new instance document version does not conform to the current JSON schema version (shown in Fig. 3). Obviously, both the temporal JSON schema (JournalsTemporalJSONSchema.json) and the temporal JSON document (JournalsTemporalJSONDocument.json) have been updated to take into account the new conventional JSON schema version (JournalsSchema_V2.json) and the new conventional JSON instance document version (Journals_V2.json), respectively.

```
{"journals":[
  {"journal":{
      "ISSN":"1234-567X",
      "name":"Non-standard Databases",
      "editor-in-chief":"Ahmad Jamil",
      "publisher":"IZI Global",
      "scopus-indexed":true,
      "currentQuartile":"Q2",
      "periodicity":"quarterly"
} } ] }
```

**Fig. 4.** The second conventional JSON instance document version for journals ("Journals_V2.json"), on April 15, 2019

```
{"$schema":"http://json-schema.org/draft-04/schema#",
 "id":"http://jsonschema.net",
 "type":"object",
 "properties":
   {"journals":
       {"id": "http://jsonschema.net/journals",
        "type":"array",
        "items":
          {"type":"object",
           "properties":
              {"journal":
                 { "type":"object",
                   "properties":
                      { "ISSN":{"type":"string"},
                        "name":{"type":"string"},
                        "editor-in-chief":{"type":"string" },
                        "publisher":{"type":"string"},
                        "scopus-indexed":{"type":"boolean"},
                        "currentQuartile":{"type":"string"},
                        "periodicity":{"type" "string"}
                    },
                    "required":["name","editor-in-chief",
                                "periodicity"]
              } },
              "required":["journal"]
   } } },
   "required":["journals"]}
```

**Fig. 5.** The second conventional JSON schema version of journals ("JournalsSchema_V2.json"), defining the structure of "Journals_V2.json", on April 15, 2019

The necessary sequence of JSON schema change operations that have been implicitly generated by the system and executed (transparently to the user, on April 15, 2019) in order to create the new conventional JSON schema version (shown in Fig. 5) is listed in the following:

```
1.  AddSliceToTemporalJSONSchema(
    "JournalsTemporalSchema.json", conventionalJSONSchema,
    "JournalsSchema_V1.json", "JournalsSchema_V2.json");
2.  RenamePropertyInConventionalJSONSchema(
    "JournalsSchema_V2.json", "$..journal.properties.editor",
    "editor-in-chief");
3.  RenamePropertyInConventionalJSONSchema(
    "JournalsSchema_V2.json", "$..journal.properties.quartile",
    "currentQuartile");
4.  DropPropertyFromConventionalJSONSchema(
    "JournalsSchema_V2.json",
    "$..journal.properties.periodicity");
5.  AddPropertyToConventionalJSONSchema(
    "JournalsSchema_V2.json", "$..journal.properties", last,
    "periodicity", "string");
6.  AddPropertyToConventionalJSONSchema(
    "JournalsSchema_V2.json", "$..journal.properties.name",
    before, "ISSN", "string");
7.  AddPropertyToConventionalJSONSchema(
    "JournalsSchema_V2.json", "$..journal.properties.editor-in-
    chief", after, "publisher", "string");
8.  AddPropertyToConventionalJSONSchema(
    "JournalsSchema_V2.json", $..journal.properties.publisher",
    after, "scopus-indexed", "boolean");
```

## 5 Conclusion

In this paper, we have proposed an approach for managing implicit JSON schema versioning based on updates to JSON-based Big Data instance documents. Indeed, some specific JSON instance updates produce a new JSON instance document version that is no longer conformant to its JSON schema version. In a traditional environment, such an update would be rejected. However, in many cases, the document produced by the update has to be anyway accepted. Our approach allows accepting this updated JSON instance document by implicitly generating the necessary sequence of JSON schema change operations that must be executed on the old JSON schema version to produce a new schema version to which the new instance document version conforms.

Moreover, we have integrated this approach into our τJSchema framework in order to extend it for the support of implicit temporal JSON schema versioning driven by Big Data evolution. Notice that τJSchema was initially designed to only support schema versioning caused by schema change operations which are explicitly specified by the user.

We think that our proposal makes the management of JSON schema versioning more flexible and more user-friendly, as it allows performing JSON instance updates whose effects do not comply with the current JSON schema version. Furthermore, our approach helps defining, progressively and guided by instance updates, the correct JSON schema of some JSON instance document(s) whose exact structure is not fully known at its creation time.

To show the feasibility of our approach, we plan to implement it through an extension of our τJSchema-Manager tool, initially developed to only support explicit temporal JSON schema versioning, to also support implicit JSON schema changes triggered by JSON instance document updates.

Moreover, since in this work we have focused only on instance update operations acting on properties, we plan to extend the approach by also dealing with the update operations involving the other components of a JSON instance document not considered here, that is objects and arrays.

Last but not least, we also intend to address another aspect related to JSON instance updates triggering JSON schema changes. Indeed, we think that the user could have a requirement of adding, to the τJSchema-based NoSQL data store, an entire JSON instance document version, even created outside of the environment, as a new instance version of the current JSON schema version. The satisfaction of such a requirement could lead to the implicit creation of a new JSON schema version, if the new instance version is not valid with respect to the current schema version.

## References


1. Brahmia, Z., Grandi, F., Oliboni, B., Bouaziz, R.: Schema versioning. In: Khosrow-Pour, M. (ed.) Encyclopedia of Information Science and Technology, 3rd edn, pp. 7651–7661. IGI Global, Hershey (2015)
2. Brahmia, S., Brahmia, Z., Grandi, F., Bouaziz, R.: τJSchema: a framework for managing temporal JSON-based NoSQL databases. In: Proceedings of the 27th International Conference on Database and Expert Systems Applications (DEXA 2016), Part 2, Porto, Portugal, pp. 167–181 (2016)
3. Brahmia, S., Brahmia, Z., Grandi, F., Bouaziz, R.: Temporal JSON schema versioning in the τJSchema framework. J. Digit. Inf. Manag. **15**(4), 179–202 (2017)
4. Brahmia, Z., Grandi, F., Oliboni, B., Bouaziz, R.: Schema versioning in conventional and emerging databases. In: Khosrow-Pour, M. (ed.) Encyclopedia of Information Science and Technology, 4th edn, pp. 2054–2063. IGI Global, Hershey (2018)
5. Brahmia, S., Brahmia, Z., Grandi, F., Bouaziz, R.: Managing temporal and versioning aspects of JSON-based big data via the τJSchema framework. In: Proceedings of the International Conference on Big Data and Smart Digital Environment (ICBDSDE 2018). Studies in Big Data, Casablanca, Morocco, vol. 53, pp. 27–39. Springer (2018)
6. Corbellini, A., Mateos, C., Zunino, A., Godoy, D., Schiaffino, S.N.: Persisting big-data: the NoSQL landscape. Inf. Syst. **63**, 1–23 (2017)
7. Currim, F., Currim, S., Dyreson, C.E., Joshi, S., Snodgrass, R.T., Thomas, S.W., Roeder, E.: τXSchema: Support for Data- and Schema-Versioned XML Documents. Technical Report TR-91, TimeCenter (2009). http://timecenter.cs.aau.dk/TimeCenterPublications/TR-91.pdf

8. Cuzzocrea, A.: Temporal aspects of big data management: state-of-the-art analysis and future research directions. In: Proceedings of the 22nd International Symposium on Temporal Representation and Reasoning (TIME 2015), Kassel, Germany, pp. 180–185 (2015)
9. Davoudian, A., Chen, L., Liu, M.: A survey on NoSQL stores. ACM Comput. Surv. **51**(2), Article 40 (2018)
10. Dyreson, C.E., Snodgrass, R.T., Currim, F., Currim, S., Joshi, S.: Validating quicksand: schema versioning in τXSchema. In: Proceedings of the 22nd International Conference on Data Engineering Workshops (ICDE Workshops 2006), Atlanta, GA, USA, p. 82 (2006)
11. IETF (Internet Engineering Task Force): The JavaScript Object Notation (JSON) Data Interchange Format, Internet Standards Track document, December 2017 (2017). https://tools.ietf.org/html/rfc8259
12. IETF (Internet Engineering Task Force): JSON Schema: A Media Type for Describing JSON Documents, Internet-Draft, 19 March 2018 (2018). https://json-schema.org/latest/json-schema-core.html
13. Information Resources Management Association (IRMA): Big Data: Concepts, Methodologies, Tools, and Applications. IGI Global, Hershey (2016)
14. Khosla, P.K., Kaur, A.: Big data technologies. In: Mittal, M., Balas, V.E., Hemanth, D.J., Kumar, R. (eds.) Data Intensive Computing Applications for Big Data, pp. 28–55. IOS Press, Amsterdam (2018)
15. NoSQL Databases. www.nosql-database.org
16. Sharma, S., Tim, U.S., Gadia, S.K., Wong, J., Shandilya, R., Peddoju, S.K.: Classification and comparison of NoSQL big data models. Int. J. Big Data Intell. **2**(3), 201–221 (2015)
17. Snodgrass, R.T., Dyreson, C.E., Currim, F., Currim, S., Joshi, S.: Validating quicksand: schema versioning in τXSchema. Data Knowl. Eng. **65**(2), 223–242 (2008)

# A Novel Tool DSMOTE to Handel Imbalance Customer Churn Problem in Telecommunication Industry

Samaher Al_Janabi(✉) and Fatma Razaq

Department of Computer Science, Faculty of Science for Women (SCIW), University of Babylon, Babylon, Iraq
samaher@uobabylon.edu.iq, razaqfatima200@gmail.com

**Abstract.** This paper presents new methodology for handling imbalance data customer churn problem in telecommunication industry, it based on developed SMOTE algorithm, where traditional SMOTE algorithm has many problems. Therefore, this paper aiming to develop SMOTE algorithm to satisfy the law of prediction says "The values generated from any predictor is true if and only if the predictor building from real samples otherwise the prediction model can lead to false results if the predictor builds from virtual values". DSMOTE will be developed SMOTE using the principle of ten sampling instead of arbitrary percentage (e.g. 100%, 200%, etc.) to split the data. And the Quadratic Principle (e.g. $2^1$, $2^2$,... etc.) to detect the numbers of neighbors, instead of depending on the number of samples, The evaluation of DSMOTE is based on both error and correct rates. We get the best results when the total dataset split into 90% as training dataset and 10% as a testing dataset with a number of neighbors equal to $2^4$ based on both measures. By comparing between the results of the SMOTE and DSOMTE, the later showed better performance and need less time and space in implementation.

**Keywords:** SMOTE · Sampling · Churn customer · Quadratic · Imbalance problem

## 1 Introduction

Telecommunication industry is evolving rapidly with the passage of time, and they are facing serious revenue losses because of churning with the growing competition in the telecommunication market. Many telecommunication companies are discovering the cause of losing clients by enumerating their loyalty. Here, the word "churn" refers to a customer with a tendency of leaving the company and move to another competitor, for various reasons [1].

According to several reviews [2–6], customer churn is divided into three categories: Voluntary Churners, Non-voluntary Churners, and Silent Churners. Voluntary churners are those customers who want to quit the contract and move to another service provider. In such case, the situation can be considered in more details that are centered on numerous churn motives, such as the technology change, regulation, contract expiration, handset change, service quality, competition. Non-voluntary churners are customers


Y. Farhaoui (Ed.): BDNT 2019, LNNS 81, pp. 36–50, 2020.
https://doi.org/10.1007/978-3-030-23672-4_4

who have their service retract by their service provider. For instance, the service provider can choose to revoke the service with the customer for several reasons such as abuse of service and not disbursing the bill. Silent churners are those customers who discontinue the contract without any prior knowledge or notifications of either the company or the customer. Figure 1 shows these types of customer churn.

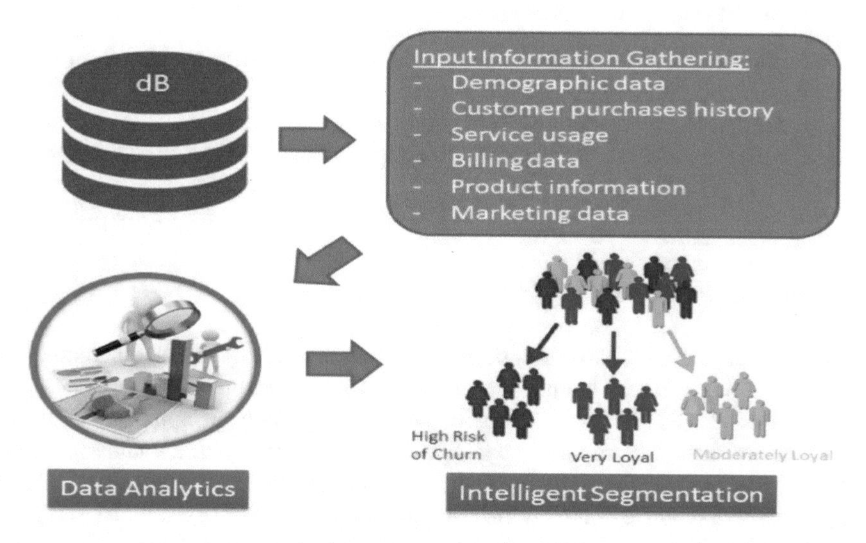


**Fig. 1.** Categories of customer churn

In the telecommunication sector, customers seek good quality service, competitive pricing, and value for money. However, when they do not find what they are expecting, they can easily terminate using the service and change from a service provider to another without restrictions [7]. This has led telecommunication companies to offer customers with great inducements in an attempt to encourage them to shift to their services [8].

The phenomenon related to customer abandonment is called customer churn, while the process of calculating the probability of future churning behaviors in the database based on the past prior behavior using a predictive model is usually called customer churn prediction [9]. Customer churn is typically computed as a relative number in percentage (i.e. the churn rate) and in telecommunication sector, the churn rates are often reported monthly, which might be misleading since some customers can leave the service at any time, whereas others are locked in longer contracts that are not due to renew on a monthly basis [10].

Imbalance class occurs when the number of instances from one class in a dataset is less than another class. This problem has significant influences on the classification accuracy in most data mining and ML techniques. In dataset of customer churn, the number of negative instances is higher than the positive instances which reflect imbalanced class distribution. Feature selection main goal is to select the most important features without changing the original data representation, and therefore, selecting a subset of the features which are relevant for the task to achieve the highest classification accuracy.

The main goal of this proposal is How can develop Synthetic Minority Over-Sampling Technique (SMOTE) be used to handle imbalance problem. The suggest methodology is divided into three stages:

- Collected dataset (e.g. collected dataset is dataset aggregation process from various sources of the Internet).
- joint the dataset of customer with calling database
- handle the imbalance problem by develop Synthetic Minority Over-Sampling Technique (DSMOTE).

## 2 Literature Survey

During the last decade, a worldwide range of studies has applied data mining and machine learning (ML) techniques for customer churn modeling.

Adnan et al. (2017) [11] The main problem is finding an approach to handle imbalance problem, the proposed approach to solve this problem, the under sampling method based on reduce majority instances, Orange dataset comprises of 50,000 sampling where only 3276 are churner sample but when PSO based under sampling, the dataset has balanced (churn and non-churn is equal in distribution) The results display the improve prediction performance when the dataset is handle based on PSO under sampling, and the improve sensitivity and AUC lead to the fact GP-AdaBoost is able to efficiently predict churners when PSO balanced training set is used. The similarity in handling imbalance but in different way from our work, the difference in this paper use PSO under sampling to handle imbalance and use for evaluation performance AUC (Sensitivity and Specificity).

Bing et al. 2017 [12] The main problem is imbalance problem, the many from solutions to handle imbalance problem, which can be nearly split into three groups: data-level, algorithm-level and ensemble solutions. he experimental results show that different types of solutions explained different behaviors, the empiricist results offer that the evaluation metric has a big effect on the execution of machines, the similarity in this paper with our work is handled imbalance but by state and art techniques while, and evaluation metrics (Expected Profit Measure (EMP) measure with the classical AUC and top-decile lift measure).

Vijayal et al. 2017 [13] The main problem is orange data-set contains imbalance ratio, The majority of the customers about 46,328 are non-churn customers and about 3672 customers are churn customers, also the approach for reduce size of data and to eliminate imbalance levels that utilizes Random under sampling, when the Random under sampling used, then the data set size was reduced to 7344 records, with number of records in each of the classes is equal, results: showing in three cases Performance on Orange-1000, Performance on Orange-5000 and Performance on Orange-7344 and found The performance of the Orange 7344 has better compared to the Orange 5000 and Orange 1000 because. The similarity in this paper with our work is handle imbalance problem, while Differently in this paper different in use the algorithm and set of evaluation metric.

# 3 Main Tools

## 3.1 Knowledge Discovery in Data Base (KDD)

Data mining is a synonym for the discovery of knowledge from the data, and it is widely used by the term. On the other hand, other people see that data mining is only an abstract step in the process of knowledge discovery [18]. Knowledge discovery consists of several steps:

- Data cleaning (to reduce noise and some inappropriate data) and Data integration (data collection from several different sources).
- Data selection (data retrieval is associated with the task of analysis of the database).
- Data conversion (the process of converting or merging data into appropriate mining models through aggregation).
- Data extraction (the most important stage where the data is extracted from different patterns), Evaluation of patterns to identify very important patterns that represent knowledge based on some common metrics.

Finally, knowledge is introduced (using visual and knowledge techniques to transfer information extracted in the previous steps to the user) [21, 22].

## 3.2 Data Mining

Data mining has received clear attention in all fields, especially in recent years due to the existence of very large amounts of data and the need to convert them into useful information or useful data. This useful information is used in maintaining customers, controlling science, and detecting fraud [21].

Data mining is the process of choosing, exploration, and modeling a large amount of data to unveil previously unheralded data patterns for business advantage. It also can be defined as the exploration and analyses of expansive amounts of information in order to distinguish significant designs. And it includes selecting, investigating, and modeling huge amounts of information to reveal already obscure designs, and at last comprehensible data, from huge databases pointing at solving commerce issues, data mining can be utilized to construct the taking after sorts of models, Classification clustering, Regression, Forecasting, Clustering Association run the show discovery, Arrangement design discovery, and deviation detection [21, 22].

## 3.3 Predictive or Perspective Data Mining

Prediction model refers to the task of building a show for the target variable as a work of the informative variable and the objective of these tasks is to predict the value of a particular attribute based on the values of other attributes [17]. The attribute to be predicated is commonly known as the target or dependent variable, while the attributes used for making the prediction are known as the explanatory or independent variables. There are two types of predicative modeling tasks:

- Classification which is utilized to watchful target variable.
- Regression which is utilized for nonstop target variable.

Prediction methods in data mining are broadly utilized to back optimizing of future decision-making in numerous diverse fields such as marketing, broadcast communications, healthcare, and Restorative Determination.

### 3.4 Imbalance Data

The problem of imbalance has an unfavorable impact on the standard learning of the classification. But most of them tend towards the larger class. Unbalance problem is a common problem in many applied data science and machine learning problems. A dataset is called imbalanced if it contains many more samples from one class than the other classes. When one class is represented by a small training data samples (i.e. minority class), and the other classes represent the remaining of the training data samples (i.e. the majority classes), this kind of datasets are considered as unbalanced dataset. Many solutions have been proposed to solve this problem, These solutions can be categorized into three sets data-level, algorithm-level, and gathering solutions. Data-level arrangements apply resampling as a preprocessing step to decrease the negative impact caused by un-balance problem. Algorithm level solutions aim to create modern algorithms or adjust existing ones to bias learning towards the minority class. Also there are two approaches to handle imbalance problem: under-sampling to remove some majority class. Over-sampling by add some minority class that has useful information. And hyper-sampling which is a mix of both over-sampling and under-sampling [12, 13].

The main concern is that the problem of data imbalance cannot be detected early. This problem varies from one company to another. Therefore, it is necessary to detect and determine the lack of imbalance problem and to find suitable methods for making the data balanced. For example, the classification in which the data is sufficiently trained in all classes. The main disadvantages of churn data are its high-volume size, and they hold the imbalance problem these disadvantages effect the accuracy of classifiers. The best solution to the former two problems is the PSO algorithm and its types (e.g. Simulated Annealing, and Feature Selection and Simulated Annealing) [13].

### 3.5 Churn Prediction

Churn prediction is giving estimations about customer are most likely to leave the organization in the close future, there are numerous distinctive reasons for the customer to churn like relocating, lake of service quality, sudden death, and unknown reasons [14].

Customer churn is the term utilized by a variety of commerce to denote the movement of a customer from a benefit provider to another. There are different reasons for this movement, such as benefit-cost quality, and loyalty to the benefit provider In hone, the cost of attracting an unused customer is e.g. 5 to 6 times higher than keeping a new loyal customer. As a result, companies are always Use another term. e.g. encouraged to reduce customer churn [14–16].

Customer churn prediction is important and has received more attention from a variety of businesses such as telecommunication (telecom) operators and as smartphone producers. Advances in these businesses such as modern services and technologies, as well as advances in the field of machine learning and data mining increased the competition in the market [14–16].

In previous years, Churn prediction attracted the attention of different researchers, since one of the main goals of data analysis is to provide a high-quality telecommunication services, in 2009, the ACM Conference on Knowledge Discovery and Data Mining (KDD) competed in predicting customers by using large sets of data from orange factors. Decision trees, neural networks, and K-mean were used to build a forecasting and customer division model. Based on the analysis of these techniques [22].

## 4 Proposed Method

Our goal of this paper is to handle the imbalance problem of customer churn in telecommunication companies. The block diagram that shows the proposed method explained in Fig. 2. We can summarize the steps of proposed method as follow:

- Collection dataset that described in Table 1.
- Join between the behavioral data (i.e. customer dataset), and the contract dataset (i.e. company dataset) as explained in pseudo #1.
- Split dataset into training and test dataset
- Apply DSMOT on training dataset as explained in pseudo #3. While pseudo #2 show the traditional SMOTE.,
- Evaluate the results based on two measures (i.e. error and correct rate) to determine:
  - Best number of neighbors.
  - Best split based on 10-cross validation
  - Suitable values to handle imbalance problem.

Feature selection, where, the main goal of this step is to reduce the search space and time, by using in-post processing stage.

The data given to us contains 3,333 observations extracted from a data warehouse. The dataset contains demographic as well as usage data of various customers.

In Table 1 shows that the database of contract and behavior contain 21 features.

In Table 2 shows Imbalance Rate based on different sampling split of original dataset. when train 90% and test 10% Continue to arrive train 10% and test 90%.

**Table 1.** Descriptive the dataset attributes

| No. | Dataset | Dataset desorption |
|---|---|---|
| 1 | Account Length | Its value is real. The period in which the account was active |
| 2 | Area Code | Categorical |
| 3 | Int'l Plan | The international plan is operational (YES, NO) |
| 4 | Vail Plan | Voice Mail plan is operational (YES, NO) |
| 5 | Vail Message | Number of voicemail messages |
| 6 | Day Mins | Overall minutes daily that utilize |
| 7 | Day Calls | Overall daily phoning |
| 8 | Day Charge | Overall day charge |
| 9 | Eve Mins | Overall evening minutes |
| 10 | Eve Calls | Overall evening calls |
| 11 | Eve Charge | Overall evening charge |
| 12 | Night Mins | Nighttime minutes |
| 13 | Night Calls | Overall nighttime phoning |
| 14 | Night Charge | Overall night-time |
| 15 | Intl Mins | The international utilized minutes |
| 16 | Intl Calls | Overall phoning internationally |
| 17 | Intl Charge | Overall cost of the international |
| 18 | Custer Calls | Number of customer service calls provided |
| 19 | Churn | Customer Churn (Target Variable yes = churn, no = not churned) |
| 20 | State | The place which that customer resides |
| 21 | Phone Number | Customer's telephone number |

**Table 2.** Imbalance rate based on different rate

| Sampling | Split dataset | | Imbalance rate |
|---|---|---|---|
| | Testing rate | Training rate | |
| Sample #1 | 90% | 10% | 0.131 |
| Sample #2 | 80% | 20% | 0.140 |
| Sample #3 | 70% | 30% | 0.134 |
| Sample #4 | 60% | 40% | 0.136 |
| Sample #5 | 50% | 50% | 0.12 |
| Sample #6 | 40% | 60% | 0.126 |
| Sample #7 | 30% | 70% | 0.130 |
| Sample #8 | 20% | 80% | 0.140 |
| Sample #9 | 10% | 90% | 0.135 |

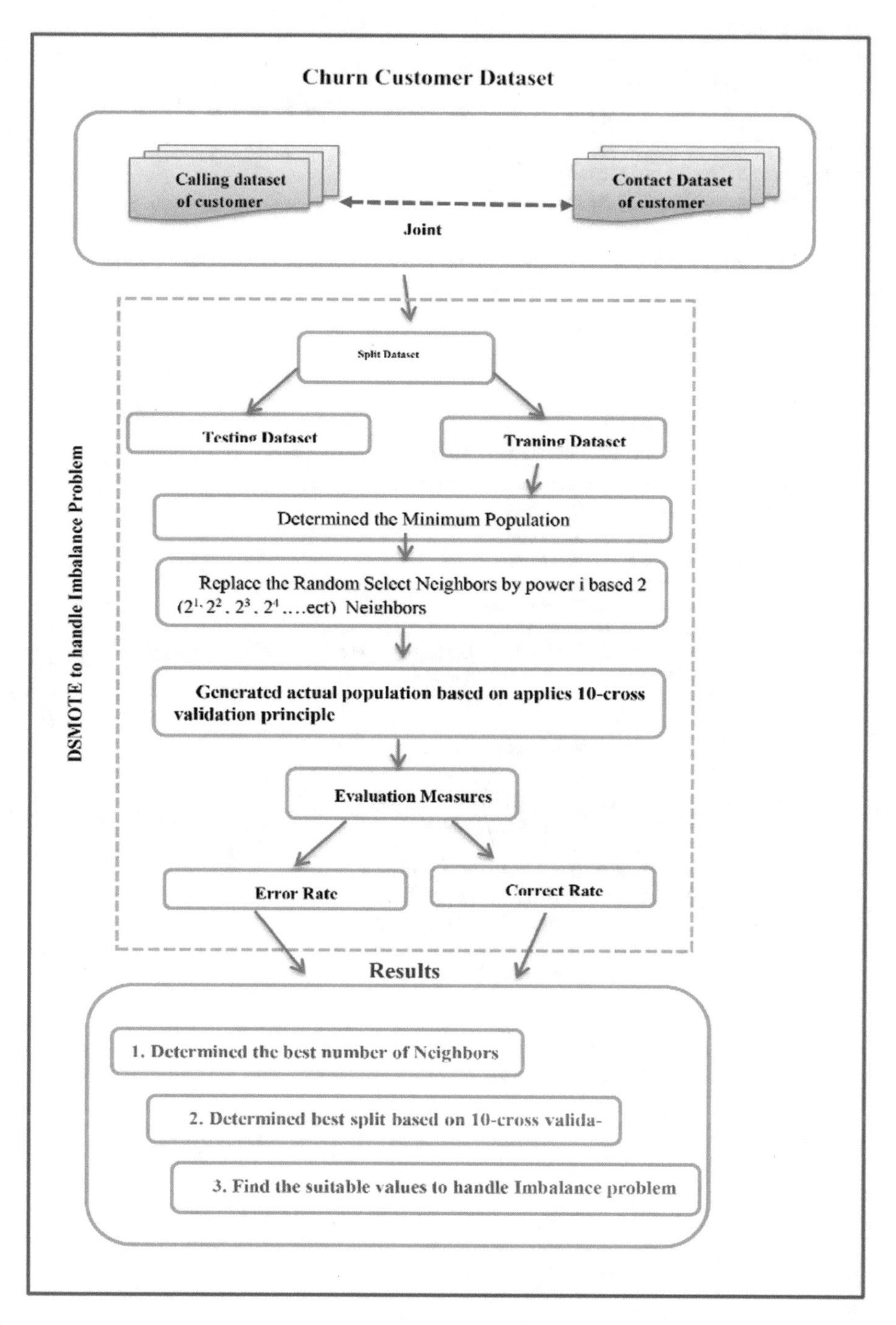


**Fig. 2.** Block diagram of DSMOTE

**Pseudo #1: Joiner**

***Input:*** *Two files from dataset*

***Output:*** *Joiner the Two File*

***Set:*** *F1:First File, F2:Scound File, F3:Thrid File, diff_index: array for keeping index of columns not repeated*

***Step1:*** *Read First file and Second File*

***Step2:*** *Compare Header Of Two Files*

***Step2.1:*** *Fetch Content Header Of Two File*

***Step 2.2:*** *Compare header length of first file with header length of second file*

- *If header_f1.length>=header_f2.length*
- *For int i=0 To header_f1.length*
- *For int j=0 To header_f2.length*
- *If header_f1[i]. equals header_f2[j]*
- *Bool=1*
- *If Bool==0*
- *Diff_index.add=i*

***Step 2.3:*** *repeat step2.2 but when header of second file is larger*

***Step 3:*** *Generated array of merge content the two file in single file*

***Step 3.1*** *: If content_f1.size()<content_f2.size()*

- *For int i=0 To content_f1.size()*
- *Line1=fetch of first file feature*
- *Line2=fetch of second file feature*
- *//from line2, some data is required*
- *For int j=0 To diff_index.size()*
- *line1[line1.length-1] = line1[line1.length-1].concat(ine2[diff_index.get(j)]);*
- *writer(line1);*
- *index=i*

***Step 3.2:*** *// now complete from the second file*

- *For int i=index To content_f2.size();*
- *For int j=0 To header_f1.length*
- *line_1[j]=" "*
- *from line2, some data is required*
- *for int k=0 To diff_index.size()*

***Step 3.3*** *: Joint the two array line 1 and line2 in line1*

- *line_1[line_1.length-1]=line_1[line_1.length-1].concat(","+line_2[diff_index.get(k)])*
- *Writer(line_1)*
- *// end if*

***Step 3.4*** *repeat step (3.1,3.2, and 3.3) but, If content_f2.size<content_f1.size*

*End*

## Pseudo #2: Traditional SMOTE

***Input***: *Collection of amount Databases related to customer and company(N1), neighbor's quadratics (K1), number of minatory class samples(T1).*

***Output***: *balance data for churn customer attribute by actual generated proportion,*

***Set:*** *NF: number of attributer, T1: Array for minatory class samples, New: counter for save number of S generated , S: Array for new synthetic samples .*

***Step 1: Generated Minority of Populate***

- *If N1 < 100% generated initialization randomization of minatory class samples (T1)*
- *T1=N1/100 *T1*
- *N1=100*

***Step2***: *Compute amount oversampling*

- *N1=(int)(N1/100)*

***Step3: compute*** *nearest neighbors:*

- *For i=1 To T1*
- ***compute*** *nearest neighbors for I, and save indices in NN*
- *end for*

*Step4:* ***generated minority of populate***

- *While N1 <> 0*

***Step4.1:*** *Selected random number between 1 and k*

- *N2=Rand ((*$Max_I$ *-* $Min_i$*) +* $Min_i$*)*

***Step4.2 :*** *build loop for number of attributer : For attrib=1 To NF*

- *Different=T[NN[nn]][attrib]-T[i[attrib]*
- *Gap=generated random number between [0,1]*
- *S[new][attrib]=T1[i][attrib]+Gap*different.*
- *End for*
- *New++*
- *N1=N1-1*
- *End while*

*End*

## Pseudo #3: DSMOTE

***Input:*** *N1:List represent collection databases to customer and company, K1:List represent of neighbors quadratics , T1:List of number of minatory class samples*

***Output:*** *S[new][attrib]: array of new synthetic*

***Set:*** *NF: number of attributer, T1:Array for minatory class samples, New: counter for save number of S generated ,S: Array for new synthetic samples, CT : Array contains cases training data set based on 10_cross validation*

***Step 1:*** *Split the selected database into Training and Testing Dataset using 10-Fold Cross*

*Validation techniques for Generated Minority Of Populate.*

- *Tindex=0*
- *For L=1 To 9*
- *For J=0 To CT. Length*
- *If CT. Length <>0 && CT[J]<T1*
- *T1=TC[J]*
- *Tindex=J*

***Step2**: compute amount oversampling*
- *N1=(int)(N1/100).*

***Step3: compute** neighbors based on quadratics*
- *$[2^0,....2^i]$,i=[1,2,3,4,5,6].*

***Step4 :** generated new synthetic samples*
- *While N1 <> 0*

***Step4.1:** Selected random number of neighbors quadratics .*
- *N2=Rand(Max $_I$ --- $Min_i$) + $Min_i$*
- *$NN=2^{N2}$*

***Step4.2 :** build loop for number of attributer :*
- *For attrib=1 To NF*
- *Different=T[NN[nn]][attrib]-T[i[attrib]*
- *Gap=generated random number between[0,1]*
- *S[new][attrib]=T1[][attrib]+Gap*different.*
- *New++*
- *End while*
- *End for*

*End*

# 5 Results

In this section, we will explain the total imbalance rate: Total imbalance Rate = Number Minatory Samples\Total number of Samples in Database

Total Imbalance Rate = 484/3334

Total Imbalance Rate = 0.145

Also, we will explain the results of two approaches to handle imbalance of the dataset (traditional SMOTE and Developed SMOTE).

## 5.1 Traditional SMOTE

When we choose neighbors randomly (K1 = 1, N1 = 100%, K1 = 2 when N1 = 200%. And K1 = 3 when N1 = 300%). When the algorithm is evaluated based on Error and Correct Rate, we will obtain these results (Table 3).

$$\text{Error Rate} = \text{False Positive (FP)} + \text{False Negative (FN)}/\text{Total}. \quad (1)$$

$$\text{Correct rate} = 1 - \text{Error Rate} \quad (2)$$

**Table 3.** Error Rate SMOTE based on neighbors randomly (K1 = 1 N1 = 100%, K1 = 2 when N1 = 200% and K1 = 3 when N1 = 300%)

| # Neighbors | Sample #1 | Sample #2 | Sample #3 | Sample #4 | Sample #5 | Sample #6 | Sample #7 | Sample #8 | Sample #9 |
|---|---|---|---|---|---|---|---|---|---|
| 1 | 0.191 | 0.212 | 0.192 | 0.182 | ***0.172*** | 0.222 | 0.223 | 0.204 | 0.230 |
| 2 | 0.194 | 0.196 | 0.18 | 0.174 | ***0.172*** | 0.224 | 0.228 | 0.200 | 0.220 |
| 3 | 0.182 | 0.188 | 0.179 | 0.169 | ***0.169*** | 0.222 | 0.212 | 0.203 | 0.222 |
| 4 | 0.179 | 0.196 | 0.182 | ***0.168*** | 0.169 | 0.227 | 0.215 | 0.195 | 0.228 |
| 5 | ***0.170*** | 0.200 | 0.183 | 0.172 | 0.169 | 0.224 | 0.231 | 0.187 | 0.237 |

The above table explains the error rate of SMOTE based on choosing different values of neighbors and different split of dataset using sampling principle. The best value of error rate is (0.168), The database is divided into 60% for training and 40% for testing and number of neighbors is 4 (Table 4).

The above table explains the correct rate of smote based on choose different values of neighbors and different split of dataset using sampling principle. The best value of correct rate is (0.832). The database is divided into 60% for train and 40% for test and number of neighbors is 4.

**Table 4.** Correct Rate of SMOTE based on neighbors randomly (K1 = 1 N1 = 100%, k = 2 when N1 = 200% and K1 = 3 when N1 = 300%)

| # Neighbors | Sample #1 | Sample #2 | Sample #3 | Sample #4 | *Sample #5* | Sample #6 | Sample #7 | Sample #8 | Sample #9 |
|---|---|---|---|---|---|---|---|---|---|
| 1 | 0.808 | 0.787 | 0.808 | 0.817 | ***0.827*** | 0.777 | 0.776 | 0.795 | 0.769 |
| 2 | 0.805 | 0.803 | 0.82 | 0.825 | ***0.827*** | 0.775 | 0.771 | 0.799 | 0.779 |
| 3 | 0.817 | 0.811 | 0.821 | 0.830 | ***0.830*** | 0.777 | 0.787 | 0.796 | 0.777 |
| 4 | 0.820 | 0.803 | 0.818 | ***0.832*** | 0.830 | 0.772 | 0.784 | 0.804 | 0.771 |
| 5 | ***0.830*** | 0.799 | 0.817 | 0.827 | 0.830 | 0.775 | 0.768 | 0.812 | 0.762 |

## 5.2 DSMOTE

**A. Evaluation the Results of DSMOTE Based on Error Rate**

In this case study, DSMOTE used the 10-fold Cross Validation [21] and estimation the number of neighbors based on Quadratic principle, we will evaluate the results of DSMOTE based on two measures (i.e. Error Rate, Correct Rate and Correct [19, 20]) (Table 5).

**Table 5.** Error Rate of DSMOTE based on the 10-fold Cross Validation and quadratic neighbors

| # Neighbors | Sample #1 | *Sample #2* | Sample #3 | Sample #4 | Sample #5 | Sample #6 | Sample #7 | Sample #8 | Sample #9 |
|---|---|---|---|---|---|---|---|---|---|
| $2^1$ | 0.143 | ***0.128*** | 0.149 | 0.153 | 0.144 | 0.148 | 0.151 | 0.151 | 0.171 |
| $2^2$ | ***0.122*** | 0.151 | 0.164 | 0.172 | 0.145 | 0.16 | 0.154 | 0.158 | 0.195 |
| $2^3$ | ***0.140*** | 0.182 | 0.168 | 0.167 | 0.169 | 0.161 | 0.152 | 0.181 | 0.178 |
| $2^4$ | 0.188 | 0.173 | 0.168 | 0.173 | 0.167 | 0.147 | 0.161 | 0.152 | ***0.143*** |
| $2^5$ | 0.155 | 0.176 | 0.168 | 0.158 | 0.168 | ***0.153*** | 0.169 | 0.178 | 0.159 |

The above table explains the error rate of smote based on choosing different values of quadratic neighbors and different split of dataset using nine samples. The best value of error rate is (0.122). The database is divided into 90%for train and 10% for test and number of neighbors is 22.

**B. Evaluation the Results of DSMOTE Based on a Correct Rate Measure**

In this case, DSMOTE uses the 10-Fold Cross Validation. And uses the Quadratic Principle to estimate the number of neighbors (Table 6).

**Table 6.** Correct Rate of DSMOTE based on the 10-fold Cross Validation and quadric neighbors

| # Neighbors | Sample #1 | Sample #2 | Sample #3 | Sample #4 | Sample #5 | Sample #6 | Sample #7 | Sample #8 | Sample #9 |
|---|---|---|---|---|---|---|---|---|---|
| $2^1$ | 0.856 | ***0.871*** | 0.851 | 0.846 | 0.855 | 0.851 | 0.848 | 0.848 | 0.828 |
| $2^2$ | ***0.878*** | 0.848 | 0.836 | 0.827 | 0.854 | 0.839 | 0.845 | 0.841 | 0.804 |
| $2^3$ | ***0.859*** | 0.817 | 0.832 | 0.832 | 0.830 | 0.839 | 0.847 | 0.818 | 0.821 |
| $2^4$ | 0.811 | 0.826 | 0.832 | 0.826 | 0.832 | 0.852 | 0.838 | 0.847 | ***0.856*** |
| $2^5$ | 0.844 | 0.823 | 0.832 | 0.841 | 0.831 | ***0.846*** | 0.830 | 0.821 | 0.840 |

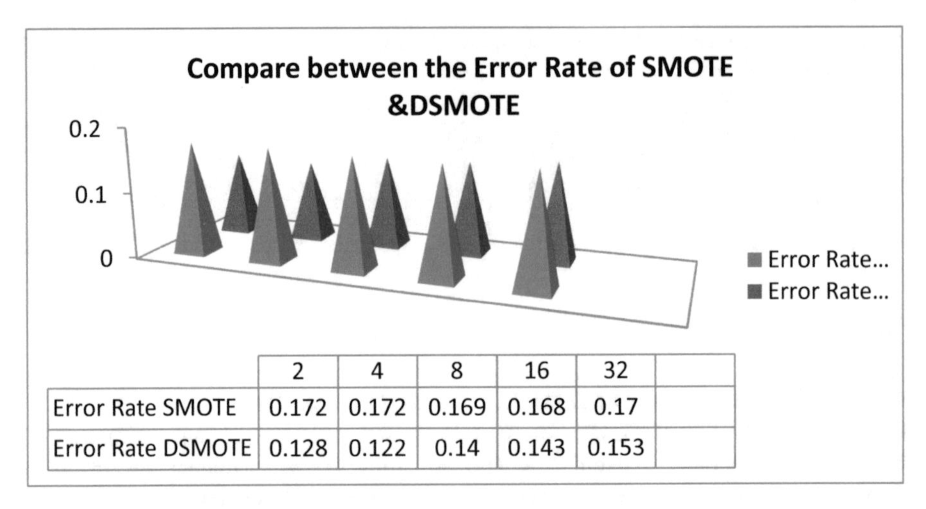


**Fig. 3.** Compare between the Error Rate of SMOTE & DSMOTE for best value for each test

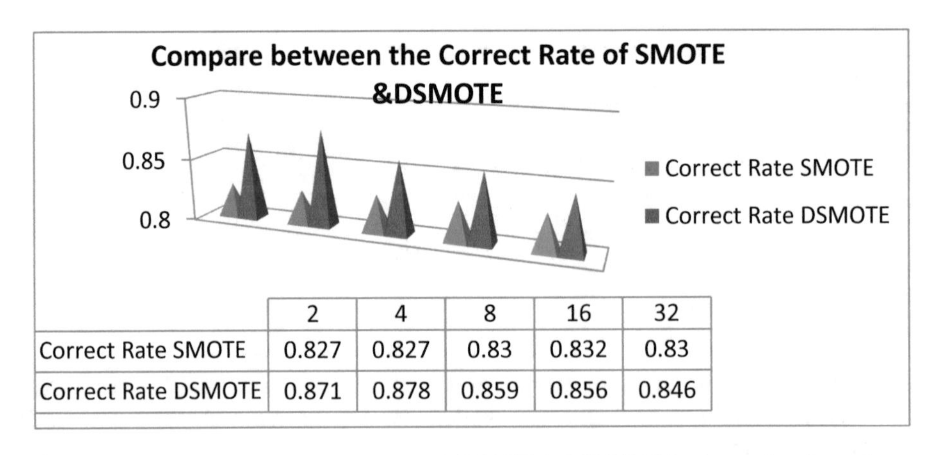


**Fig. 4.** Compare between the correct rate of SMOTE & DSMOTE for best value for each test

The above table explains the correct rate of DSMOTE based on choosing different values of quadratic neighbors and different split of dataset using nine samples. The best value of correct rate is (0.878). The database is divided into 90%for train and 10% for test and number of neighbors is 22.

## 6 Conclusion

In this paper, we used two principles to develop SMOTE (i.e., the 10-Fold Cross and Quadratic Principle); the main purpose of *the 10-Fold Cross Principle* to generated the truth values by sampling data rather than using duplication of minority sampling based on Traditional SMOTE that lead to generated virtual sampling. And *the Quadratic Principle* to determined number of neighbors, because the Traditional SMOTE need to duplication sampling 1600% to take number of neighbors = 16 and this lead to generation big virtual population, full memory quickly and can't use it's results in the any post-processing method such as "classification, prediction or optimization". Therefore, this point is considered one of the main challenges that handle by proposed method (DSMOTE).

As a result, DSMOTE technique is a powerful tool for solving imbalance problems compare with SMOTE as explained in Figs. 3 and 4 based on two measures correct and error rate. From experiments analysis of this technique, we observed the following: it's held the advantage among SMOTE since it's not randomization. Therefore, techniques with mathematical basis like DSMOTE were more powerful and faster. We can say DSMOTE give an optimal solution because it utilized features of mathematics such as linear combination, simplification, derivatives, and integration. Moreover, from the analysis, we found similar parameters shared between the imbalances handle techniques like the target variable and interest variable and specified them as a higher priority. In general, this paper considers as the first step for any prediction techniques that can be used in the next post-processing stage. Because any prediction model can lead to true prediction if it is based on true samples such as DSMOTE, but it leads to false prediction if the prediction construction is from virtual samples such as (SMOTE).

## References

1. Sharma, A., Panigrahi, D., Kumar, P.: A neural network based approach for predicting customer churn in cellular network services. arXiv preprint arXiv:1309.3945 (2013)
2. Yu-Teng, C.: Measuring the impact of data mining on churn management. Internet Res. Electron. Netw. Appl. Policy **11**(5), 375–387 (2015)
3. Hashmi, N., Butt, N.A., Iqbal, M.: Customer churn prediction in telecommunication a decade review and classification. Int. J. Comput. Sci. **10**(5), 271–282 (2013)
4. Saradhi, V.V., Palshikar, G.K.: Employee churn prediction. Expert Syst. Appl. **38**(3), 1999–2006 (2011)
5. AlOmari, D., Hassan, M.M.: Predicting telecommunication customer churn using data mining techniques. In: International Conference on Internet and Distributed Computing Systems, pp. 167–178. Springer, September 2016

6. Coussement, K., Van den Poel, D.: Improving customer attrition prediction by integrating emotions from client/company interaction emails and evaluating multiple classifiers. Expert Syst. Appl. **36**, 6127–6134 (2013)
7. Owczarczuk, M.: Churn models for prepaid customers in the cellular telecommunication industry using large data marts. Expert Syst. Appl. **37**(6), 4710–4712 (2010)
8. Rygielski, C., Wang, J.C., Yen, D.C.: Data mining techniques for customer relationship management. Technol. Soc. **24**(4), 483–502 (2002)
9. Coussement, K., De Bock, K.W.: Customer churn prediction in the online gambling industry: the beneficial effect of ensemble learning. J. Bus. Res. **66**(9), 1629–1636 (2013)
10. Statista: Average Monthly Churn Rate for Wireless Carriers in the United States from 1st Quarter 2013 to 1st Quarter 2016 (2016). https://www.statista.com/statistics/283511/average-monthly-churn-rate-top-wireless-carriers-us/. Accessed 14 Oct 2017
11. Idris, A., Iftikhar, A., ur Rehman, Z.: Intelligent churn prediction for telecommunication using GP-AdaBoost learning and PSO undersampling. Clust. Comput., 1–15 (2017)
12. Zhu, B., Baesens, B., vanden Broucke, S.K.L.M.: An empirical comparison of techniques for the class imbalance problem in churn prediction. J. Inf. Sci. **408**, 84–99 (2017)
13. Vijaya, J., Sivasankar, E.: An efficient system for customer churn prediction through particle swarm optimization based feature selection model with simulated annealing. Clust. Comput. (2017)
14. Sousse: pp. 951–961 (2012). https://doi.org/10.1109/setit.2012.6482042
15. Wang, H.F.: Intelligent Data Analysis: Developing New Methodologies through Pattern Discovery and Recovery, Developing New Methodologies through Pattern Discovery and Recovery. IGI Global (2009)
16. María, O., Cristián, B., Wouter, V., Carlos, S., Bart, B., Jan, V.: Social network analytics for churn prediction in telco: model building, evaluation and network architecture. Expert Syst. Appl. **85**, 204–220 (2017)
17. Al_Janabi, S., Mahdi, M.A.: Evaluation prediction techniques to achievement an optimal biomedical analysis. Int. J. Grid Util. Comput. (2019)
18. Al-Janabi, S., Alkaim, A.F.: A nifty collaborative analysis to predicting a novel tool (DRFLLS) for missing values estimation. J. Soft Comput. (2019). https://doi.org/10.1007/s00500-019-03972-x
19. Al-Janabi, S., Al-Shourbaji, I.: A smart and effective method for digital video compression. 2016 7th International Conference on Sciences of Electronics, Technologies of Information and Telecommunications (SETIT), Hammamet, pp. 532–538 (2016). https://doi.org/10.1109/setit.2016.7939927
20. Al-Janabi, S., Al-Shourbaji, I.: A hybrid Image steganography method based on genetic algorithm. In: 2016 7th International Conference on Sciences of Electronics, Technologies of Information and Telecommunications (SETIT), Hammamet, pp. 398–404 (2016). https://doi.org/10.1109/setit.2016.7939903
21. Ali, S.H.: Miner for OACCR: case of medical data analysis in knowledge discovery. In: IEEE, 2012 6th International Conference on Sciences of Electronics, Technologies of Information and Telecommunications (SETIT), Sousse, pp. 962–975 (2012). https://doi.org/10.1109/setit.2012.6482043
22. Ali, S.H.: A novel tool (FP-KC) for handle the three main dimensions reduction and association rule mining. In: IEEE, 2012 6th International Conference on Sciences of Electronics, Technologies of Information and Telecommunications (SETIT), Sousse, pp. 951–961 (2012). https://doi.org/10.1109/setit.2012.6482042

# Quality and Reliability Data Fusion for Improving Decision Making by Means of Influence Diagram: Case Study

Abdelaziz Lakehal[1(✉)], Tarek Khoualdia[1,2], and Zoubir Chelli[3]

[1] Department of Mechanical Engineering, Mohamed Chérif Messaadia University, P.O. Box 1553, 41000 Souk-Ahras, Algeria
lakehal21@yahoo.fr

[2] LGMREIU Laboratory, Mohamed-Cherif Messaadia University, Souk Ahras, Algeria

[3] Departement of Electrical Engineering, Mohamed Chérif Messaadia University, P.O. Box 1553, 41000 Souk-Ahras, Algeria

**Abstract.** The article focuses on the decision making in fault diagnosis of lubrication systems. In these systems, the diagnosis covers, in the majority of cases, the mechanical reliability or analysis of lubricating oils, but in a separate manner. In this section, the mechanical reliability is considered in combination with the lubricant quality, but the diagnosis process is always infected by uncertainties. Bayesian network (BN) model is developed and used as a decision-making tool. From this one, it is possible to quantify the probability of failure of this system. The diagnosis of failures is based on using Fault-Tree (FT) and Bayesian Network (BN). Firstly, a conversion from FT to BN is presented to establish a quick and accurate diagnosis. Secondly, the diagnosis is optimized by means of Influence Diagram (ID) which measures the preference.

**Keywords:** Influence diagram · Fault tree · Bayesian network · Lubrication system · Decision making

## 1 Introduction

Fault diagnosis, safety, maintenance, and dependability are the main areas of research where BNs have made a strong contribution [1]. Diagnosis always ends with decision making, which is not easy. In the situations where information is incomplete and uncertainty exists, it is possible to call on BNs to facilitating diagnosis and decision-making. Other techniques may be also used in fault diagnosis alongside BNs such as fault tree, fuzzy logic, and decision tree [2]. Hybridization between these techniques gives promising results and facilitates the decision-making.

Bayesian modeling for reliability studies has also been widely exposed in recent years [3]. Whatever the system, static or dynamic, BN can model their reliability using dynamic Bayesian network. Several studies have been carried out whose objective and field are varied. The rehabilitation of water networks, the evaluation of the availability of electrical power, and the optimization of emergency plan in gas network operation


Y. Farhaoui (Ed.): BDNT 2019, LNNS 81, pp. 51–58, 2020.
https://doi.org/10.1007/978-3-030-23672-4_5

[4–6] are some applications of reliability modeling by dynamic Bayesian network that link rehabilitation and reliability, availability and reliability, and safety and reliability.

Beside the dynamic Bayesian networks exists other extensions of the BN. Object-oriented Bayesian networks (OOBN) and influence diagrams (ID) are BNs developed to address issues related mainly to system complexity and decision making. This work exposes a new application of the BN to alleviate decision-making in fault diagnosis. ID that measure the modeler's preferences will be used to simulate the consequences of the taken decisions. The application presented in this paper is new, as it deals with the relationship between quality and reliability in the same model. The reliability of a lubrication system will be presented in a case study alongside the quality of the oil. This type of system is found in the most industrial systems.

## 2 Research Tools

### 2.1 Fault Tree

FT is a method of dependability. It is widely used to determine the combinations of scenarios that could lead to a dreaded event. In the quantitative analysis by FT, variables are considered as independent. The FT allows a qualitative and quantitative analysis of faults. The top event represents the dreaded event. The logic gates "AND" and "OR" are the basis of the connection between the studied events.

### 2.2 Bayesian Networks

BNs have made a strong contribution to solving various problems related to fault diagnosis and prediction [7]. Uncertainty, lack of information or incomplete information; are the main addressed problematic. Causality and interdependencies between variables have also been modeled by BNs. The structure of the BN is simple and intuitive. Their logic is easily understandable by a non-specialist in modeling. Bayesian reasoning in fault diagnosis allows the definition of the fault probabilities of a complex system while defining the most fragile elements leading to failure and the most probable causes of this failure.

The parameters of a BN are given as probabilities. They are calculated from the Bayes theorem, and the relation between causes and effect is given by the following formula:

$$P(V_t/V_{t-1}) = \prod_{i=1}^{N} P\left(V_{i,t}/C(V_i)_{i,t}\right) \tag{1}$$

Where C ($V_i$) is the set of parents (or causes) of $V_i$ in DAG (Fig. 1).

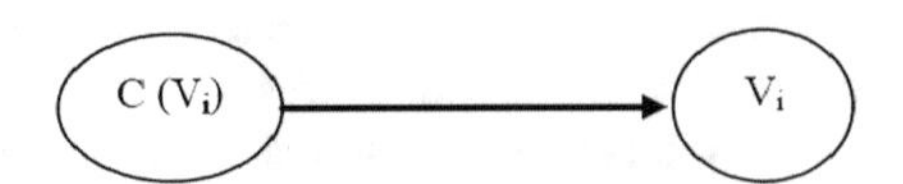


**Fig. 1.** Basic BN

### 2.3 Influence Diagrams

IDs are efficient modeling tools for representation and analysis of decision-making under uncertainty. IDs provide a natural representation for decision-making with minimal uncertainty and confusion to the decision-maker. The solution to a decision problem consists in determining an optimal strategy that maximizes expected utility for the decision-maker and calculates the maximum expected utility to adhere to this strategy [8].

## 3 Bayesian Modeling

### 3.1 Description of the Installation

The illustrated installation in Fig. 2, consists of a turbo-compressor comprising three cascaded machines: a steam turbine coupled to a compressor, itself connected to a gas turbine. The oil system is very complex since it does not only feed bearings and other machine lubrication points and its accessories, but also regulators and positioners. Mainly, lubrication is provided by a principal oil pump P1 driven by a steam turbine T. However, there exist two other auxiliary oil pumps: P2 driven by the three-phase motor M1, and P3 driven by a direct current electrical motor M2. All three pumps are designed to ensure redundancy; if the turbo-pump TP1 fails the motor-pump M1P2 starts automatically, and if this last fails lubrication is provided by M2P3.

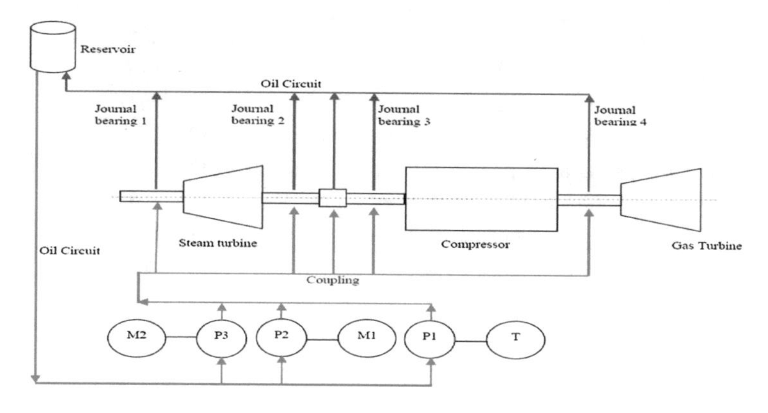


**Fig. 2.** The installation scheme

### 3.2 Fault Tree Analysis

In this case-study, unavailability of lubrication is taken as the dreaded event (Fig. 3). In practice, in order to build the FT, the necessary and sufficient causes leading to unavailability of lubrication are needed. The solution, given in this contribution, allows identifying bad oil quality and system failure as intermediate events, which will constitute the first level of the FT. Following the same methodology, the second level will be obtained.

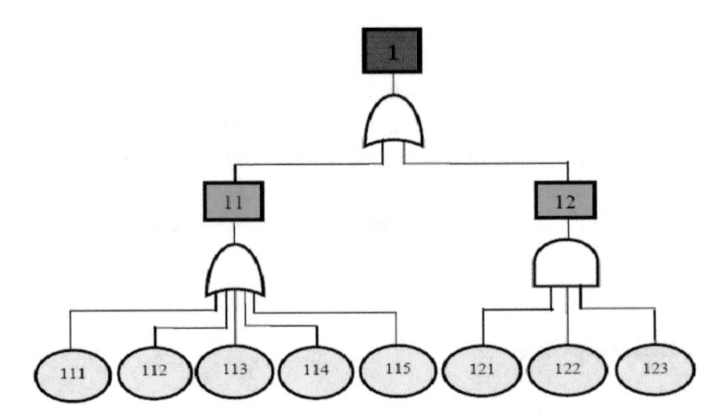


**Fig. 3.** Analysis of unavailability of lubrication system by FT

The causes of the unavailability of lubrication are either intrinsic or extrinsic. Intrinsic related to the components reliability of the lubrication system, or/and extrinsic related to the oil quality (Fig. 3). The analysis conducted on 45 samples as well as the judgments made by the experts' operators have allowed defining the occurrence probabilities of the basic events (see Table 1).

**Table 1.** The basic events probabilities

| N° | Event | Code | Probability |
|---|---|---|---|
| 01 | **Unavailability Lubrication** | **1** | **OR gate** |
| 02 | **Bad quality of oil** | **11** | **OR gate** |
| 03 | Passing the limit value of the viscosity | 111 | 0.0222 |
| 04 | Passing the limit value of the density | 112 | 0 |
| 05 | Passing the limit value of the water content | 113 | 0.8667 |
| 06 | Passing the limit value of the acidity | 114 | 0.0444 |
| 07 | Passing the limit value of the flash point | 115 | 0 |
| 08 | **Faulty lubrication system** | **12** | **AND gate** |
| 09 | TP1 does not operate | 121 | 0.06 |
| 10 | M1P2 does not operate | 122 | 0.03 |
| 11 | M2P3 does not operate | 123 | 0.02 |

### 3.3 Modeling by Bayesian Network

From Fig. 3 a conversion of the FT into BN is carried out. The result of Fig. 4 is generated by Netica software [9]. Inference in this BN gives a probability of lubrication system unavailability witch equal to that calculated by the FT, i.e. P(1) = 0.8753. By analyzing the probabilities of events (11) and (12), we conclude that the unavailability of the lubrication system is mainly due to the bad quality of the oil, itself mainly due to a high water content.

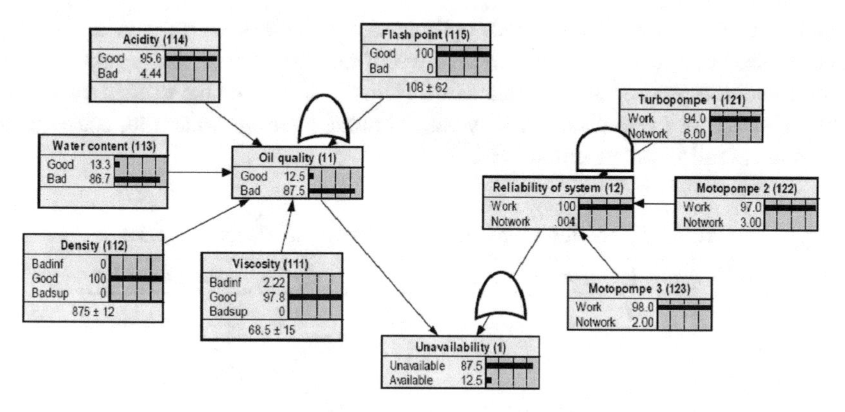


**Fig. 4.** Modeling by Bayesian network.

# 4 Decision-Making by Using Influence Diagram

In an FT, from the review of the probabilities of intermediate events leading to the top event, it is possible to prioritize system changes by identifying the most likely causes of an adverse event. Reducing the probability of this leading event can be considered by removing or reducing the occurrence probability of basic events and/or improving the reliability of the system by adding an "AND" gate between the top event and basic events.

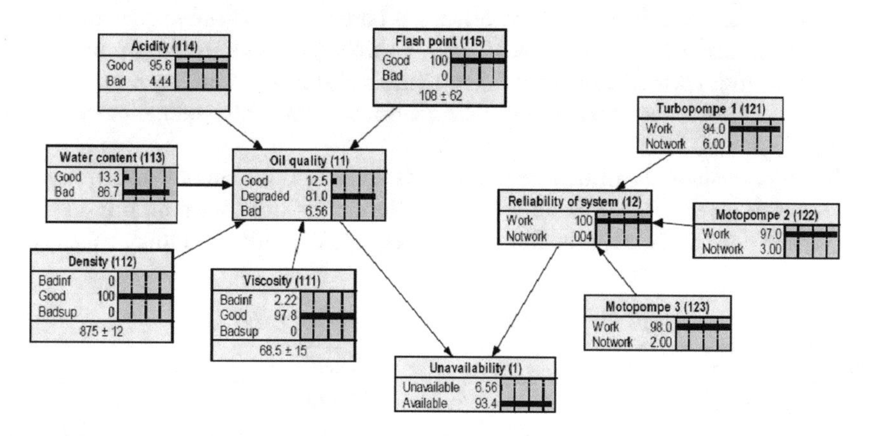


**Fig. 5.** BN with three states of the "quality" variable

By means of the FT of Fig. 3 and the BN of Fig. 4, we can easily notice that a corrective action on oil quality significantly reduces the occurrence probability of the dreaded event (1). Suppressing the event (113) is technically possible (by centrifugation) without oil change. On the contrary, suppressing events (111) and (114) is possible only by changing oil. Through this analysis we note that an aspect and an economic utility are required, which optimization coincides with the prioritization of changes presented in Table 2. But in the field of maintenance of vital equipment, such as the turbo-compressor of our installation (Fig. 2), safety is also needed as a second utility. A decrease in oil viscosity may cause journal bearings wear and, consequently, serious degradation of the equipment.

**Table 2.** The top event probability based on corrective actions

| N° | Suppression of the basic event | P(1) Top event probability |
|---|---|---|
| 1 | P(113) = 0 | P(1) = 0.066 |
| 2 | P(114) = 0 | P(1) = 0.870 |
| 3 | P(111) = 0 | P(1) = 0.873 |
| 4 | P(113) = 0 et P(114) = 0 | P(1) = 0.022 |
| 5 | P(113) = 0 et P(111) = 0 | P(1) = 0.044 |
| 6 | P(114) = 0 et P(111) = 0 | P(1) = 0.867 |
| 7 | P(113) = 0 et P(114) = 0 et P(111) = 0 | P(1) = 0.001 |

From this, it is clear that the diagnosis process does not only involve an analysis of the different failure scenarios and a reduction of the probability of the dreaded event, but also a decision-making related to the hierarchization of corrective actions priorities while respecting the utility function. It should be noticed that this last task is not achievable with a FT.

Preference is a key element common to all decision-making problems, without which no decision can be made. Very often, it is based on an objective quantity, such as material use, performance, or financial gain. However, decision-making problems can involve quantities which have no evident numerical measure, such as health or client satisfaction. Another complication is a set of attributes probably contradictory, such as price and quality.

In this case study, there are two preferences: the first is the suppression of the event (113), which gives P (1) = 0.066 with a significant risk due to the drop in viscosity. The second one is the suppression of the event (111), which minimizes the risk of

**Table 3.** Input data and results of the ID of Fig. 5

| Decision | Centrifugation (True) | | Centrifugation (False) | |
|---|---|---|---|---|
| Event | P(113) True = 0.867 | P(113) False = 0.133 | P(113) True = 0.867 | P(113) False = 0.133 |
| Utility | 0 | −1000 euro | 1000 euro | 0 |
| Results | −133.3 euro | | 866.7 euro | |
| Decision | Oil Change (True) | | Oil Change (False) | |
| Event | P(111) True = 0.022 | P(111) False = 0.978 | P(111) True = 0.022 | P(111) False = 0.978 |
| Utility | 24000 euro | 10714 euro | −24000 euro | 0 |
| Results | 11008.9 euro | | −532.8 euro | |

journal bearings wear with a high occurrence probability P(1) = 0.873. A measure of desirability is quantified by the utility represented, in this case, by the cost of the consequence. Input data and results are shown in Table 3, and the ID of Fig. 6 supports the decision-making.

The total cost price includes 24,000 euro for the journal bearings, 10,714 euro for the quantity of oil ensuring the lubrication and operation of the installation, and 1000 euro for centrifuge operation. Results related to the centrifugation are quite contradictory in terms of benefit, because the operation of the system is possible even in the presence of water in oil, but this can cause failure over time. However, from results of Table 3 it is clear that oil change provides a significant gain in the case of a drop in viscosity. The direction of the arc connecting the two decision nodes of Fig. 6 indicates the decision-making order or preference, which is not feasible with a FT. From this figure the priority in the decision-making process is given to oil change.

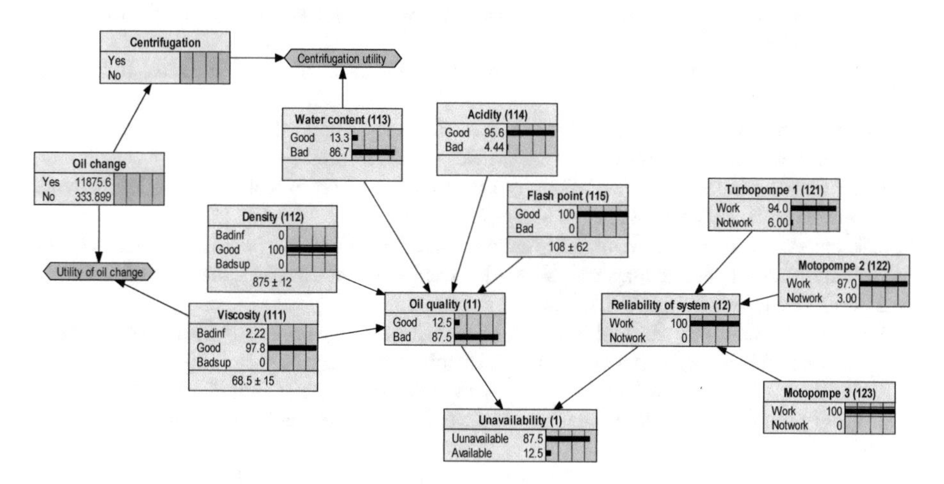


**Fig. 6.** Decisions modelling by ID.

This decision-making is always possible by using a FT to reduce the probability of occurrence of the dreaded event, but it does not model the risk that may occur as a result of this decision. It is also important to note that the conversion of a BN into FT is not possible.

## 5 Conclusion

This paper discusses and demonstrates a new decision making tool in maintenance that passes through two steps: firstly, a FT is constructed, and secondly a conversion of this tree into a BN is carried out. BN presented here in its hybrid form eases the representation of different oil analyzes used in preventive maintenance. Also and through the conditional probability table, it allows the addition of a state and even a variable with high flexibly. The BN, shown in the case study, has allowed analyzing the causes of unavailability of the lubrication system of a strategic machine whose shutdown causes significant economic losses. Marginalizing oil quality evaluation and concentrating the maintenance activities on improving the reliability of the system components increases the unavailability probability. Also, from the ID presented in Sect. 4 allowed prioritizing different decisions based on the associated utilities, and determining the consequences of each decision.

**Acknowledgment.** The authors like to thank the Algerian general direction of research (DGRSDT) for their financial support.

## References

1. Weber, P., Medina-Oliva, G., Simon, C., Iung, B.: Overview on Bayesian networks applications for dependability, risk analysis and maintenance areas. Eng. Appl. Artif. Intell. **25**, 671–682 (2012)
2. Cai, B., Huang, L., Xie, M.: Bayesian networks in fault diagnosis. IEEE Trans. Industr. Inf. **13**(5), 2227–2240 (2017)
3. Cai, B. Kong, X. Liu, Y. Lin, J. Yuan, X. Xu, H., Ji, R.: Application of bayesian networks in reliability evaluation. IEEE Trans. Industr. Inf. (2018, in press)
4. Lakehal, A., Laouacheria, F.: Reliability based rehabilitation of water distribution networks by means of Bayesian networks. J. Water Land Dev. **34**(1), 163–172 (2017)
5. Lakehal, A., Ghemari, Z.: Availability assessment of electric power based on switch reliability modelling with dynamic Bayesian networks: case study of electrical distribution networks. J. Math. Syst. Sci. **5**(7), 289–295 (2015)
6. Lakehal, A., Ghemari, Z.: Optimisation of an emergency plan in gas distribution network operations with Bayesian networks. Int. J. Reliab. Saf. **10**(3), 227–242 (2016)
7. Williamson, J.: Foundations for bayesian networks. In: Corfield, D., Williamson, J. (eds) Foundations of Bayesianism. Applied Logic Series, vol. 24. Springer, Dordrecht (2001)
8. Kjærulff, U.B., Madsen, A.L.: Bayesian Networks and Influence Diagrams: A Guide to Construction and Analysis. Springer-Verlag, New York (2008)
9. Norsys Software Corp. https://www.norsys.com/

# Data Mining and Machine Learning Approaches and Technologies for Diagnosing Diabetes in Women

Safae Sossi Alaoui[1](✉), Brahim Aksasse[1](✉), and Yousef Farhaoui[2](✉)

[1] Faculty of Sciences and Techniques, Department of Computer Science, M2I Laboratory, ASIA Team, Moulay Ismail University, Errachidia, Morocco
sossialaouisafae@gmail.com, baksasse@yahoo.com

[2] Faculty of Sciences and Techniques, Department of Computer Science, M2I Laboratory, IDMS Team, Moulay Ismail University, Errachidia, Morocco
youseffarhaoui@gmail.com

**Abstract.** Data mining and machine learning are two interesting areas of computer science that go hand in hand in identifying hidden patterns and extracting valuable information from data. Indeed, Data mining covers the entire process of data analysis including machine learning which aims at constructing programs that learn automatically from experiences. The main purpose of this paper is to make a comparative study of four well-known classification algorithms namely Naive Bayes, Neural network, Support vector machines and Decision tree in order to categorize female patients into two groups; having diabetes or not. Therefore, after adopting well-chosen criteria based on confusion matrix, we run the selected algorithms in two different data mining technologies Weka and Orange. Indeed, the results obtained demonstrate that support vector machines; implemented in Weka toolkit as SMO, is the best technique in terms of accuracy, sensitivity and precision when handling diabetes in women dataset.

**Keywords:** Data mining · Machine learning · Classification algorithms · Weka · Orange

## 1 Introduction

Diabetes is a set of diseases in which human body has high levels of blood sugar. This issue can risk for many health problems such as skin infections, eye problems, nerve damage… etc. Diabetes is too danger for women because it can affect both mothers and their babies during pregnancy [1]. Therefore, the need for predicting this disease from data mining and machine learning techniques becomes extremely crucial.

Data mining [2] refers to the process of discovering interesting knowledge from a large amount of data. It has its great application in diverse areas such as telecommunication industry, marketing, business, biological data analysis and other scientific applications.


Y. Farhaoui (Ed.): BDNT 2019, LNNS 81, pp. 59–72, 2020.
https://doi.org/10.1007/978-3-030-23672-4_6

When dealing with machine learning, there exist several types of techniques. Each type gives a different impact or result depending on the problem that we try to solve. For our case, we focus on classification task which arrange the data into predefined classes. It provides many algorithms and techniques, but the common widely used are Naive Bayes, Neural network; Support vector machines (SVM) and Decision tree.

The rest of the paper is organized as follows; in the second section we present the research background which depicts diabetes in women, machine learning concepts and the different works related to data mining especially the latest comparative studies of classification algorithms. In the third section, we describe the methodology followed and the tools used. In the fourth section, we present the different results obtained. In the last section, we conclude our paper.

## 2 Research Background

### 2.1 Diabetes in Women

Diabetes; generally known as diabetes mellitus, is a set of metabolic diseases which describes the insufficient or lack of insulin hormone in the body produced by the pancreas organ. Both women and men can be affected by this health problem, however, diabetes in women is more complex than men. Certainly according to several studies [3], women has the higher levels of getting heart attack because she usually receives less aggressive treatment for cardiovascular risk factors. Also, the complications of diabetes in women are so difficult to diagnose and she mostly can be victim of blindness, depression and so on [1].

Many symptoms of diabetes in women are similar to men, although, there are some which are unique for women including urinary infections, female sexual dysfunction, polycystic ovary syndrome, vaginal and oral yeast infections as well as vaginal thrush [1]. Diabetes during pregnancy has probably negative effects on the health of the baby namely its great growth than average and the possibility of being premature birth or even stillborn. However, it can be significantly reduced if the blood sugars are under control.

### 2.2 Machine Learning Concepts

Machine Learning (ML) [4] is a scientific field related to the commonly used term “Artificial intelligence”; which is a simulation of human intelligence processes by machines. The core objective of machine learning is to make a prediction based on samples of previous observations; in other words to learn from experience like humans do. Machine learning tasks can be divided into two main categories: supervised learning and unsupervised learning.

**Supervised learning** requires that both the inputs and outputs are already known; which means that the data used to train the algorithm is already labeled with correct answers in order to generate reasonable predictions for the response to new data. When the outputs are discrete, the **Classification** task is used to categorize the data into specific groups or classes. Otherwise, when the outputs are continuous, the **Regression** task is employed to work with a data range or if the output is a real number.

**Unsupervised learning** aims at identifying complex processes and hidden patterns from input data without labeled responses. **Clustering** is the most common unsupervised learning technique. It is based on finding similarities between objects in input data in order to separate them into groups and to assign a new label to each group.

In our work, we want to categorize the data into two classes; tested positive and tested negative; which means being diabetes or not. Also, the data has previously known inputs and outputs with discrete values. Therefore, we emphasize to handle supervised learning approaches especially classification algorithms shown in Fig. 1.

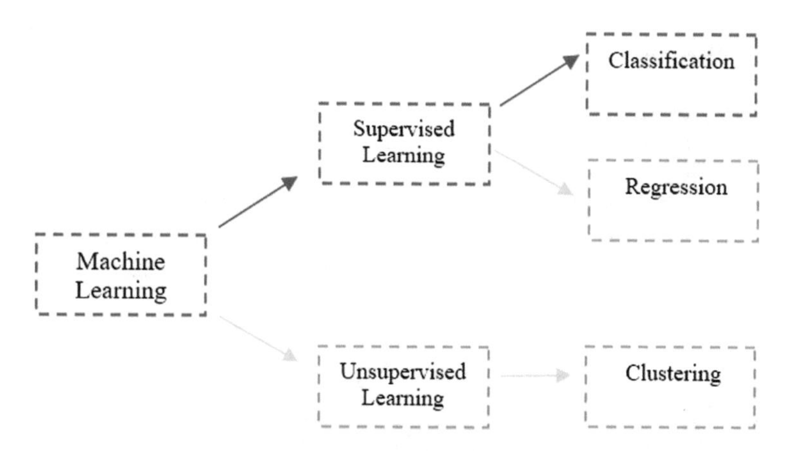


**Fig. 1.** Machine learning tasks

## 2.3 Related Works

Today, data mining community is interested to make comparative studies for several classification algorithms.

For example, Dogan and Tanrikulu [5] have made a comparison of Fourteen classification algorithms representing the different types of classification models (the AIRS2P, C4.5, CART, CSCA, Ex-CHAID, IBk, Logistics, LogitBoost, MLP, MLVQ, NaiveBayesian, QUEST, RSES, and SVM) for various datasets; exactly ten; and their implementation techniques. WEKA, SPSS, and Rosetta, were the main components used to run the selected algorithms. In addition, this study was based on multiple criteria such us algorithm accuracies, speed (CPU time consumed) and robustness.

Moreover, researchers like Lim et al. [6] have made a comparison of prediction accuracy, complexity, and training time of thirty-three old and new Classification Algorithms namely twenty-two decision tree, nine statistical, and two neural network algorithms by using thirty-two datasets. In this work, experiments generally showed that the mean error rates of several algorithms are sufficiently similar that their differences are statistically insignificant. Unlike error rates, there are huge differences between the training times of these algorithms.

Also, Rashid et al. [7] searched on decision support system for Diabetes Mellitus through Machine Learning Techniques. SMO, C4.5 and ANN are three classification models designed for classification, prediction, and description purposes to offer complete knowledge about Diabetes Mellitus patients.

In addition, Gupta et al. [8] focused on the following classification algorithms; BayesNet, NaiveBayes, J48, JRip, OneR and Decision Table; in order to analyze crime and accident datasets from Denver City, USA during 2011 to 2015. The criteria used in this study, are correct classification, incorrect classification, True Positive Rate (TP), False Positive Rate (FP), Precision (P), Recall (R) and F-measure (F). They have also used two different test methods: k-fold cross-validation and percentage split.

# 3 Methodology

For achieving the goal of this research, we have based on Fig. 2 which demonstrates the methodology followed in this study. Firstly, it consists on one dataset related to women diabetes.

Secondly, it specifies the two software used among the best four open source tools available in the market. Thirdly, it compares a set of selecting classification algorithms in the two chosen toolkits in machine learning and data mining researches. Finally, it shows the different results obtained.

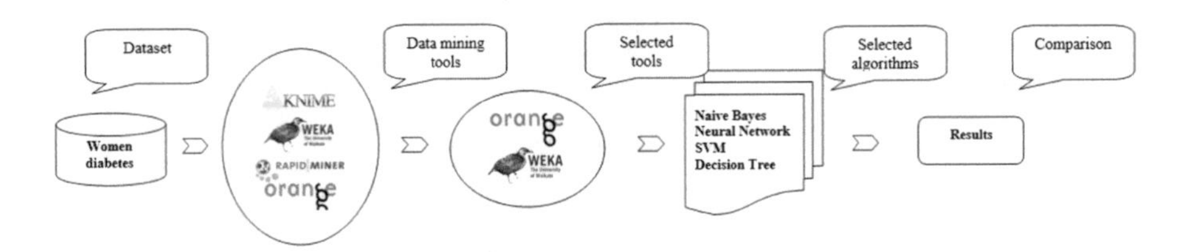


**Fig. 2.** Methodology of study

## 3.1 Dataset Description

In this paper, we have used one dataset related to women diabetes; at least 21 years old of Pima Indian heritage; taken from the UC Irvine Machine Learning Repository which is a collection of databases, domain theories, and data generators that are used by the Data mining community for the empirical analysis of machine learning algorithms [9] (Tables 1 and 2).

**Table 1.** Description of the dataset

| Name of dataset | Pima Indians diabetes database |
|---|---|
| Original owners | National Institute of Diabetes and Digestive and Kidney Diseases |
| Date received | 9 May 1990 |
| Number of instances | 768 |
| Number of attributes | 9 |
| Type of attributes | Numeric |
| Number of classes | 2 |

**Table 2.** Description of attributes

| Name of attribute | Details of attribute |
|---|---|
| preg | Number of times pregnant |
| plas | Plasma glucose concentration a 2 h in an oral glucose tolerance test |
| pres | Diastolic blood pressure (mm Hg) |
| skin | Triceps skin fold thickness (mm) |
| insu | 2-Hour serum insulin (mu U/ml) |
| mass | Body mass index (weight in kg/(height in m)^2) |
| pedi | Diabetes pedigree function |
| age | Age (years) |
| class | Class variable (tested positive, tested negative) |

### 3.2 Tools Used

Plenty of open source tools are available for data mining in the market, but the most powerful tools are Weka, KNIME, Orange, and RapidMiner. Bellow descriptions of these data mining toolkits:

**Weka** [10] (Waikato Environment for Knowledge Analysis) is a widely used suite of machine learning software written in Java and developed at the University of Waikato in New Zealand. It is very sophisticated and it supports many tasks including data preprocessing, clustering, classification, regression, visualization, and feature selection. Weka is free under the GNU General Public License and it provides access to SQL databases. It can be accessible from both the command line and the user interface named the Explorer.

**Orange** [11] is a Python-based toolkit for data mining. It is very powerful and can be used for both novices and experts. It has a complete set of components for data preprocessing, feature scoring and filtering, modeling, model evaluation, and exploration techniques.

**KNIME** [12] (Konstanz Information Miner) is a user friendly open source data analytics, reporting and integration platform written in java and based on Eclipse. It gives users the ability to add plugins which provide additional functionalities.

**RapidMiner** [13] is an environment for machine learning and data mining experiments written in the java programming language. It provides advanced analytics through template-based frameworks. It can be utilized for both research and real-world data mining tasks such as data preprocessing and visualization, predictive analytics and statistical modeling, evaluation, and deployment.

According to a recent paper [14], which has conducted a comparison study between four machine learning toolkits namely; Weka, Orange, Tanagra, and KNIME over some classification methods. They proved that Weka was the best tool in terms of the ability to run the selected classifier, followed by Orange, then KNIME and finally Tanagra. Also, the two first toolkits have achieved the highest performance improvements when moving from the Percentage Split test mode to the Cross Validation test. Therefore, thanks to their various advantages we had chosen the two open source software Weka and Orange shown in Table 3 to evaluate our classification algorithms.

**Table 3.** Details of selected tools

| Technology | Weka | Orange |
|---|---|---|
| Full name | Waikato environment for knowledge analysis | Orange, Data mining fruitful & Fun |
| Developers | The University of Waikato | The University of Ljubljana |
| Year | 1997 | 1997 |
| Programing language | Java | Python, Cython, C++, C |
| License | GNU General Public License | GNU General Public License |
| Operating system | Windows, OS X, Linux | Cross-platform |
| Version | Weka 3.8 | Orange 3.11 |

### 3.3 Classification Algorithms

Overall, Classification in data mining is an outstanding task that can be used to portion the data into classes that are already predefined. Numerous classification algorithms are available. Yet, in this research, we had selected four well known classification algorithms; Naive Bayes, Neural Network, SVM and Decision tree.

Each of these algorithms is implemented in the two machine learning technologies; Weka and Orange which are outlined in Table 4.

**Table 4.** Description of selected algorithms

| Algorithm name | Weka implementation | Orange implementation | Years | Introduced by |
|---|---|---|---|---|
| Naive Bayes | Naive Bayes | Naive Bayes | 1995 | John and Langley [15] |
| Decision tree | C4.5(J48) | Tree | 1993 | Quinlan [16] |
| Support vector machines | SMO | SVM | 1992 | Boser et al. [17] |
| Neural network | MLP | Neural network | 1986 | Rumelhart et al. [18] |

#### 3.3.1 Naive Bayes

Naive Bayes is statistical method for classification based on Bayes' Theorem:

$$P(H/X) = P(X/H)P(H)/P(X). \quad (1)$$

Where X is data attribute and H is some hypothesis.

It can solve problems involving both categorical and continuous valued attributes.

Moreover, the name Naive comes from the fact that this classifier considers that the effect of the value of a predictor on a given class is independent of the values of other predictors.

Naive assumption of «class conditional independence»:

$$P(X/Ci) = P(x1, x2, \ldots\ldots, xn/Ci) = P(x1/Ci) * P(x2/Ci) * \ldots * P(xn/Ci). \quad (2)$$

Where X: (x1, x2, x3, … , xn) is the different data attributes and C: C1, C2, C3, … , Cm is a set of classes.

A Naive Bayes classifier predicts class membership probabilities such as the probability that a given feature belongs to a particular class which maximizes this probability P(X|Ci)*P(Ci).

### 3.3.2 Decision Tree

Decision tree is supervised machine learning that construct models in the form of a tree structure. One of the best decision tree algorithm is C4.5 a classification technique witch improves the ID3 algorithm by managing both continuous and discrete properties, missing values and pruning trees after construction. It is available in Weka tool as J48 and it is used to generate a decision tree based on Shannon entropy in order to pick features with the greatest information gain as nodes.

Entropy at a given node t:

$$\text{Entropy}(t) = -\Sigma\ p(j\,|\,t)\ \log 2\ p(j\,|\,t). \quad (3)$$

p(j | t) is the relative frequency of class j at node t.

Information Gain:

$$\text{GAINsplit} = \text{Entropy}(t) - \Sigma\,(ni/n\ \text{Entropy}(i)). \quad (4)$$

Parent node t with n records is split into k partitions; ni is number of records in partition (node) i.

Gain Ratio:

$$\text{GainRATIOsplit} = \text{GAINsplit}/\text{SplitINFO}. \quad (5)$$

$$\text{SplitINFO} = -\Sigma\,(ni/n\ \log\ ni/n). \quad (6)$$

Parent node p is split into k partitions, ni is the number of records in partition i.

### 3.3.3 Support Vector Machines (SVM)

Support vector machines (SVM) is among the most robust and accurate methods in all well-known data mining algorithms. SVM is implemented in Weka as SMO (Sequential minimal optimization) [19]. It is a supervised learning model its goal is to find a hyperplane that can separate two classes of given samples with a maximal margin. Intuitively, a margin refers to the amount of space, or separation, between the two classes as defined by a hyperplane. Geometrically, the margin corresponds to the shortest distance between the closest data points to any point on the hyperplane.

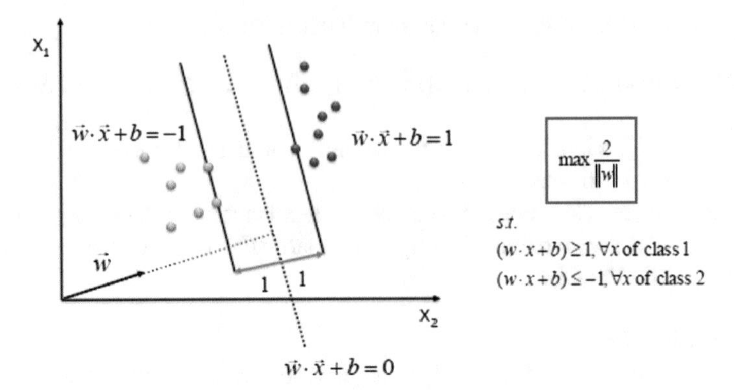


**Fig. 3.** SVM algorithm

Figure 3 illustrates the optimal hyperplane which can be defined as Eq. 7 where w and b denote the weight vector and bias respectively.

$$\mathrm{w}^T\mathrm{x} + b = 0. \tag{7}$$

### 3.3.4 Neural Network

Neural network is a mathematical model based on the emulation of biological neural system. The Neural network algorithm is implemented in Weka as Multi-Layer perceptron (MLP) which is a feedforward algorithm that consists of one or more layers between input and output layer. The term "feedforward" significates that data flows are in one direction; from input to output layer. MLP is trained with a supervised learning technique called backpropagation learning algorithm. MLP is an amelioration of the standard linear perceptron. MLPs are widely used for pattern classification, recognition, prediction and approximation. Multi-Layer Perceptron can solve problems which are not linearly separable.

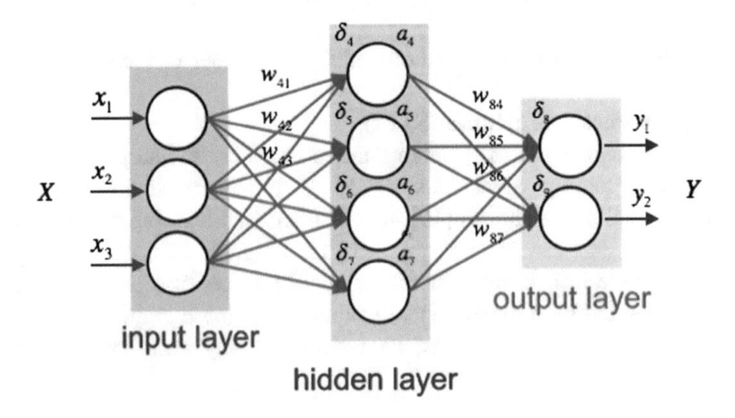


**Fig. 4.** Neural network algorithm

A Neural network algorithm can be represented graphically as follows (Fig. 4):

### 3.4 Criteria of Comparison

It is very important to evaluate classifiers in order to know which methods are the best. Indeed, there are several metrics that can be utilized to measure the performance of a classifier or predictor; among these criteria: accuracy, speed, robustness, scalability, interpretability. In our work, we utilize the first two criteria in order to compare selected algorithms:

- **Accuracy**: refers to the ability of the model to correctly predict the class label of new or previously unseen data.
- **Speed**: means the computation costs involved in generating and using the model. It depends on the algorithm complexity.
- **Robustness**: refers to the ability of the model to make correct predictions given noisy data or data with missing values.
- **Scalability**: refers to the ability to construct the model efficiently given large amount of data.
- **Interpretability**: refers to the level of understanding and insight that is provided by the model [20].

A more detailed performance description via; precision, recall, true- and false positive rate are useful measures for comparing classifiers. All these values are based on the confusion matrix and can be computed from it.

**The confusion matrix** [21] named also contingency table in the case of two classes, the confusion matrix is 2 × 2. The number of correctly classified instances is the sum of diagonals in the matrix; all others are incorrectly classified. Figure 5 shows a representation of confusion matrix, where: TP = true positive, TN = true negative, FP = false positive, FN = false negative.

| | | Predicted Class | |
|---|---|---|---|
| | | Yes | No |
| Actual Class | Yes | TP | FN |
| | No | FP | TN |

**Fig. 5.** Confusion matrix

**The True Positive (TP) rate** [21] refers to the proportion of examples which were classified as class x, among all examples which truly have class x. In the confusion matrix, this is the diagonal element divided by the sum over the relevant row. It is equivalent to Recall or sensitivity.

$$\text{True positive rate} = \text{Recall} = \text{TP}/(\text{TP} + \text{FN}). \tag{8}$$

**The False Positive (FP) rate** [21] is defined as the proportion of examples which were classified as class x, but belong to a different class, among all examples which are not of class x. In the matrix, this is the column sum of class x minus the diagonal element, divided by the rows sums of all other classes.

$$\text{False positive rate} = \text{FP}/(\text{TP} + \text{FN}). \tag{9}$$

**The Precision** [21] means the proportion of the examples which truly have class x among all those which were classified as class x. In the matrix, this is the diagonal element divided by the sum over the relevant column.

$$\text{Precision} = \text{TP}/(\text{TP} + \text{FP}). \tag{10}$$

**The F-Measure** [21] is simply a combined measure for precision and recall:

$$2 * \text{Precision} * \text{Recall}/(\text{Precision} + \text{Recall}). \tag{11}$$

## 4 Results

Empirical studies have proved that Cross validation is the most elaborate method for dataset with less than 1000 instances. It consists on a number of folds n which must be specified. The dataset is randomly reordered and then split into k folds of equal size. In each iteration, one fold is used for testing and the other k − 1 folds are used for training the classifier. For our case, we have chosen the default option: 10-fold cross-validation. Figure 6 illustrates a snapshot of our experiment, in fact, File widget reads the input data file of diabetes dataset, and send them to the Data Table widget which presents them in a spreadsheet format. In addition, the widget Test & Score tests our selected machine learning algorithms namely Naive Bayes, Neural Network, SVM and Tree, on diabetes dataset and the output is employed by the widget Confusion Matrix for analyzing the performance of our classifiers.

It is quite clear; that we can consider an algorithm is a perfect classifier if at least on the training data, all instances were classified correctly as well as all errors are zero. However, it is not the case in reality. Therefore we can admit that a best classifier is the algorithm with the maximum of Correctly Classified Instances or the minimum of Incorrectly Classified Instances. Tables 5 and 6 show the correctly and incorrectly instances for each algorithm generated by Weka and Orange respectively.

Figure 7 gives the results of Weka toolkit which outlines that SMO classifier has the highest accuracy with 76.8%, followed by Naive Bayes having correct classification rate of 75.75%, then MLP with 74.75% and finally Decision tree C4.5 has determined least correct instances with 74.49%.

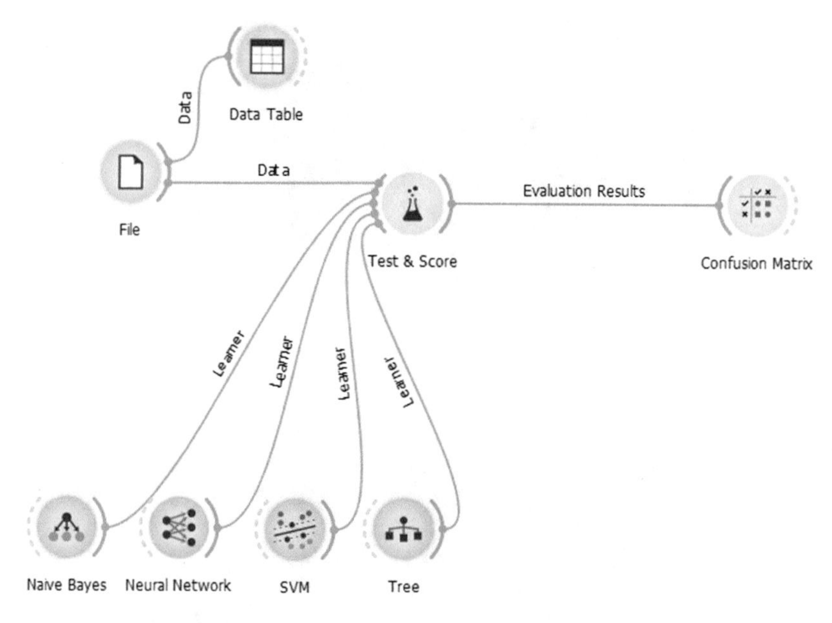


**Fig. 6.** Snapshot of experiment implemented by Orange

**Table 5.** Correctly and incorrectly classified instances for each algorithm (Weka)

| Algorithm | Correctly classified instances | Incorrectly classified instances |
|---|---|---|
| Naive Bayes | 586 | 182 |
| MLP | 579 | 189 |
| SMO | 594 | 174 |
| C4.5 (J48) | 567 | 201 |

**Table 6.** Correctly and incorrectly classified instances for each algorithm (Orange)

| Algorithm | Correctly classified instances | Incorrectly classified instances |
|---|---|---|
| Naive Bayes | 570 | 198 |
| Neural network | 589 | 179 |
| SVM | 579 | 189 |
| Tree | 536 | 232 |

On the other side, Orange results clarified in Fig. 8, show that Neural Network has the higher accuracy with 76.7%, followed by SVM with 75.4% then Naive Bayes with 74.2% and finally Tree with 69.8%.

Certainly, it is quite clear that best classifier in terms of speed is the algorithm with the minimum of Execution time. According to Weka, Naive Bayes is the fast classifier that took 0.03 s then Decision tree C4.5 with 0.05 s. While SMO time to build the model was 0.08 s, and MLP was the slowest classifier with 1.9 s (Table 7 and Fig. 9).

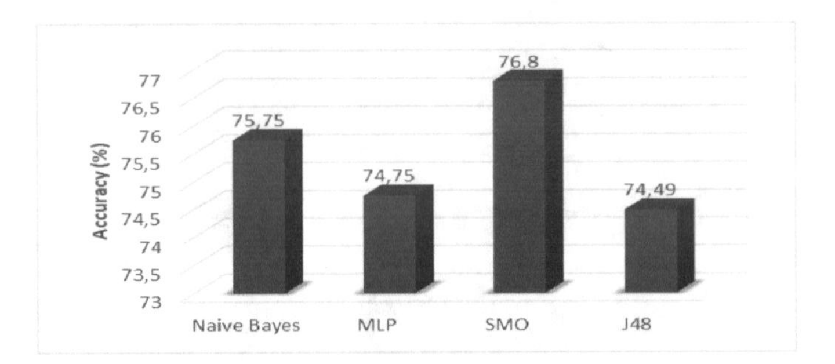


**Fig. 7.** Classifiers accuracy (Weka)

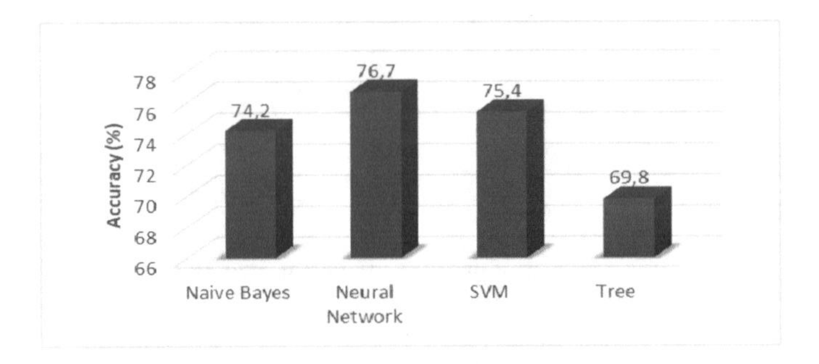


**Fig. 8.** Classifiers accuracy (Orange)

**Table 7.** Execution time of selected algorithms (Weka)

| Algorithm | Execution time (second) |
|---|---|
| Naive Bayes | 0,03 |
| MLP | 1,9 |
| SMO | 0,08 |
| C4.5 (J48) | 0,05 |

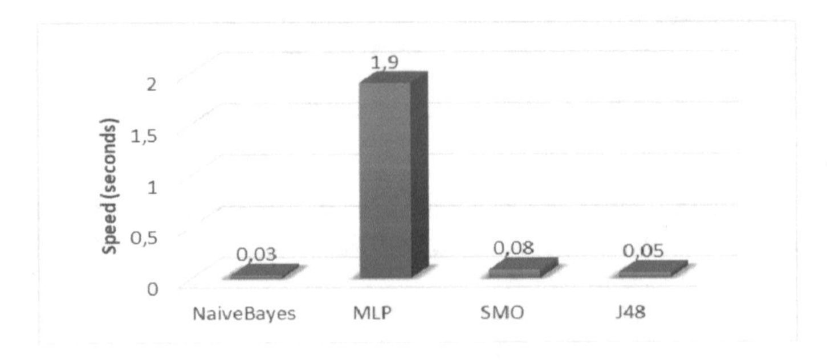


**Fig. 9.** Execution time of selected algorithms (Weka)

**Table 8.** Detailed performance description (Weka)

| Algorithm | TP rate | FP rate | Precision | Recall | F-Measure |
|---|---|---|---|---|---|
| Naive Bayes | 0,763 | 0,307 | 0,759 | 0,763 | 0,76 |
| MLP | 0,754 | 0,314 | 0,75 | 0,754 | 0,751 |
| SMO | 0,773 | 0,334 | 0,769 | 0,773 | 0,770 |
| C4.5 (J48) | 0,738 | 0,327 | 0,735 | 0,738 | 0,736 |

**Table 9.** Detailed performance description (Orange)

| Algorithm | TP rate | FP rate | Precision | Recall | F-Measure |
|---|---|---|---|---|---|
| Naive Bayes | 0,742 | 0,382 | 0,755 | 0,742 | 0,746 |
| Neural network | 0,767 | 0,328 | 0,763 | 0,767 | 0,764 |
| SVM | 0,754 | 0,324 | 0,750 | 0,754 | 0,751 |
| Tree | 0,698 | 0,282 | 0,693 | 0,698 | 0,695 |

Bellow Tables 8 and 9 show the TP and FP rate of each classifier, the weighted average of Precision, Recall and F-Measure, obtained by using the 10-fold cross-validation approach in Weka and Orange.

For Weka, SMO has the highest TP Rate and Recall with the value 0.773, it has also the greatest precision with 0.769, followed by Naive Bayes having TP rate and recall value of 0.763. Then MLP with 0.754 and finally decision tree C4.5 with 0.738.

Concerning Orange performance, Neural network has the greatest value of TP Rate and Recall with 0.767 then SVM with 0.754 then Naive Bayes with 0.742 then Tree with 0.698. While precision values are 0.763 for Neural network, 0.755 for Naive Bayes, 0.750 for SVM and 0.693 for Tree.

## 5 Conclusion

This paper has concentrated on using Data mining and machine learning techniques in order to make prediction for health problem, especially diabetes in women; a disease in which the body cannot control the level of sugar in the blood. Therefore, we have chosen the most four well-known classifiers; Naive Bayes, Neural network, SVM and Decision tree; in order to pick the best algorithm for our dataset. Also, we have made a comparison between selected classification algorithms using two data mining technologies Weka and Orange. Indeed, according to the results obtained, our analysis indicates that support vector machines classifier implemented in Weka as SMO has the highest accuracy with 76.8%, sensitivity with 77.3% and precision with 76.9%.

For the future work, we try to use a dataset with a huge number of instances. Certainly, it can lead us to manage Big data mining technologies with the aim of achieving the precise and perfect predictions.

## References

1. How Diabetes Affects Women: Symptoms, Risks, and More. http://www.healthline.com/health/diabetes/symptoms-in-women
2. Wang, Y., Zhang, J.: Exploring topics related to data mining on Wikipedia. Electron. Libr. **35**, 667–688 (2017)
3. Peters, S.A.E., Huxley, R.R., Woodward, M.: Diabetes as risk factor for incident coronary heart disease in women compared with men: a systematic review and meta-analysis of 64 cohorts including 858,507 individuals and 28,203 coronary events. Diabetologia **57**, 1542–1551 (2014)
4. Kavakiotis, I., Tsave, O., Salifoglou, A., Maglaveras, N., Vlahavas, I., Chouvarda, I.: Machine learning and data mining methods in diabetes research. Comput. Struct. Biotechnol. J. **15**, 104–116 (2017)
5. Dogan, N., Tanrikulu, Z.: A comparative analysis of classification algorithms in data mining for accuracy, speed and robustness. Inf. Technol. Manag. **14**, 105–124 (2013)
6. Lim, T.-S., Loh, W.-Y., Shih, Y.-S.: A comparison of prediction accuracy, complexity, and training time of thirty-three old and new classification algorithms. Mach. Learn. **40**, 203–228 (2000)
7. Rashid, T.A., Abdulla, S.M., Abdulla, R.M.: Decision support system for diabetes mellitus through machine learning techniques. Int. J. Adv. Comput. Sci. Appl. (IJACSA) **7** (2016)
8. Gupta, A., Mohammad, A., Syed, A., Halgamuge, M.N.: A comparative study of classification algorithms using data mining: crime and accidents in Denver City the USA. Education **7**, 374–381 (2016)
9. UCI Machine Learning Repository: Pima Indians Diabetes Data Set. https://archive.ics.uci.edu/ml/datasets/Pima+Indians+Diabetes
10. Weka 3 - Data Mining with Open Source Machine Learning Software in Java. http://www.cs.waikato.ac.nz/ml/weka/
11. Orange – Data Mining Fruitful & Fun. https://orange.biolab.si/
12. KNIME | Open for Innovation. https://www.knime.org/
13. Data Science Platform | Machine Learning. https://rapidminer.com/
14. Wahbeh, A.H., Al-Radaideh, Q.A., Al-Kabi, M.N., Al-Shawakfa, E.M.: A comparison study between data mining tools over some classification methods. IJACSA Int. J. Adv. Comput. Sci. Appl. Spec. Issue Artif. Intell. **8**, 18–26 (2011)
15. John, G., Langley, P.: Estimating continuous distributions in Bayesian classifiers (1995)
16. Salzberg, S.L.: C4. 5: Programs for Machine Learning by J. Ross Quinlan. Morgan Kaufmann Publishers, Inc. (1993). Mach. Learn. **16**, 235–240 (1994)
17. Boser, B.E., Guyon, I.M., Vapnik, V.N.: A training algorithm for optimal margin classifiers. In: Proceedings of the Fifth Annual Workshop on Computational Learning Theory, pp. 144–152. ACM (1992)
18. Rumelhart, G., Hinton, G., Williams, R.: Learning internal representations by error propagation. Presented at the (1986)
19. Platt, J.C.: Sequential Minimal Optimization: A Fast Algorithm for Training Support Vector Machine (1998)
20. Stefanowski, J.: Data Mining - Evaluation of Classifiers (2008). http://www.cs.put.poznan.pl/jstefanowski/sed/DM-4-evaluatingclassifiersnew.pdf
21. Kirkby, R., Frank, E., Reutemann, P.: Weka explorer user guide for version 3-5-8. Univ, Waikato (2007)

# Multi Dimensional Hidden Markov Model for Credit Scoring Systems in Peer-To-Peer (P2P) Lending

El Annas Monir[1](✉), Mohamed Ouzineb[1], and Badreddine Benyacoub[2]

[1] Institut National de Statistique et d'Economie Appliquée, Rabat, Morocco
elannas.mounir@gmail.com, ouzineb.insea@gmail.com
[2] Faculty of Sciences, University Mohammed-V, Rabat, Morocco
benyacoubb@gmail.com

**Abstract.** Credit scoring models become a key role for lending institutions to distinguish good applicants (likely to repay) from bad applicants (likely to default), and attracting significant attention from researchers and market participants. Many standard statistical and machine learning techniques are used in the literature to build credit scoring models from training sets consisting of people in their records who were given loans in the past.

Online peer-to-peer (P2P) lending is a new financing channel which is based on electronic business platform and electronic commerce credit. In P2P lending, borrowers and lenders can use the internet platform to achieve online transactions. There is lower transaction cost, while the loan process is simple and easy to operate. Small and micro enterprises and individual borrowers that are difficult to get loans from the bank do not need loan guarantor and collateral in P2P, so they can get financing more easily. But it means higher credit risk to lenders.

Many classification methods such as neural networks, support vector machines and random forests, have been suggested in the literature to improve credit scoring models in terms of their statistical performance. But there execution time can grow exponentially with the size of the instances, especially with the grow of Big Data sources like mobile phone data and social network data for credit scoring applications [18]. For these cases, metaheuristics like Hidden Markov Model (HMM) present a good alternative solution technique. A limitation of HMM is that it only supports one observed random variable. In case of the credit scoring, there are more variables that are relevant to the estimation. Multi dimensional Hidden Markov Model (MDHMM) extends hidden Markov model by supporting multiple observed variables. The iterative procedure of MDHMM greatly simplifies parallelized implementation and computations of innovative Big Data sources in credit scoring.

The aim of this paper is to investigate the performance of MDHMM to addresses the credit scoring problem in Peer-To-Peer (P2P) lending. The performance of the proposed MDHMM method is validated on Lending Club (Peer-to-peer lending) credit dataset.


Y. Farhaoui (Ed.): BDNT 2019, LNNS 81, pp. 73–83, 2020.
https://doi.org/10.1007/978-3-030-23672-4_7

# 1 Introduction

After the recent world financial crisis, more attention was given from banks and financial institutions to credit risk, since it can cause great cost losses to owners, managers, workers, lenders, clients, community and government. Therefore, it is very important to predict bankruptcy and decide whether to grant credit to new applicants or not. One of the primary tools used by banks is credit scoring or scorecards, it's used to estimate how likely an applicant is in default. Originally, it was conducted using subjective judgments of human expert, but it's impossible nowadays because of the large number of applicants, the vast amount of information, the great commercial competition, the slowness of the process and the frequent errors.

Credit scoring is a model developed to determine if loan customers belong to a good applicant group or a bad applicant group. Therefore, it's a classification problem where a new applicant must be categorized into one of the predefined classes based on a number of attributes that describe the economic and the socio-demographic situation of the applicant.

P2P lending platforms, a new financial intermediary between borrowers and lenders, experience an astonishing growth since their inception. For example, the biggest P2P lending platform in USA, Lending Club, almost double the amount of issued loans each year. Lending Club publishes information about all issued loans on their websites. For purpose of our paper, we have taken a data set of loans issued between June 2007 and June 2018 that contains 2,004,090 loans.

Many classification methods have been suggested in the literature to tackle the problem of credit scoring. These techniques include traditional statistical methods as discriminant analysis, logistic regression, k-nearest neighbour, decision trees, and machine learning techniques as neural networks, support vector machines and hidden markov models [1–6].

A limitation of HMM is that it only supports one observed random variable. One could increase the dimensionality of the observed variable. However, this would lead to a phenomenon called curse of dimensionality, the learning time would increase greatly up to the point where it would no longer be tractable. More training data would be necessary too. Multi dimensional HMM addresses this issue by supporting multiple observed variables. It is assumed that these variables are mutually statistically independent [12–14].

The remainder of this paper is organized as follows: Sect. 2 summarizes the first order of HMM and describes the Multi dimensional HMM. Section 3 shows the experiments setup and results. Finally, Sect. 4 concludes this paper.

# 2 Background

## 2.1 First Order Hidden Markov Model

A hidden Markov model is a statistical model in which the system being modeled is assumed to be a Markov process with unobserved (hidden) states. The state is not directly visible, but output, dependent on the state, is visible. The output is an alphabet or signs are observed or emitted. Each state has a probability distribution over the possible output tokens. Therefore the sequence of tokens generated by an HMM gives some information about the sequence of states.

*Elements of Hidden Markov Model*

An HMM model is characterized by the following elements [7]:

- $S_t$ The random variable representing the state at time t, where $0 \leq S_t \leq N-1$ and $0 \leq t \leq T-1$, N is the number of states in the model, and T being the length of the observation sequence. Let $S = (S_0, S_1, \ldots, S_{T-1})$ be the states sequence.
- $O_t$ The random variables representing the observation at time t, where $0 \leq O_t \leq M$ and $0 \leq t \leq T$, where M is the number of observation symbols, and T the length of the observation sequence. Let $O = (O_0, O_1, \ldots, O_{T-1})$ be the observation sequence.

In this paper we consider the values of $O_t$ and the values of $S_t$ to be discrete.

- A $= \{a_{ij}\}$ the state transition probabilities matrix, where A $\in R^{N \times N}$ and $a_{ij} = P(S_{t+1} = j | S_t = i)$
- B $= \{b_i(k)\}$ the observation probability matrix, where B $\in R^{N \times M}$ and $b_i(k) = P(O_t = k | S_t = i)$
- $\pi = \{\pi_i\}$ be the initial probability vector, where $\pi \in R^N$ and $\pi_i = P(S_t = i)$
- we denote an HMM as a triplet $\lambda = (\pi, A, B)$

There are three common operations that can be performed on HMM: learning, inference and scoring.

*Training with Fully Observed Data (Supervised Learning)*

Learning is used to estimate the model parameters $\lambda$ that maximize the probability that given sequences of observed symbols were generated by the model. If we observe $S$ and $O$ that will be a supervised learning, $\lambda$ can be then computed exactly by the following counts:

$$a_{ij} = \frac{Number\ of\ transitions\ from\ state\ j\ to\ state\ i}{Number\ of\ times\ in\ state\ i}$$

$$b_i(k) = \frac{Number\ of\ times\ we\ are\ in\ state\ i\ and\ we\ see\ a\ symbol\ k}{Number\ of\ times\ we\ are\ in\ state\ i}$$

$$\pi_i = \frac{Number\ of\ times\ in\ state\ i}{number\ of\ observations}$$

*Training with Baum-Welch algorithm (Unsupervised Learning)*

Baum-Welch is a learning algorithm [10] that is based on the principles of Expectation Maximization (EM) [11]. It does not require the sequences of hidden states $S$ for learning. To find the optimum model parameter $\lambda$ that maximizes $P(O|\lambda)$ we first define the following probabilities:

- $\alpha_t(i) = P(O_0, O_2, \ldots, O_t, S_t = i | \lambda)$
- $\beta_t(i) = P(O_{t+1}, O_{t+2}, \ldots, O_{T-1} | S_t = i, \lambda)$

- $\gamma_t(i) = P(S_t = i|O, \lambda)$
- $\zeta_t(i,j) = P(S_t = i, S_{t+1} = j|O, \lambda)$

Then given a random initial conditions for $\lambda$, (it can also be set using prior information about the parameters if it is available):

for $i = 0, 1 \ldots, N-1$ and $t = 1, 2 \ldots, T-1$

- $\alpha_0(i) = \pi_i b_i(O_0)$ and $\alpha_t(i) = [\sum_{j=0}^{N-1} \alpha_{t-1}(j) a_{ji}] b_i(O_t)$

for $i = 0, 1 \ldots, N-1$ and $t = T-2, T-3 \ldots, 0$

- $\beta_{T-1}(i) = 1$ and $\beta_t(i) = \sum_{j=0}^{N-1} a_{ij} b_j(O_{t+1}) \beta_{t+1}(j)$

for $i, j \in 0, \ldots, N-1$ and $t = 0, 2 \ldots, T-2$

- $\zeta_t(i,j) = \frac{\alpha_t(i) a_{ij} b_j(O_{t+1}) \beta_{t+1}(j)}{\sum_{k=0}^{N-1} \alpha_{T-1}(k)}$

for $i = 0, 1 \ldots, N-1$ and $t = 0, 2 \ldots, T-2$

- $\gamma_t(i) = \frac{\alpha_t(i) \beta_t(i)}{\sum_{k=0}^{N-1} \alpha_{T-1}(k)} = \sum_{j=0}^{N-1} \zeta_t(i,j)$

The parameter reestimation formulas, are described by the following expressions:

for $i, j \in 0, \ldots, N-1$ and $k = 0, 1 \ldots, M-1$

- $\widehat{\pi_i} = \gamma_0(i)$
- $\widehat{a_{ij}} = \frac{\sum_{t=0}^{T-2} \zeta_t(i,j)}{\sum_{t=0}^{T-2} \gamma_t(i)}$
- $\widehat{b_i(k)} = \frac{\sum_{t=0, O_t=k}^{T-1} \gamma_t(i)}{\sum_{t=0}^{T-1} \gamma_t(i)}$

The parameter $\lambda$ will be estimated iteratively by Baum-welch procedure. as follow:

1. Start with the best guess of parameters $\lambda$ for the model
2. Estimate $\alpha_t(i)$ and $\beta_t(i)$, $\zeta_t(i,j)$, $\gamma_t(i)$ using the observation sequence $O$
3. Update $\lambda$
4. Repeat steps 2 and 3 until a convergence to a stationary point of the likelihood.

*Inference*

Inference returns the most likely sequence of hidden states, given a sequence of observed symbols $O$. From the definition of $\gamma_t(i)$ it follows that the most likely state at time t is the state $S_i$ for which $\gamma_t(i)$ is maximum, where the maximum is taken over the index i.

*Scoring*

Given a sequence of observed symbols $O$ the probability that the sequence was generated by HMM with parameters $\lambda$ is called a score.

From the definition of $\alpha_t(i)$ it follows that the score is: $P(O|\lambda) = \sum_{j=0}^{N-1} \alpha_{T-1}(j)$

## 2.2 Multi Dimensional HMM

MDHMM can be considered as an extension of hidden Markov model. The difference is in the number of observed variables. HMM defines only a single observed variable, whereas the MDHMM supports multiple observed variables with one common hidden sequence. In this paper we will assumes independence between the different dimensions of the input data.

*Elements of MDHMM*

The parameters the multi-dimensional HMM can be defined as follows [12–14].

- $S_t$ The random variable representing the state at time t, where $0 \leq S_t \leq N-1$ and $0 \leq t \leq T-1$ , N is the number of states in the model, and T being the length of the observation sequence.
- $S = (S_0, S_1, \ldots, S_{T-1})$ be the state sequence.
- $O_t$ The random variables representing the observation at time t. where $O_t = \{O_t^0, O_t^1, \ldots, O_t^{M-1}\}$ and $0 \leq t \leq T-1$, M is the number of observed variables, and T is the length of the observation sequence.
- $O = (O_0, O_1, \ldots, O_{T-1})$ be the observation sequence from the M variables.
- $O^m = (O_0^m, O_1^m, \ldots, O_{T-1}^m)$ the m-th sequence of observation.

In this paper we consider the values of observed symbols $O_t^m$ and the values of $S_t$ to be discrete.

- A $= \{a_{ij}\}$ the state transition probabilities matrix, where A $\in R^{N \times N}$ and $a_{ij} = P(S_{t+1} = j | S_t = i)$
- $B^m = \{b_i^m(k)\}$ the m-th observation probability matrix, where $B^m \in R^{N \times L_m}$ and $b_i^m(k) = P(O_t^m = o_k^m | S_t = i)$ where $o_k^m$ is a possible value of the random variable of the observation $O_t^m$. $L_m$ being the number of observations symbols of m-th observation sequence.
- $\pi = \{\pi_i\}$ be the initial probability vector, where $\pi \in R^N$. and $\pi_i = P(S_t = i)$
- we denote an MDHMM as a triplet $\lambda = (\pi, A, B^{0:M-1})$
- An MDHMM has the joint probability function: $P(O_t, S_t) = P(S_{t-1})$ $P(S_t|S_{t-1}) \prod_{m=0}^{M-1} P(O_t^m|S_t)$
- $P(S_t|S_{t-1})$ is represented by the matrix A
- Each $P(O_t^m|S_t)$ is represented by the matrix $B^m$.

Similarly to HMM, there are three common operations that can be performed on MDHMM: learning, inference and scoring.

Learning is used to estimate the model parameters $\lambda = (\pi, A, B^{0:M-1})$ such that $\prod_{m=0}^{M-1} P(O^m|\lambda)$ is maximized. In other words, it is desired to maximize the product of probabilities that the given sequence of observed symbols was generated by the given model.

Inference returns the most likely sequence of hidden states, given a sequence of observations.

Given a sequence of observations, the probability that it was generated by MDHMM with parameters $\lambda$ is called a score.

*Supervised Learning*

The learning approach is similar to the one described in the case of HMM. The only difference is that there are multiple emission functions $B^m$

*Unsupervised Learning*

The parameter learning process can be performed by means of the Baum-Welch algorithm extended for MDHMM based on the probabilities $\alpha_t(i)$, $\beta_t(i)$, $\gamma_t(i)$ and $\gamma_t(i,j)$:

- $\alpha_t(i) = P(O_0, O_1, \ldots, O_t, S_t = i|\lambda)$
- $\beta_t(i) = P(O_t, O_{t+1}, \ldots, O_{T-1}|S_t = i, \lambda)$
- $\gamma_t(i) = P(S_t = i|O, \lambda)$
- $\zeta_t(i,j) = P(S_t = i, S_{t+1} = j|O, \lambda)$

The computation of these variables for the MDHMM are straightforward extensions of the ones for HMMs

- $\alpha_0(i) = \pi_i \prod_{m=0}^{M-1} b^m(O_0^m)$
- $\alpha_t(i) = [\sum_{j=0}^{N-1} \alpha_{t-1}(j) P(S_t = i, S_{t-1} = j, \lambda)] \prod_{m=0}^{M-1} P(O_t^m|S_t = i, \lambda)$
  $= [\sum_{j=0}^{N-1} \alpha_{t-1}(j) a_{ji}] \prod_{m=0}^{M-1} b^m(O_t^m)$
- $\beta_t(i) = \sum_{j=0}^{N-1} P(S_{t+1} = i, S_t = j, \lambda) \prod_{m=0}^{M-1} P(O_{t+1}^m|S_{t+1} = i, \lambda)\beta_{t+1}(i)$
  $= [\sum_{j=0}^{N-1} a_{ij}\beta_{t+1}(i)] \prod_{m=0}^{M-1} b^m(O_{t+1}^m)$
- $\beta_{T-1}(i) = 1$
- $\zeta_t(i,j) = \frac{\alpha_t(i)\beta_{t+1}(j)a_{ij} \prod_{m=0}^{M-1} b^m(O_{t+1}^m)}{\sum_{j=0}^{N-1} \alpha_{T-1}(j)}$
- $\gamma_t(i) = \frac{\alpha_t(i)\beta_t(i)}{\sum_{j=0}^{N-1} \alpha_{T-1}(j)} = \sum_{j=0}^{N-1} \zeta_t(i,j)$

The variables $\alpha_t(i), \beta_t(i), \zeta_t(i,j), \gamma_t(i)$ are calculated for each training sequence $O^m$ and then, re-estimation is performed on the accumulated values. To generate the new better model $\widehat{\lambda}$, the parametre $\lambda$ can be updated as follows:

- $\widehat{\pi_i} = \gamma_0(i)$
- $\widehat{a_{ij}} = \frac{\sum_{t=0}^{T-2} \zeta_t(i,j)}{\sum_{t=0}^{T-2} \gamma_t(i)}$

- $\widehat{b_i^m(k)} = \frac{\sum_{t=0/O_t^m=o_k^m}^{T-1} \gamma_t(i)}{\sum_{t=0}^{T-1} \gamma_t(i)}$

The iterative process described above is performed until some standard stopping criterion is met.

*Inference*

In the similar way as in the case of HMM. It is extended to support multiple observed variables. Inference returns the most likely sequence of hidden states $S$, given a sequence of observed symbols $O$.

From the definition of $\gamma_t(i)$ it follows that the most likely state at time t is the state $S_i$ for which $\gamma_t(i)$ is maximum, where the maximum is taken over the index i.

*Scoring*

Scoring of a sequence of observations $O$ is as usual performed the same way as in case of HMM.
From the definition of $\alpha_t(i)$ it follows that The score is:

$$P(O|\lambda) = \sum_{j=0}^{N-1} \alpha_{T-1}(j)$$

### 2.3 Credit Scoring Problem

Considering the problem of Credit scoring. we define the set of hidden states to be $S_0$ = good applicant, $S_1$ = bad applicants. At every time t, we will consider the hidden state inconditionally depends on the hidden state at time t − 1. Our work focuses on learning the emission structure not the transition structure.

A set of observed variables: Age, Salary and Housing..., assumed to be conditionally independent of each other. The observed variables at time t are conditionally depend on the hidden state at time t.

The Data used to applied in our proposed model contains numerical, categorical and continuous types. In order to obtain a suitable data for model, we dissect all continuous attributes, and we reorganized all data as sequence of numerical symbols for the training.

## 3 Experimental Setup

### 3.1 Data Description

The P2P dataset we used is from Lending Clubs dataset [17]. Excluding records containing obvious errors and the characteristics, with missing information, and by keeping accepted records, with both good/bad statuses observed, we got a dataset consisting of 799,443 issued loans including 158,592 defaults, with 8 attributes, which 4 numerical, and 4 categorical.

### 3.2 K Fold Corss Validation

To minimize the impact of data dependency and improve the reliability of the estimates, k-fold cross validation is used to create random partitions of the data sets. The procedure of k-fold cross-validation is as follow:

1. The data set is split into k mutually folds of nearly equal size.
2. Choose the first subset for training set and the k-1 remainder for training set.
3. Build the model on the training set.
4. Evaluate the model on the testing set by calculating the evaluation metrics.
5. Alternately choose the following subset for testing set and the k-1 remainder for training set.
6. The structure of the model is then trained k times each time using $k-1$ subsets (training set) for training and the performance of the model is evaluated k-1 on the remaining subset (testing set).
7. The predictive power of classifier is obtained by averaging the k validation fold estimates found during the k runs of the cross validation process.

The common values for k are sometimes 5 and 10. The Cross validation method is used in this work to assess the performance of classification techniques and we choose 10 as value for k for our experiments evaluation method. This approach can be computationally expensive, but does not waste too much data (as it is the case when fixing an arbitrary test set), and lower the variance of the estimate.

### 3.3 Handling the Imbalanced Dataset

We will handle the imbalanced data set by SMOTE (Synthetic Minority Over-sampling Technique). SMOTE is an over-sampling method, it creates synthetic samples of the minority class. Hence making the minority class equal to the majority class. SMOTE does this by selecting similar records and altering that record one column at a time by a random amount within the difference to the neighbouring records.

### 3.4 Evaluation Metrics

There are many metrics that can be used to measure the performance of a classifier different fields have different preferences for specific metrics due to different goals. In this paper the performance of the models used was first measured in terms of average, accuracy, precision, recall, and F1-score. We will also calculate the area under the curve (AUC) created from the K-fold cross-validation.

The following equations show the process to calculate the accuracy, precision, recalls and F1-score:

$$Accuracy = \frac{TP+TN}{TP+TN+FP+FN}$$

$$Precision = \frac{TP}{TP+FP}$$

$$Recall = \frac{TP}{TP+FN}$$

$$F1 = \frac{2(Precision*Recall)}{Precision+Recall} = \frac{2TP}{2TP+FP+FN}$$

Where true positive (TP) is the number of instances that actually belong to the good group that were correctly classified as good by the classifier, true negative (TN) is the number of instances that belong to the bad group and correctly classified as bad, false positive (FP) is the number of instances that are of the bad group but mistakenly classified as good, and, finally, false negative (FN) is the number of instances that are actually of good but incorrectly classified as bad

### 3.5 Computation

We conduct the experimental analysis for the proposed method and compared the results with other models, using Scikit-learn an open source Python library that provides a range of machine learning algorithm, and we used GridSearchCV for parameters setting and combinations.

### 3.6 Experiments Results

Table 1 compare the accuracy, accuracy, precision, recall, F1-score, and AUC for methods, namely LR, SGDC, RandomForest, SVM, KNN, and MDHMM, for the Lending Clubs dataset without over-sampling, performed with 10 fold cross validation.

**Table 1.**

| | Mesures of performance | | | | |
|---|---|---|---|---|---|
| | *Accuracy* | *Precision* | *Recall* | *AUC* | *F1-score* |
| SGDC | 79.94 | 44.03 | 1.87 | 64.13 | 3.48 |
| RandomForest | 75.97 | 27.65 | 13.06 | 56.82 | 17.70 |
| SVM | 77.85 | 4.09 | 5.30 | 53.32 | 3.29 |
| KNN | 79.86 | 36.92 | 2.10 | 60.73 | 3.97 |
| MDHMM | 73.29 | 86.70 | 58.13 | 59.35 | 62.41 |

The results arising from Lending Clubs dataset, show that, the best accuracy is obtained by SGDC and KNN 79% which is followed by SVM 77.85%, RandomForest 75.97% and MDHMM 73.29%. However, the Precision for MDHMM is highest than other classification models with 86.70% followed by SGDC and KNN with 44.03% and 36.92%. It is also interesting to note the recall for MDHMM gives the highest value 58.13% followed by RandomForest with just 13.06% and lower for the remaining models. For the AUC value obtained by the modles, the best value is obtained by SGDC and KNN with 64.13% 60.73% followed by MDHMM with 59.35%.

Table 2 compare the accuracy, accuracy, precision, recall, F1-score, and AUC for methods, namely LR, SGDC, RandomForest, SVM, KNN, and MDHMM, for the Lending Clubs dataset with over-sampling, performed with 10 fold cross validation.

**Table 2.**

| | Mesures of performance | | | | |
|---|---|---|---|---|---|
| | *Accuracy* | *Precision* | *Recall* | *AUC* | *F1-score* |
| SGDC | 53.06 | 31.35 | 59.97 | 64.42 | 40.00 |
| RandomForest | 84.15 | 85.68 | 79.07 | 90.06 | 79.11 |
| SVM | 51.51 | 59.54 | 65.93 | 55.96 | 46.45 |
| KNN | 53.44 | 50.73 | 81.29 | 62.79 | 62.58 |
| MDHMM | 61.15 | 72.53 | 58.65 | 62.08 | 64.85 |

The results after the over-sampling of the dataset, show that, the best accuracy and precision values is obtained by RandomForest followed by MDHMM, its noted that KNN, SVM, and SGDC have decreased in accuracy. However, its noted that those models shows significant improvement in precision and recall values.

From the results presented above, we can consider MDHMM model as an alternative way in credit scoring.

## 4 Conclusion

In this paper, implemented a new approach using MDHMM for credit scoring problem, and compared the classification performance with the baseline methods. The method was validated on real-world peer-to-peer lending datasets. The empirical results show that the proposed method have a high prediction rate in one hand, and on the other hand, by examining the precision and recall of MDHMM model also shows better performance than other methods widely used in credit scoring. A key advantage of the proposed system over the existing systems is that, we can build a set of models based on iterative procedure and performed in parallel, providing a efficient way for credit scoring systems with big data sources.

## References


1. Baesens, B., Van Gestel, T., Viaene, S., Stepanova, M., Suykens, J., Vanthienen, J.: Benchmarking state-of-the-art classification algorithms for credit scoring. J. Oper. Res. Soc. **54**(6), 627–635 (2003)
2. Henley, W.E., Hand, D.J.: Construction of a k-nearest neighbour credit-scoring system. IMA J. Math. Appl. Bus Ind. **8**, 305–321 (1997)
3. Sahin, Y., Bulkan, S., Duman, E.: A cost-sensitive decision tree approach for fraud detection. Expert. Syst. Appl. **40**, 5916–5923 (2013)
4. West, D.: Neural network credit scoring models. Comput. Oper. Res. **27**, 1131–1152 (2000)
5. Huang, C.L., Chen, M.C., Wang, C.J.: Credit scoring with a data mining approach based on support vector machines. Expert. Syst. Appl. **33**, 847–856 (2007)
6. Teng, G.-E., He, C.-Z., Xiao, J., Jiang, X.-Y.: Customer credit scoring based on HMM/GMDH hybrid model. Knowl. Inf. Syst. **36**(3), 731–747 (2013)
7. Rabiner, L.R.: A tutorial on Hidden Markov Models and selected applications in speech recognition. Proc. IEEE **77**, 257–286 (1989)
8. Chen, M.-Y., Kundu, A., et al.: Off-line handwritten word recognition using a hidden Markov model type stochastic network. IEEE Trans. Pattern Anal. Mach. Intell. **16**(5), 481–496 (1994). ISSN 0162-8828
9. Khadr, M.: Forecasting of meteorological drought using Hidden Markov Model (case study: the upper Blue Nile river basin, Ethiopia). Ain Shams Eng. J. **7**(1), 47–56 (2016). ISSN 2090-4479
10. Baum, L.E., Petrie, T., Soules, G., Weiss, N.: A maximization technique occurring in the statistical analysis of probabilistic functions of Markov chains. Ann. Math. Stat. **41**(1), 164–171 (1970)
11. Bilmes J.A.: A gentle tutorial of the EM algorithm and its application to parameter estimation for Gaussian mixture and Hidden Markov Models, U.C. Berkeley, TR-97-021 (1998). http://citeseer.ist.psu.edu/1570.html
12. Li, X., Parizeau, M., Plamondon, R.: Training Hidden Markov Models with multiple observations-a combinatorial method? IEEE Trans. Pattern Anal. Mach. Intell. **22**(4), 371–377 (2000)
13. Ye, F., Yi, N., Wang, Y.: EM algorithm for training high-order Hidden Markov Model with multiple observation sequences. J. Inf. Comput. Sci. **8**(10), 1761–1777 (2011)
14. Hadar, U., Messer, H.: High-order Hidden Markov Models? Estimation and implementation. In: Proceedings of the IEEE/SP 15th Workshop on Statistical Signal Processing, pp. 249–252 (2009)
15. Badreddine, B., Souad, B., Abdelhak, Z., Ismail, E.: Classification with Hidden Markov Model. Appl. Math. Sci. **8**(50), 2483–2496 (2014)
16. UCI machine learning repository. http://archive.ics.uci.edu/ml
17. https://www.lendingclub.com/info/download-data.action
18. Óskarsdóttir, M., et al.: The value of big data for credit scoring: enhancing financial inclusion using mobile phone data and social network analytics. Appl. Soft Comput. J. (2018)

# Pragmatic Method Based on Intelligent Big Data Analytics to Prediction Air Pollution

Samaher Al_Janabi(✉), Ali Yaqoob, and Mustafa Mohammad

Department of Computer Science, Faculty of Science for Women (SCIW), University of Babylon, Babylon, Iraq
samaher@itnet.uobabylon.edu.iq

**Abstract.** Deep learning, as one of the most popular techniques, is able to efficiently train a model on big data by using large-scale optimization algorithms. Although there exist some works applying machine learning to air quality prediction, most of the prior studies are restricted to several-year data and simply train standard regression models (linear or nonlinear) to predict the hourly air pollution concentration. The main purpose of this proposal is design predictor to accurately forecast air quality indices (AQIs) of the future 48 h. Accurate predictions of AQIs can bring enormous value to governments, enterprises, and the general public -and help them make informed decisions. We Will Build Model Consist of four Steps: (A) Determine the Main Rules (contractions) of avoiding emission (B) Obtaining and pre-processing reliable database from (KDD CUP 2018) (C) Building Predator have multi-level based on Long Short-term Memory network corporative with one of optimization algorithm called (Partial Swarm) to predict the PM2.5, PM10, and O3 concentration levels over the coming 48 h for every measurement station. (D) To evaluate the predictions, on each day, SMAPE scores will be calculated for each station, each hour of the day (48 h overall), and each pollutant (PM2.5, PM10, $SO_{x,}$ CO, O3 and $NO_x$). The daily SMAPE score will then be the average of all the individual SMAPE scores.

**Keywords:** Air pollutant prediction · Big data · Prediction · LSTM · PSO

## 1 Introduction

The proposed title deals with the study of gases that cause air pollution. The proposed title deals with the study of gases that cause air pollution. Air pollution has become a serious threat to large parts of the Earth's population [1]. Air pollution control has attracted the attention of governments, industrial companies and scientific communities. There are two sources of air pollution: the first source is natural sources such as volcanoes, forest fires, radioactive materials, etc. The second source is industrial sources produced by human activities such as factories, vehicles, remnants of war and others. These pollutants produce different types of gases, including sulfur oxides (SOx), carbon monoxide (CO), nitrogen oxides ($NO_x$) and ozone (O3). Another type of contaminant is $PM_{2.5}$ (particulate matter), a mixture of compounds (solids and liquid droplets), the most dangerous types of contaminants cause cardiovascular disease plus


Y. Farhaoui (Ed.): BDNT 2019, LNNS 81, pp. 84–109, 2020.
https://doi.org/10.1007/978-3-030-23672-4_8

pollutant $PM_{10}$. This work calculates concentrations of air pollutants (6 types of air pollution above) using a predictive model capable of handling large data efficiently and producing very accurate results. Therefore, a long short-term memory model using the swarm algorithm was used and developed. So that the network is trained to concentrate this polluted air on a number of stations and every hour and the results are evaluated using Root Mean error the scale of assessment so that this model can predict within the next 48 h. These data are analyses, conversion, testing, modelling, accurate information mining, restructuring and storage. This information helps us make decisions about these data and takes them across several stages before they are entered into the network. It is the data collection phase of the different version. This is the data problem to be expected. Are the determinants and stage of understanding the nature of these data in themselves? This data is a very important stage where the data we deal with contains missing data processed by each column and then these data are entered after analysis in a model developed and evaluated by SMAPE.

The remind of this paper organization as follow. Section 2 show the related work. Section 3 explain the main concept used to handle this problem. Section 4 show the prototype of pragmatic system while Sect. 5 show the results. Finally, Sect. 6 description the discussion and conclusion of the problem statement.

## 2 Related Works

The issue of air quality prediction is one of the vital topics that are directly related to the lives of people and the continuation of a healthy life in general. Since the subject of this thesis is to find a modern predictive way to deal with this type of data that is huge and operates within the field of data series. Therefore, in this part of the thesis, we will try to review the works of the former researchers in the same area of our problem and compare these works in terms of five basic points, namely the database used. The methods to assess the results, the advantages of the method, and their limitations.

Bun et al. [1] used "deep recurrent neural network (DRNN)" that is reinforced with the novel pre-training system using auto-encoder principally design for time series prediction. Moreover, sensors chosen is performed within DRNN without harming the accuracy of the predictions by taking of the sparsity found in the system. The method used to solve particulate matter ($PM_{2.5}$) that is one of the air populations. this method reveals results in more accuracy than the poor performance of the "Noise reduction (AE)". Evaluation the results based on four measures: "Root mean square error (RMSE), precision (P), Recall (R) and F-measure (F)". Our work is similar that work by using the same technique RNN while we predicate based on "long short-term memory".

Samaher et al. [2] the researcher found in this work the hybrid system uses "genetic neural computing (GNC)" to analyze and understand the resulting data from the concentration of dissolved gases. Where it is used in four subgroups for the purpose of analysis and assembly based on the C57.104 Specified by "IEEE" using "GA". Is inserted the clustering data into the neural network for the purpose of predicting different types of errors. This hybrid system generates decision rules which identify the error accurately. There are two measures used in this work are: "The Davies–Bouldin

(DB) index and MSE". This work has proven to provide less cost to solve the problem. And through this method facilitates the process of prediction and identify more accurate ideas through the analysis of errors and ways to address them. This work is similar to our work in terms of using neural networks but the difference is using a" Swarm" algorithm with "LSTM".

Xiang et al. [3] used model "spatiotemporal deep learning (STDL)"-based on air quality prediction method. It used a "stacked autoencoder (SAE)" method to extract inherent air quality characteristics, in addition, it is trained in a greedy layer-wise method. it compared your' model with traditional time series prediction models, the model can predict the air quality of all stations at the same time and shows the temporal stability in all seasons. In addition, a comparison with the "spatiotemporal artificial neural network (STANN)", "autoregression moving average (ARMA)", and "support vector regression (SVR)" models demonstrates. The results of that model evaluation by three measures are "RMSE", "MAE" "MAPE". Our work is similar that work by using the same technique RNN to prediction the air quality index but we deal with huge data, "long-short-term memory" ("LSTM") to enhance network work.

Xiang et al. [4] used "long short-term memory neural network" extended ("LSTME") model which is the association of spatial-temporal links to predict the concentration of air pollutants. Long-term memory layers (LSTM) have been used to automatically extract potential intrinsic properties from historical air pollutants data, and assistive data, Contains meteorological data and timestamp data, has been incorporated into the proposed model to improve performance. This technique evaluation by three measures "RMSE", "MAE" and "MAPE". A comparison with the "spatiotemporal artificial neural network (STANN)", "autoregression moving average (ARMA)", and "support vector regression (SVR)" models demonstrates. Our work is similar that work by using long-short-term memory ("LSTM") as part of the repeated neural network structure. While differ by using another evaluation measure.

Osama et al. [5] "shown a new deep learning-based ozone level prediction model, which considers the pollution and weather correlations integrally. This deep learning model is used to learn ozone level features, and it is trained using a grid search technique. A deep architecture model is utilized to represent ozone level features for prediction. Moreover, experiments demonstrate that the proposed method for ozone level prediction has superior performance. The outcome of this study can be helpful in predicting the ozone level pollution in Aarhus city as a model of smart cities for improving the accuracy of ozone forecasting tools". Results of that model evaluation based on "RMSE", "MAE", "MAPE", squared "$R^2$" and correlation coefficient. Our work will also use memory in ("LSTM") for the processing large data, but differ by finding the optimal structure of that neural network by partial swarm algorithm.

Lifeng et al. [6] the author found that obtaining the best air quality prediction uses the GM model (1.1) the fractional order accumulation (FGM (1.1)) to find the Expected the average annual concentrations of "PM2.5, PM10, SO2, NO2, and 8-h O3 And O-24 h". The measure used in this search is "MAPE". Using the method of "FGM (1.1)" obtained much higher than the traditional GM model (1.1), the average annual concentrations of "PM2.5, PM10, SO2, NO2, O8-O3, and O3 24-h" will decrease from 2017 to 2020. This work is similar to our work in terms of predicting the concentration of air pollutants and finding ways to address them, but the difference in terms of the method of prediction using "LSTM".

Olalekan et al. [7] in this work the method was used Sensor measurement: SNAQ boxes and network deployment, Sensor measurement validation, Source apportionment to create a predictive model for modeling (ADMS-Airport), the concentration of pollutants to determine the air quality model. The results showed in this study can be applied in many environments that suffer from air pollution. Which will reduce the potential health effects of air quality and the lowest cost, as well as monitor greenhouse gas emissions? This work is similar to our work to determine the concentration of air pollutants but the method used for our work is "LSTM –RNN".

Congcong et al. [8] in order to extract high temporal-spatial features have been used by the Merge of the "convolutional neural network (CNN)" and "long short-term memory neural network (LSTM-NN), and meteorological data and aerosol data were also Merge, in order to refinement model prediction performance. The data collected from 1233 air quality monitoring stations in Beijing and the whole of China were used to verify the effectiveness of the proposed C-LSTME model. The results show that the present model has achieved better performance than current state-of-the-art Technologies for different time predictions at various regional and environmental scales. This technique evaluation by three measures "RMSE", "MAE" and "MAPE". Our work used "LSTM" through RNN but after fined the best structure of that network. It differs by using another evaluation measures.

Zhigen et al. [9] the prediction method was based on "classification and regression tree ("CART")" and was combined with the "ensemble extreme learning machine (EELM)" method. Subgroups were created by dividing datasets by creating a shallow hierarchy tree by dividing the data set through "CART". Where at each node of the tree "EEL Models" are done using the training samples of the node, to minimize the verification errors sequentially to all the tree sub-trees by identifying the numbers of hidden intestines where each node is considered as root. Finally, EEL models for each path of the leaf is compared to the root on each leaf and only one path is selected with the smallest error checking the leaf. The measures that used are RMSE and MAPE. This method proved that the experimental results of the measurements used: can address the global-local duplication of the method of prediction in each leaf, "CART-EELM" work better than the models "RF, v-SVR, and EELM", "CART-EELM" also shows superior performance compared "EELM and K-means-EELM seasonal". Our work is similar to this work using the same set of six data for air pollution "(PM2.5, O3, PM10, SO2, NO2, CO)", but we differ in terms of the mechanism of reducing air pollutants where we use the RNN method.

Hongmin et al. [10] using a new air quality forecasting method, and proposing a new positive analysis mechanism that includes complex analysis, improved prediction units, pre-treatment data, and air quality control problems. This system analyzes the original series using the entropy model and data processing process. The "MOMVO" algorithm was used to achieve the required standards and "LSSVM" to achieve the best accuracy in addition to stable prediction. There are three ratings used in this work are: "RMSE", "MAE" and "MAPE". The result of the application of the proposed method on the data set showed a good performance in the analysis and control of air quality, in addition to the approximation of values with high precision. Our work used the same

**Table 1.** Compare among the previous works

| Name | Dataset/database | Method | Evaluation | Advantage |
|---|---|---|---|---|
| Bun et al. [1] | Air quality index "(AQI)" http://uk-air.defra.gov.uk | "(DRNN)" that is enhanced with a novel pre-training method using auto-encoder | • "(RMSE)"<br>• "(P)"<br>• "(R)"<br>• "(F)" | 1- The numerical experiments show that DRNN with our proposed pre-training method is superior to when using a canonical and a state-of-the-art auto-encoder training method when applied to time series prediction. The experiments confirm that when compared against the PM2:5 prediction system VENUS<br>2- NN was known as "(RNN)", in contrast with "(FNN)", has been shown to exhibit very good performance in modeling temporal structures and has been successfully applied to many real-world problems |
| Samaher et al. [2] | | • "(GNC)"<br>• "BPNN" | • "MSE"<br>• "(DB)" | This work has proven to provide less cost to solve the problem. And through this method facilitates the process of prediction and identify more accurate ideas through the analysis of errors and ways to address them |
| Xiang et al. [3] | Air quality using PM2.5 (http://datacenter.mep.gov.cn/). | 1- "(STDL)" based air quality prediction method<br>2- "(SAE)" | • "(RMSE)"<br>• "(MAE)"<br>• "(MAPE)" | 3- Compared with traditional time series air quality prediction models, our model was able to predict the air quality of all monitoring stations simultaneously, and it showed satisfactory seasonal stability. We evaluated the performance of the proposed method and compared it with the performance of the "STANN", "ARMA", and "SVR" models and the results showed that the proposed method was effective and outperformed the competitors |
| Xiang et al. [4] | Air quality using PM2.5 (http://datacenter.mep.gov.cn/) | • "(LSTM)" | • "(RMSE)"<br>• "(MAE)"<br>• "(MAPE)" | The "LSTME" model is capable of modeling time series with longtime dependencies and can automatically determine the optimum time lags. To evaluate the performance of our proposed model, Compared with Six different models, including our "LSTME", the traditional "LSTM" "NN", the "STDL", the "TDNN", the "ARMA" and the "SVR" models |
| Osama et al. [5] | Citypulse dataset. (ftp://http://iot.ee.surrey.ac.uk:) | 1-Deep learning approach<br>2-In-Memory computing | • "(RMSE)"<br>• "(MAE)"<br>• "(MAPE)"<br>• "($R^2$)"<br>• "(r)" | The proposed method is evaluated on citypluse dataset and compared with SVM, NN, and GLM models. the comparison results show that the proposed model is efficient and superior as compared to already existing models |

(*continued*)

**Table 1.** (*continued*)

| Name | Dataset/database | Method | Evaluation | Advantage |
|---|---|---|---|---|
| Lifeng et al. [6] | BBC, 2013. Beijing Smog: when Growth Trumps Life in China. www.bbc.com/news/magazine-21198265 | "FGM(1,1)" | • "RMSE"<br>• "MAPE" | Using the method of "FGM (1.1)" obtained much higher than the traditional GM model (1.1), the average annual concentrations of "PM2.5, PM10, SO2, NO2, O8-O3, and O3 24-h" will decrease |
| Olalekan et al. [7] | | • "SNAQ boxes and network deployment"<br>• Sensor measurement validation.<br>Source apportionment | | The results showed in this study can be applied in many environments that suffer from air pollution |
| Congcong et al. [8] | Hourly PM2.5 concentration data from 1233 air quality monitoring stations in China collected from January 1, 2016, to December 31, 2017, were acquired from the Ministry of Environmental Protection of China (http://datacenter.mep.gov.cn/). | combination of the "(CNN)" and "(LSTM-NN)" | • "(RMSE)"<br>• "(MAE)"<br>• "(MAPE)" | (1) The addition of PM2.5 information from neighboring stations, which contributes to the spatiality of the data, can considerably improve the prediction accuracy of the model<br>(2) The supplement of auxiliary data can help predict sudden changes in air quality, thereby improving the prediction performance of the model. Moreover, compared to meteorological data, the "AOD" data contributes more to the accuracy of the model<br>(3) The present model can efficiently extract more essential spatiotemporal correlation features through the combination of "3D-CNN" and stateful "LSTM", thereby yielding higher accuracy and stability for air quality prediction of different spatiotemporal scales |

(*continued*)

**Table 1.** (*continued*)

| Name | Dataset/database | Method | Evaluation | Advantage |
|---|---|---|---|---|
| Zhigen et al. [9] | Yancheng city, which is one of the 13 cities under the direct administration of Jiangsu Province, China. Yancheng city spans between northern latitude 32°34′–34°28′, eastern longitude 119°27′–120°54′ | "CART" and "EELM" | • "(RMSE)"<br>• "(MAPE") | The experimental results of the measurements used: can address the global-local duplication of the method of prediction in each leaf, "CART-EELM" work better than the models "RF, v-SVR, and EELM", "CART-EELM" also shows superior performance compared "EELM and K-means-EELM seasonal |
| Hongmin et al. [10] | The datasets from eight sites in China. (https://data.epmap.org) | • "LSSVM"<br>• "MOMVO" | • "(RMSE)"<br>• "(MAE)"<br>• "(MAPE)" | The application of the proposed method on the data set showed a good performance in the analysis and control of air quality, in addition to the approximation of values with high precision |

evaluation measures but it differs by using "LSTM" through RNN but after fined the best structure of that network.

Table 1 Shown the compare among the previous works based on five points are types of dataset, methodology used, evaluation measures, and advantages.

# 3 Main Concept

## 3.1 Big Data

There are three types of data. The first type is small data: these data are not subject to normal distribution and therefore cannot be determined and difficult to predict due to their small size and less than 30 samples, the second type is the normal data is the most common data that is organized in the table consists of columns and rows, For large data that we will deal with in our work. These big data cannot be handled in the traditional way due to their large size. From TB to ZB, these big data are either structural, semi-structured or unorganized. In large data, there are three conditions that must be met: First, the existence of available data called the data domain has three characteristics of volume, Velocity and Variety. The second condition: There is a statistic for this data called (statistical domain), which has three characteristics are Veracity, Variability and Validity. The third condition is the existence of the beneficiary of this data called (intelligence analysis domain) which has three characteristics of Visibility, value and Verdict [11].

## 3.2 Big Data Analysis Stages

The attention to large data has become of great importance to many organizations and companies over the last few years so many technologies have been developed to meet the challenges of dealing with huge data The process of analysis of the statement through several stages: the data deployment stage is the stage of data collection from different sources, the stage of understanding the problem: the phase of understanding the problem of data and the problem of this proposal are the gases and vehicles that result from air pollution, data exploration stage: After understanding the data problem, Itself. Does this data contain missing data? Is this data of an integer type, type of characters, or mixture between them?, data pre-processing stage This is a very important stage because the data we deal with contains missing data to be processed by each column in addition to the filtering of data. After processing the data, we need to build a predictive model for this data, that called the data modelling stage and then evaluate the data using one of the different measurements called the data assessment stage [12].

## 3.3 Deep Learning

It is a new research field that examines the creation of theories and algorithms that allow the machine to learn by itself by simulating neurons in the human body. And one of the branches of science that deals with artificial intelligence science. From an

Automated Learning Science branch, most in-depth learning research focuses on finding methods to derive a high degree of strippers by analyzing a large data set using linear and nonlinear mutants. "Discoveries in this area have proven significant, rapid and effective progress in many areas, including face recognition, speech recognition, computer vision, and natural language processing." Deep learning is a branch of learning, which is a branch of artificial intelligence. Where deep learning is divided into three basic categories: Deep networks under supervision or deep discrimination Models, deep unsupervised networks or giant models, deep hybrid networks" [13].

## 3.4 Prediction

Prediction is the process of studying and making guesses about future time events using specific mechanisms across time periods and different spatial intervals. It is a way for policymakers to develop plans to address future risks. The process of prediction goes through several stages, the most important of which are: determining the purpose of forecasting, gathering the data needed for a predictable phenomenon, analyzing and extracting data, and selecting the appropriate model or mechanism to predict the phenomenon under consideration and to take the appropriate decision. The forecasting process is capable of predicting the state of data based on advance data, a necessary step to give very low error rates [14] (Fig. 1).

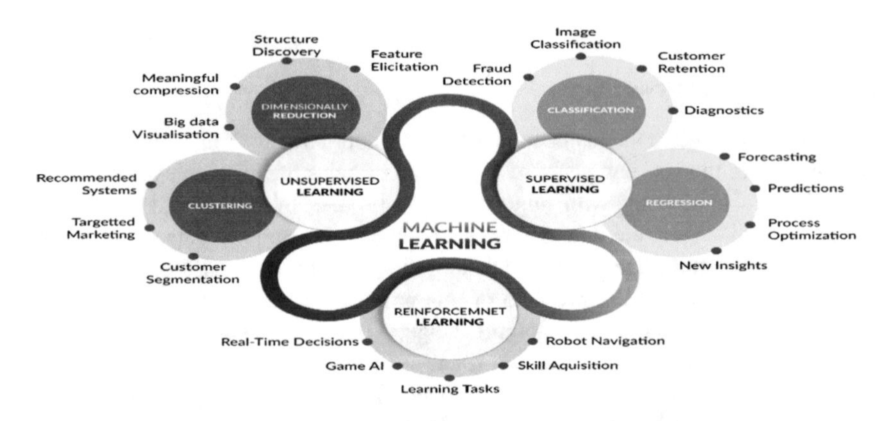


**Fig. 1.** Main Types of Machine Learning Techniques [19]

## 3.5 Air Pollution

Pollution is a serious concern that has attracted the attention of industrial companies, the international community and scientific and health communities because it contains substances harmful to the lives of humans and other living organisms. These pollutants produce different types of concentrations that cause air pollution, including sulfur dioxide $SO_2$, Nitrogen Dioxide gas $NO_2$, Carbon monoxide gas CO, Ozone gas $O_3$, Vehicles with mixed composition $PM_{2.5}$ and $PM_{10}$. Predicting concentrations that cause air pollution is very important for health community organizations and to launch an early warning and decide on these concentrations [15].

# 4 A Prototype of Smart Air Quality Prediction Model (SAQPM)

## 4.1 Problem Statement

The problem of air pollution has serious health aspects to humans and other living organisms because of the presence of harmful concentrations and very dangerous. So the prediction of air pollution attracted many companies, governments, and the scientific community to study these concentrations that cause air pollution. Where many techniques were used but did not produce results at the required level for several reasons, including Most of the techniques used cannot deal with very large data in addition to some techniques lose data because they do not contain a memory able to save this data until deep learning techniques emerged.

Deep learning, as one of the most popular techniques, is able to efficiently train a model on big data by using large-scale optimization algorithms. Although there exist some works applying machine learning to air quality prediction, most of the prior studies are restricted to several-year data and simply train standard regression models (linear or nonlinear) to predict the hourly air pollution concentration. The main purpose of this LSTM-RNN model is to predict future concentrations of air pollution 48. The challenge in this work is how to choose an algorithm that can predict the large data set at high resolution, taking into account that previous readings are not ignored, that is, they maintain all readings.

Objectives:

We will review the main objectives of this research (Fig. 2)

- Identify key constraints to avoid emissions.
- To give a specific definition of air pollution.
- Build LSTM-RNN prediction model in conjunction with a swarm algorithm to predict. The air-polluting concentration of each station and each hour to predict the concentrations of air pollution during future 48.
- Validating the suggested LSTM-RNN force using SMAPE to measure results.

In this paper, we will attempt to find answer for the following questions

- How particle swarm can be useful in building a recurrent neural network (RNN)?
- How to build a multi-layer model with a combination of two technologies) LSTM-RNN with particle swarm) so that this model can predict the concentrations of air pollution accurately and efficiently over the next 48 h?
- Is SMAPE measure enough to evaluate the results of suggesting predictor?
- How to install sensors and get data related to concentrations of air pollution?
- How can you combine two deep learning techniques that lead to reduced time and the complexity of the training phase?

## 4.2 Particle Swarm Optimization (PSO)

Is a scientific technique developed to improve solutions problems and solutions to solve the best solution to these problems and is one of the latest areas of evolution in

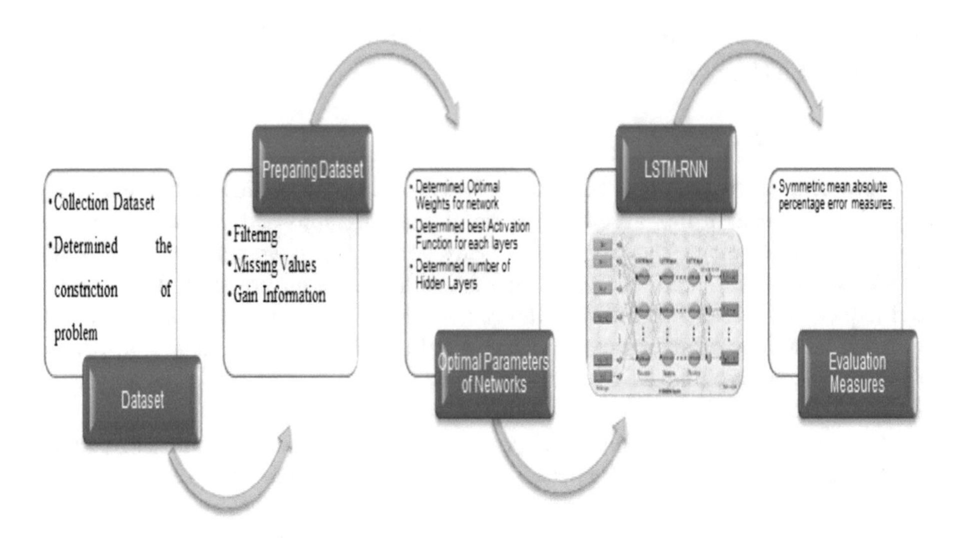


**Fig. 2.** Block Diagram of Proposed DLSTM-RNN

the field of artificial intelligence, developed by the world (Kennedy and Eberhart) in 1990, and the idea of PSO of the behavioral and social behavior of bird droppings through the idea of research About the food, where the bird squadron are looking for food from one place to another and that some birds in the squadron have the ability to distinguish the smell of food in a strong and effective with information for these birds about the best place to eat because some birds send information among themselves during the search process and Inspection of the best place for food, when the bird flock is explored for a good place for food quality, these birds use this place to get better food. Thus, the work of the swarm algorithm is the search process and replication process of the best solutions within the specific research area. The PSO algorithm can be used to solve optimization problems and problems that change by the time [16, 17].

### 4.2.1 Basic Components of the Bird Swarm Algorithm (PSO)

The swarm algorithm consists of the numbers of the population of the swarm called particles. Symbolizes them (n) which consists of n = (n1, n2, …, ni) which moves within the swarm of the search space determined by the type of problem which is multi-dimensional and the search for good initial solutions. Particles depend on their own expert and also rely on the experts and experiences of neighboring particles within the swarm, The PSO algorithm is randomly assigned to the number of particles of the squadron in the search area. The use of the squadron particles when creating the PSO algorithm depends on the velocity of the particles that comprise $V_i^t = (V_1^t, V_2^t, \ldots, V_i^t)$ the location of the particles that are composed of $X_i^t = \left(X_1^t, X_2^t, \ldots, X_i^t\right)$, where it is determined based on the previous cases of the best location of the particle itself and symbolizes it $P_{best,i}^t$, The best location of the particles in the entire swarm symbolizes it

$G^t_{best,i}$, Depending on the dimensions of problem d, consisting of (d1, d2,…dj), The speed and location of each particle are adjusted according to the following equations:

$$V_i^{t+1} = (V_1^t + c_1 r_1^t \left(P^t_{best,i} - X_i^t\right) + c_2 r_2^t \left(G^t_{best,i} - X_i^t\right) \quad (1)$$

$$X_i^{t+1} = X_i^t + V_i^t \quad (2)$$

Where:

$V_i^t$: Particle velocity i in swarm in dimension j and frequency t.
$X_i^t$: The location of the particle i in a swarm in dimension j and frequency t.
$c_1$: acceleration factor related to Pbest.
$c_2$: Acceleration factor related to gbest.
$r_1^t$, $r_2^t$: random number between 0 and 1.
t: Number of occurrences specified by type of problem.
$G^t_{best,i}$: gbest position of swarm
$P^t_{best,i}$: pbest position of particle.

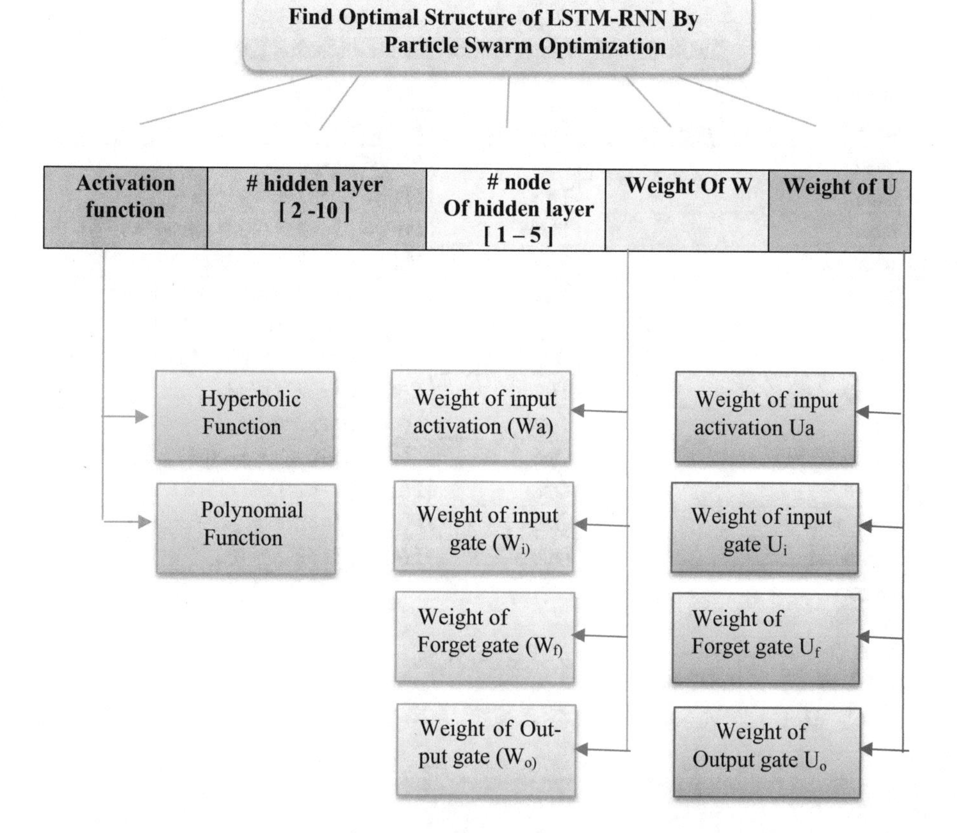


**Fig. 3.** PSO algorithm is to optimize the LSTM-RNN

The aim of the PSO algorithm is to optimize the structure of LSTM-RNN from determined the best activation function for each layer, determined number of hidden layers in network, number of neurons in each hidden layer, and the optimal weights between each layer and the layer next it as described in the diagram (Fig. 3 and Tables 2 and 3):

**Table 2.** Hyperbolic Functions with discretion [20]

| Hyperbolic function | | |
|---|---|---|
| Name of hyperbolic function | #variable | Function |
| Sinh | One | F(x) = $\frac{e^x - e^{-x}}{2}$ |
| Cosh | One | F(x) = $\frac{e^x + e^{-x}}{2}$ |
| Tanh | One | F(x) = $\frac{e^x - e^{-x}}{e^x + e^{-x}}$ |
| $\text{Sinh}^{-1}$ | One | F(x) = $\frac{2}{e^x - e^{-x}}$ |
| $\text{Cosh}^{-1}$ | One | F(x) = $\frac{2}{e^x + e^{-x}}$ |
| $\text{Tanh}^{-1}$ | One | F(x) = $\frac{e^x + e^{-x}}{e^x - e^{-x}}$ |

Where: x is input and F(x) is output.

**Table 3.** Polynomial Functions with discretion [20]

| Polynomial functions | | |
|---|---|---|
| Name of polynomial function | #variable | Function |
| Linear | One | F(x) = p1+p2*x1 |
| Linear | Tow | F(x) = p1+p2*x1+p3*x2 |
| Linear | Three | F(x) = p1+p2*x1+p3*x2+p4*x3 |
| Quadratic | One | F(x) = p1+p2*x1+p3*x1^2 |
| Quadratic | Tow | F(x) = 1+p2*x1+p3*x1^2+p4*x2 +p5*x2^2+p6*x1*x2 |
| Cubic | One | F(x) = p1+p2*x1+p3*x1^2 +p4*x1^3 |
| Product | Tow | F(x) = p1+p2*x1*x2 |
| Ratio | Tow | F(x) = p1+p2*(x1/x2) |
| Logistic | One | F(x) = p1+p2/(1+ exp(p3*(x1-p4))) |
| Log | One | F(x) = p1+p2*Log(x1+p3) |
| Exponential | One | F(x) = p1+p2*exp((p3*(x1+p4)) |
| Asymptotic | One | F(x) = p1+p2/(x1+p3) |

Where:
x is input
f(x) is output
p1, p2, p3, p4: is constants

Algorithm #1: PSO [21]

1. Set of parameters
2. ***A***: Population of agents, $p_i$: Position of agent $a_i$ in the solution space, $f$: Objective function
3. $v_i$ : Velocity of agent's $a_i$ , $V(a_i)$: Neighborhood of agent $a_i$ (fixed)
4. [x*] = PSO()
5. P = Particle_Initialization()
6. For *i*=1 to *it_max*
7. | For each particle *p* in *P* do
8. | | *fp* = f(*p*)
9. | | If *fp* is better than f(*pBest*)
10. | | | *pBest* = *p*
11. | | End
12. | End
13. | *gBest* = best *p* in *P*
14. | For each particle *p* in *P* do
15. | | # $v = v + c1*rand*(pBest - p) + c2*rand*(gBest - p)$
16. | | # $p = p + v$
17. | End
18. End

## 4.3 Long Short-Term Memory (LSTM)

LSTM was proposed in 1997 by Sepp Hochreiter and Jürgen Schmidhuber. By introducing Crossover Fixed Error (CEC) modules, LSTM deals with gradient and burst problems. The initial version of the block included LSTM cells, input and output gateways. LSTM achieved record results in natural language text compression, unauthorized handwriting recognition and won the Handwriting Competition (2009). LSTM networks were a key component of the network, which achieved a standard audio error rate of 17.7% over the traditional natural speech data set (2013). (LSTM) are units of the Recurrent Neural Network (RNN). The RNN is often called LSTM (or LSTM only). The common LSTM module consists of a cell, an input port, an output port, and a forgotten gateway. The cell remembers values at random intervals and the three gates regulate the flow of information inside and outside the cell. LSTM networks are well suited for classifying, processing and predicting predictions based on time series data, where there may be an unknown delay for important events in a time series. LSTMs have been developed to deal with fading and fading problems that can be encountered when training traditional RNNs. The relative lack of sense of gap length is the LSTM feature on RNNs, Hidden Markov models and other sequential learning methods in many applications [18].

### 4.3.1 LSTM Architecture and Algorithm

There are many structures for LSTM modules. The common structure of a cell (the memory part of the LSTM module) and three "organizations", usually called portals, consist of the flow of information within the LSTM module: an input gateway, an output gateway, and a forgotten gateway. Some differences in the LSTM module do not contain one or more of these portals or may have other portals. For example, duplicate units do not contain portals (GRUs) on the output portal.

Intuitively, the cell is responsible for tracking dependencies between elements in the input sequence. The input gateway controls the extent to which a new value is flowing into the cell. The ny gate controls how long a value in the cell and the output gateway controls how much the value in the cell is used to calculate the output activate the LSTM module. LSTM port activation function is often the logistics function.

There are connections to and from LSTM portals, a few of which are frequent. The weights of these links, which must be learned during training, determine how the gates work.

### 4.3.2 The Variables in LSTM – RNN

This algorithm required setup multi variables at the beginning then through it work will update these variables by apply computation operations. as shown below (Fig. 4):

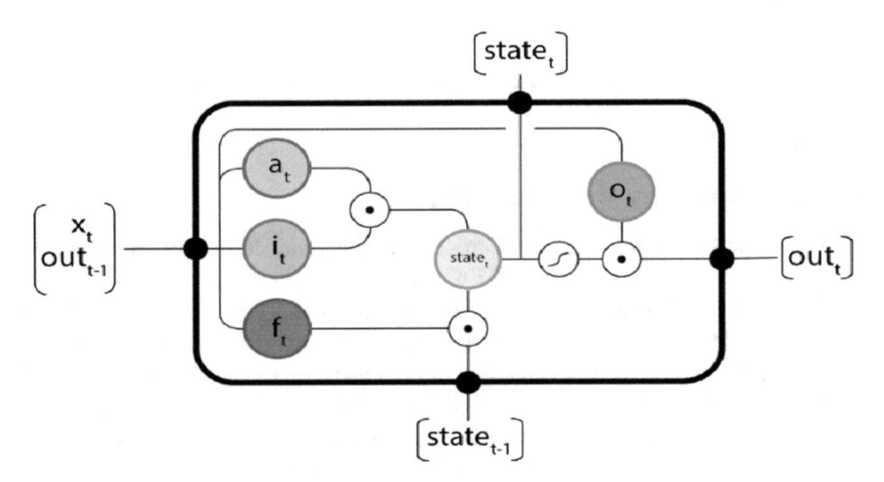


**Fig. 4.** LSTM cell

**Step 1: The forward components**
***Step 1.1***: *Compute the gates:*
Input activation:

$$\mathrm{a_t} = \tanh(\mathrm{W_a}.\ \mathrm{X_t} + \ \mathrm{U_a}.\ \mathrm{out_{t-1}} + \mathrm{b_a}) \tag{3}$$

Input gate:

$$\mathrm{i_t} \ = \sigma(\mathrm{WI.}\ \mathrm{X_t} + \ \mathrm{U_i}.\ \mathrm{out_{t-1}} + \mathrm{b_i}) \tag{4}$$

Forget gate:

$$\mathrm{f_t}\sigma\,(\mathrm{Wf.X_t} + \ \mathrm{U_f}.\ \mathrm{out_{t-1}} + \mathrm{b_f}) \tag{5}$$

Output gate:

$$\mathrm{o_t}\sigma(\mathrm{Wo.}\ \mathrm{X_t} + \ \mathrm{U_o}.\ \mathrm{out_{t-1}} + \mathrm{b_o}) \tag{6}$$

Then fined:
Internal state:

$$\mathrm{State} = \ \mathrm{a_t} \odot \mathrm{i_t} + \ \mathrm{f_t} \odot \ \mathrm{state_{t-1}} \tag{7}$$

Output:

$$\mathrm{outt} = \tanh(\mathrm{state}) \ \odot \ \mathrm{ot} \tag{8}$$

Where

$$\text{Gate } \mathrm{S_t} = \begin{bmatrix} a_t \\ i_t \\ f_t \\ o_t \end{bmatrix}, \mathrm{W} = \begin{bmatrix} W_a \\ W_i \\ W_f \\ W_o \end{bmatrix}, \mathrm{U} = \begin{bmatrix} U_a \\ U_i \\ U_f \\ U_o \end{bmatrix}, \mathrm{b} = \begin{bmatrix} b_a \\ b_i \\ b_f \\ b_o \end{bmatrix}$$

**Step. 2: The backward components:**
**Step 2.1.** Find
$\Delta t$ the output difference as computed by any subsequent.
$\Delta OUT$ the output difference as computed by the next time-step

$$\delta \mathrm{out_t} = \Delta_\mathrm{t} + \Delta \mathrm{out_t} \tag{9}$$

$$\delta state_t = \delta out_t \odot o_t \odot \left(1 - tanh^2(state_t)\right) + \delta state_{t+1} \odot f_{t+1} \tag{10}$$

**Step 2.2:** Gives:

$$\delta a_t = \delta state_t \odot i_t \odot (1 - a_t^2) \tag{11}$$

$$\delta i_t = \delta state_t \odot a_t \odot i_t \odot (1 - i_t) \tag{12}$$

$$\delta f_t = \delta state_t \odot state_{t-1} \odot f_t \odot (1 - f_t) \tag{13}$$

$$\delta o_t = \delta out_t \odot \tanh(state_t) \odot o_t \odot (1 - o_t) \tag{14}$$

$$\delta x_t = W^t.\delta state_t \tag{15}$$

$$\delta out_{t-1} = U^t.\delta state_t \tag{16}$$

**Step 3:** update to the internal parameter

$$\delta W = \sum_{t=0}^{T} \delta gates_t \otimes x_t \tag{17}$$

$$\delta U = \sum_{t=0}^{T} \delta gates_{t+1} \otimes out_t \tag{18}$$

$$\delta b = \sum_{t=0}^{T} \delta gates_{t+1} \tag{19}$$

Algorithm #2: DLSTM-RNN

---

Input: X: Dataset of Air Pollution
Output: prediction value of $PM_{2.5}$, $PM_{10}$, $O_3$, $SO_2$, CO, $NO_2$

1. Set of Parameters: $X_t$, $OUT_t$, $W_a$, $W_i$, $W_f$, $W_o$, $U_a$, $U_i$, $U_f$, $U_o$, $b_a$, $b_i$, $f_t$, $b_o$, $STEAT_t$.
2. The forward components
3. Call PSO
4. For each Time (t) in X and OUT, apply Dataset by:
5. Compute: $a_t$, $i_t$, $f_t$, $o_t$.
6. # Input activation: $a_t = \tanh(W_a . X_t + U_a. out_{t-1} + b_a)$
7. # Input gate: $i_t = \sigma(WI. X_t + U_i. out_{t-1} + b_i)$
8. # forgte gate: $f_t = \sigma(Wf. X_t + U_f . out_{t-1} + b_f)$
9. # Output gate: $O_t = \sigma(Wo. X_t + U_o . out_{t-1} + b_o)$
10. Compute $STATE_t$, $OUT_t$
11. # internal state : $state_t = a_t \odot i_t + f_t \odot state_{t-1}$
12. # output : $out_t = \tanh(state_t) \odot o_t$
13. The Backward components.
14. Given $\Delta t$ the output difference as computed by any subsequent.
15. Given $\Delta OUT$ the output difference as computed by the next time-step
16. For each time (t) to update OUTt,$STATEt$
17. # $\delta out_t = \Delta_t + \Delta out_t$
18. # $\delta state_t = \delta out_t \odot o_t \odot (1 - tanh^2(state_t)) + \delta state_{t+1} \odot f_{t+1}$
19. For each time (t) to update of input Xt and ΔOUTt.
20. Update of Input activation, Input gate, Forget gate and Output gate.
21. # $\delta a_t = \delta state_t \odot i_t \odot (1 - a_t^2)$
22. # $\delta i_t = \delta state_t \odot a_t \odot i_t \odot (1 - i_t)$
23. # $\delta f_t = \delta state_t \odot state_{t-1} \odot f_t \odot (1 - f_t)$
24. #$\delta o_t = \delta out_t \odot \tanh(state_t) \odot o_t \odot (1 - o_t)$
25. # $\delta x_t = W^t . \delta state_t$
26. # $\delta out_{t-1} = U^t . \delta state_t$
27. END
28. END
29. The final updates to the internal parameters is compute:
30. # $\delta W = \sum_{t=0}^{T} \delta gates_t \otimes x_t$
31. # $\delta U = \sum_{t=0}^{T} \delta gates_{t+1} \otimes out_t$
32. # $\delta b = \sum_{t=0}^{T} \delta gates_{t+1}$
33. Using SMAPE to evaluate the resulted.
34. # SAMPE $= \frac{1}{n} \sum_{t=1}^{n} \frac{|F_t - A_t|}{|A_t + F_t|/2}$ .

END.

---

## 5 Experiment

Using (DLSTM) networks in Python and how you can use them to make appropriate with concentrations of air pollution predictions! In this network, we will see how you can use a time-series model known as Long Short-term Memory. LSTM models are powerful, especially for retaining a long-term memory, by design, as you will see later.

### 5.1 Data and Method

We will be using data from KDD cup 2018, where contain the name of the station and Pollution time for each of the following concentrations per hour (Table 4).

**Table 4.** Before handle the missing values

| No | Utc_Time | PM2.5 | PM10 | NO2 | O3 | CO | SO2 |
|---|---|---|---|---|---|---|---|
| 6 | 1/1/2017 19:00 | 429.0 | 141.0 | 6.5 | 3.0 | 9.0 | NaN |
| 7 | 1/1/2017 20:00 | 211.0 | 110.0 | 3.3 | 11.0 | NaN | NaN |
| …<br>… | | | | | | | |
| 309327 | 11/24/2017 1:00 | 19.0 | 20.0 | 19.0 | 0.3 | 28.0 | 2.0 |
| 309329 | 11/24/2017 2:00 | 9.0 | 21.0 | 27.0 | 28.0 | 0.4 | 2.0 |

The concentrations are: $PM_{2.5,}$ $PM_{10,}$ Sox, CO, NOx, O3.

The table above shows that there are incomplete values that will be preprocessed by a MEAN equation.

Then we will preprocess the missing values through each column (Table 5):

**Table 5.** After handle the missing values

| No | Utc_Time | PM2.5 | PM10 | NO2 | O3 | CO | SO2 |
|---|---|---|---|---|---|---|---|
| 6 | 1/1/2017 19:00 | 429.0 | 141.0 | 6.5 | 3.0 | 9.0 | 11.212 |
| 7 | 1/1/2017 20:00 | 211.0 | 110.0 | 3.3 | 11.0 | 15.78 | 11.212 |
| …<br>… | | | | | | | |
| 309327 | 11/24/2017 1:00 | 19.0 | 20.0 | 19.0 | 0.3 | 28.0 | 2.0 |
| 309329 | 11/24/2017 2:00 | 9.0 | 21.0 | 27.0 | 28.0 | 0.4 | 2.0 |

After processing for each column using MEAN the following table shows the description of the results in the previous table (Table 6).

**Table 6.** The description of the data after the preprocessing

| | PM2.5 | PM10 | NO2 | CO | O3 | SO2 |
|---|---|---|---|---|---|---|
| Count | 200.000 | 200.000 | 200.000 | 200.000 | 200.000 | 200.000 |
| Mean | 179.949 | 134.376 | 27.180 | 17.205 | 15.788 | 11.212 |
| Std | 131.835 | 123.790 | 56.373 | 20.671 | 11.056 | 2.788 |
| Min | 5.000 | 4.600 | 0.200 | 0.200 | 2.000 | 2.000 |
| Max | 500.000 | 561.000 | 208.000 | 79.000 | 61.000 | 37.000 |

## 5.2 Data Visualization

Now let's see what sort of data you have. You want data with various patterns occurring over time (Fig. 5).

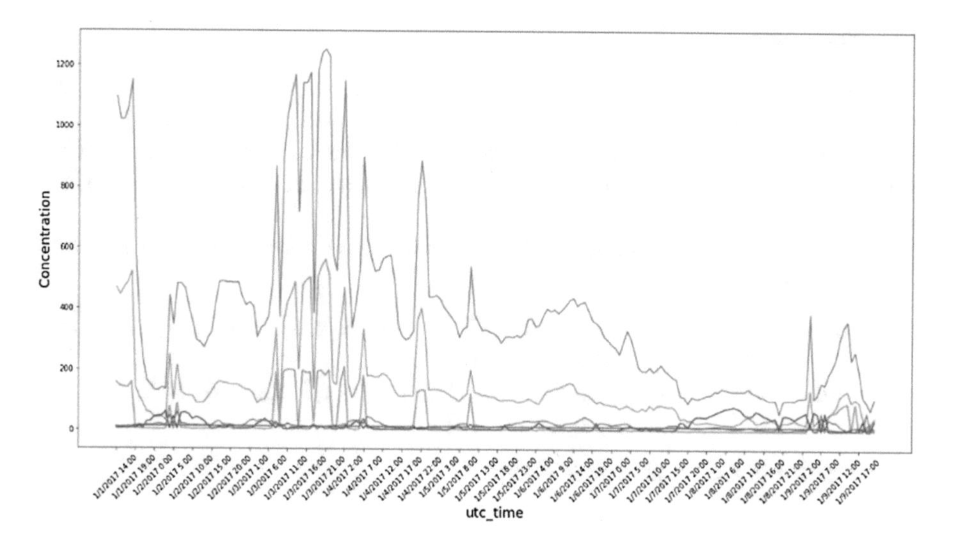


**Fig. 5.** Data Visualization

This graph already says a lot of things. The specific reason I picked this data that this graph is bursting with different behaviors of for concentrations of air pollution over time. This will make the learning more robust as well as give you a chance to test how good the predictions are for a variety of situations.

Another thing to notice is that the values at the beginning 2017 are much higher and fluctuate more than the values close to the last days. Therefore, you need to make sure that the data behaves in similar value ranges throughout the time frame. You will take care of this during the data normalization phase.

## 5.3 Normalizing the Data

Before the normalizing, we will the Data process Splitting Data into a Training set and a Test set, where we use the 70% for training and 30 for testing.

Now we need to define a scaler to normalize the data. Min Max Scalar scales all the data to be in the region of 0 and 1. You can also reshape the training and test data to be in the shape [data_size, num_features].

Due to the observation, we made earlier, that is, different time periods of data have different value ranges, and we normalize the data by splitting the full series into windows. If we don't do this, the earlier data will be close to 0 and will not add much value to the learning process. Here you choose a window size of 2500.

We can now smooth the data using the exponential moving average. This helps us to get rid of the inherent raggedness of the data concentrations and produce a smoother curve. Note that we should only smooth training data.

## 5.4 Data Generator

We are first going to implement a data generator to train our model. This data generator will have a method called. unroll batches (…) which will output a set of num_unrollings batches of input data obtained sequentially, where a batch of data is of size [batch_size, 1]. Then each batch of input data will have a corresponding output batch of data.

For example, if num_unrollings=3 and batch_size=4 a set of unrolled batches It might look like

- **input data**: [$x_0$, $x_1 0$, $x_2 0$, $x_3 0$, $x_4 0$, $x_5 0$], [$x_1$, $x_1 1$, $x_2 1$, $x_3 1$, $x_4 1$, $x_5 1$], [$x_2$, $x_1 2$, $x_2 2$, $x_3 2$, $x_4 2$, $x_3 5$]
- **output data:** [$x_1$, $x_1 1$, $x_2 1$, $x_3 1$, $x_4 1$, $x_5 1$], [$x_2$, $x_1 2$, $x_2 2$, $x_3 2$, $x_4 2$, $x_5 2$], [$x_3$, $x_1 3$, $x_2 3$, $x_3 3$, $x_4 3$, $x_5 3$]

As shown in the results below (Table 7):

**Table 7.** Data generator to train our model

| *Unrolled index 0:* | |
|---|---|
| Inputs: | [0.86032474 0.79311657 0.79409707 0.8310883 0.90970576 0.79311657] |
| Output: | [0.86032474 0.90970576 0.90970576 0.09216607 0.90970576 0.90970576] |
| *Unrolled index 1:* | |
| Inputs: | [0.79311657 0.79409707 0.8310883 0.90970576 0.90970576 0.79409707] |
| Output: | [0.8310883 0.41866928 0.8310883 0.09216607 0.41866928 0.90970576] |
| … | |
| … | |

(*continued*)

**Table 7.** (*continued*)

| *Unrolled index 132:* | |
|---|---|
| *Inputs:* | [0.79409707 0.86032474 0.8310883 0.90970576 0.79409707 0.90970576] |
| *Output:* | [0.2070513 0.79409707 0.90970576 0.2070513 0.2070513 0.02807767] |
| *Unrolled index 133:* | |
| *Inputs:* | [0.8310883 0.79311657 0.90970576 0.8310883 0.8310883 0.79311657] |
| *Output:* | [0.2070513 0.41866928 0.02807767 0.90970576 0.2070513 0.79409707] |

# 6 Discussions and Conclusions

In this paper, we used PSO algorithm to find the parameter of best structure related to RNN, these parameters include activation function (i.e., one from six types of Hyperbolic, or one from twelve types of Polynomial Functions), and then from the Experiment we find the best Hyperbolic function is Tanh and the best Polynomial Functions is Linear (Three parameter).

Through the PSO we find the best number of hidden layers is three and the best number of nodes is four. Also, find the best weights of the input between the input layer and hidden layer are (Table 8):

**Table 8.** The weights of the input between input and first hidden layers

| Node 1 | | | | Node 2 | | | | Node 3 | | | |
|---|---|---|---|---|---|---|---|---|---|---|---|
| $W_a$ | $W_i$ | $W_f$ | $W_o$ | $W_a$ | $W_i$ | $W_f$ | $W_o$ | $W_a$ | $W_i$ | $W_f$ | $W_o$ |
| 0.378 | 0.199 | 0.305 | 0.506 | 0.163 | 0.263 | 0.979 | 0.992 | 0.545 | 0.576 | 0.376 | 0.889 |
| 0.059 | 0.716 | 0.352 | 0.521 | 0.411 | 0.986 | 0.836 | 0.793 | 0.928 | 0.174 | 0.177 | 0.901 |
| 0.399 | 0.228 | 0.628 | 0.243 | 0.728 | 0.440 | 0.519 | 0.826 | 0.204 | 0.838 | 0.954 | 0.411 |
| 0.663 | 0.708 | 0.069 | 0.910 | 0.725 | 0.698 | 0.181 | 0.902 | 0.522 | 0.695 | 0.121 | 0.136 |
| Node 4 | | | | Node5 | | | | Node6 | | | |
| $W_a$ | $W_i$ | $W_f$ | $W_o$ | $W_a$ | $W_i$ | $W_f$ | $W_o$ | $W_a$ | $W_i$ | $W_f$ | $W_o$ |
| 0.486 | 0.219 | 0.191 | 0.990 | 0.091 | 0.648 | 0.422 | 0.156 | 0.515 | 0.797 | 0.680 | 0.883 |
| 0.018 | 0.968 | 0.928 | 0.564 | 0.352 | 0.405 | 0.386 | 0.865 | 0.940 | 0.938 | 0.253 | 0.580 |
| 0.062 | 0.893 | 0.098 | 0.950 | 0.159 | 0.136 | 0.356 | 0.475 | 0.951 | 0.447 | 0.096 | 0.827 |
| 0.725 | 0.823 | 0.400 | 0.291 | 0.862 | 0.312 | 0.074 | 0.475 | 0.303 | 0.544 | 0.842 | 0.279 |

Then find the best weights of input between hidden layers as explained in Tables 9, 10 and 11. While, we explained the weights of recurrent connections in Tables 12 and 13:

After building and developing the LSTM model by the PSO algorithm, this model consists of several layers capable of predicting concentrations of air pollutants. We have (32 stations) products six types of concentrations that cause air pollution

**Table 9.** The optimal weights of the input **between** first and second layer

| Node1 | | | | Node2 | | | | Node3 | | | | Node4 | | | |
|---|---|---|---|---|---|---|---|---|---|---|---|---|---|---|---|
| $W_a$ | $W_i$ | $W_f$ | $W_o$ | $W_a$ | $W_i$ | $W_f$ | $W_o$ | $W_a$ | $W_i$ | $W_f$ | $W_o$ | $W_a$ | $W_i$ | $W_f$ | $W_o$ |
| 0.882 | 0.256 | 0.604 | 0.403 | 0.373 | 0.999 | 0.252 | 0.875 | 0.350 | 0.841 | 0.158 | 0.912 | 0.113 | 0.814 | 0.287 | 0.782 |
| 0.447 | 0.412 | 0.067 | 0.900 | 0.223 | 0.860 | 0.594 | 0.913 | 0.087 | 0.273 | 0.174 | 0.537 | 0.300 | 0.895 | 0.082 | 0.636 |
| 0.338 | 0.467 | 0.832 | 0.209 | 0.039 | 0.091 | 0.547 | 0.717 | 0.286 | 0.918 | 0.394 | 0.215 | 0.603 | 0.608 | 0.930 | 0.460 |
| 0.661 | 0.362 | 0.657 | 0.592 | 0.900 | 0.763 | 0.934 | 0.029 | 0.542 | 0.767 | 0.998 | 0.794 | 0.619 | 0.623 | 0.899 | 0.483 |

**Table 10.** The optimal weights of input between second and third layer

| Node1 | | | | **Node2** | | | | **Node3** | | | | **Node4** | | | |
|---|---|---|---|---|---|---|---|---|---|---|---|---|---|---|---|
| $W_a$ | $W_i$ | $W_f$ | $W_o$ | $W_a$ | $W_i$ | $W_f$ | $W_o$ | $W_a$ | $W_i$ | $W_f$ | $W_o$ | $W_a$ | $W_i$ | $W_f$ | $W_o$ |
| 0.067 | 0.724 | 0.363 | 0.178 | 0.709 | 0.755 | 0.743 | 0.843 | 0.757 | 0.989 | 0.547 | 0.671 | 0.311 | 0.930 | 0.457 | 0.718 |
| 0.422 | 0.321 | 0.590 | 0.597 | 0.707 | 0.120 | 0.699 | 0.269 | 0.991 | 0.030 | 0.576 | 0.247 | 0.238 | 0.367 | 0.716 | 0.740 |
| 0.161 | 0.665 | 0.109 | 0.467 | 0.305 | 0.992 | 0.343 | 0.327 | 0.053 | 0.084 | 0.889 | 0.092 | 0.049 | 0.719 | 0.816 | 0.687 |
| 0.783 | 0.388 | 0.733 | 0.263 | 0.130 | 0.879 | 0.799 | 0.906 | 0.731 | 0.674 | 0.420 | 0.765 | 0.154 | 0.443 | 0.198 | 0.323 |

**Table 11.** The weights of input between third and output layer

| Node1 | | | | **Node2** | | | | **Node3** | | | | **Node4** | | | |
|---|---|---|---|---|---|---|---|---|---|---|---|---|---|---|---|
| $W_a$ | $W_i$ | $W_f$ | $W_o$ | $W_a$ | $W_i$ | $W_f$ | $W_o$ | $W_a$ | $W_i$ | $W_f$ | $W_o$ | $W_a$ | $W_i$ | $W_f$ | $W_o$ |
| 0.410 | 0.492 | 0.598 | 0.842 | 0.651 | 0.777 | 0.599 | 0.033 | 0.456 | 0.983 | 0.647 | 0.569 | 0.608 | 0.006 | 0.192 | 0.476 |
| 0.137 | 0.681 | 0.510 | 0.839 | 0.741 | 0.934 | 0.680 | 0.817 | 0.346 | 0.848 | 0.984 | 0.007 | 0.041 | 0.557 | 0.350 | 0.329 |
| 0.801 | 0.171 | 0.317 | 0.079 | 0.099 | 0.613 | 0.227 | 0.671 | 0.385 | 0.218 | 0.903 | 0.890 | 0.053 | 0.579 | 0.625 | 0.936 |
| 0.605 | 0.418 | 0.610 | 0.850 | 0.351 | 0.303 | 0.095 | 0.008 | 0.152 | 0.601 | 0.087 | 0.990 | 0.549 | 0.913 | 0.731 | 0.373 |
| 0.132 | 0.830 | 0.711 | 0.139 | 0.092 | 0.545 | 0.181 | 0.717 | 0.291 | 0.376 | 0.722 | 0.297 | 0.036 | 0.084 | 0.776 | 0.853 |
| 0.456 | 0.033 | 0.562 | 0.703 | 0.981 | 0.003 | 0.254 | 0.185 | 0.839 | 0.431 | 0.363 | 0.470 | 0.792 | 0.674 | 0.376 | 0.420 |

**Table 12.** The weight of recurrent connections

| $U_a$ | $U_i$ | $U_f$ | $U_o$ |
|---|---|---|---|
| 0.390 | 0.279 | 0.435 | 0.622 |

**Table 13.** The result of DLSTM-RNN and SMAPE

| | PM2.5 | PM10 | NO2 | CO | O3 | SO2 |
|---|---|---|---|---|---|---|
| DLSTM-RNN | 79.29 | 78.07 | 73.48 | 11.60 | 12.25 | 14.00 |
| | .. | ... | ... | ... | ... | ... |
| | 67.88 | 71.18 | 66.94 | 7.01 | 9.37 | 10.44 |
| SMAPE | 13.61 | 12.25 | 11.83 | 4.70 | 4.52 | 4.46 |
| | ... | ... | ... | ... | ... | ... |
| | 12.25 | 11.93 | 11.61 | 4.38 | 4.41 | 4.22 |

(PM2.5, PM10, Sox, CO, NOx, O3) so within one hour we have (192) reading, within one day (4608) and within 30 days after the training process the network has become our (138240) Read. After the training, DLSTM-RNN can predict air pollution concentrations over the next 48 h based on previous training.

Then we used the SMAPE error rate scale to evaluate the results from the DLSTM-RNN network for the least or the nearest error.

The combination of LSTM and SWARM has reduced training time on the network because the SWARM algorithm has provided the best function for activation and has identified the number of hidden layers and the number of nodes in each hidden layer, adding that they provide better weights, but at the same time complicate the network for the above reason.

## Compliance with Ethical Standards.

**Conflict of Interest:** The authors declare that they have no conflict of interest.

**Ethical Approval:** This article does not contain any studies with human participants or animals performed by any of the author.

## Appendix

See Table 14.

**Table 14.** Terms and meaning

| DLSTM | Developed long short – term memory |
|---|---|
| LSTM | Long short-term memory |
| PSO | Particle Swarm Optimization |
| SMAPE | Symmetric mean absolute percentage error |
| PM2.5 | Particulate matter that have a diameter of less than 2.5 μm |
| PM10 | particulate matter 10 μm or less in diameter |
| O3 | Ozone is the unstable triatomic form of oxygen |
| Sox | sulfur oxides |
| CO | carbon monoxide |
| $NO_x$ | nitrogen oxides |
| $\Theta$ | is the element-wise product or Hadamard product |
| $\otimes$ | Outer products will be represented |
| $\sigma$ | represents the sigmoid function |
| $a_t$ | Input activation |
| $i_t$ | Input gate |
| $f_t$ | Forget gate |
| $o_t$ | Output gate |
| $State_t$ | Internal state |
| $Out_t$ | Output |
| W | the weights of the input |
| U | the weights of recurrent connections |

## References


1. Ong, B.T., Sugiura, K., Zettsu, K.: Dynamically Pre-Trained Deep Recurrent Neural Networks using environmental monitoring Data for Predicting PM2.5. https://doi.org/10.1007/s00521-015-1955-3
2. Al-Janabi, S., Rawat, S., Patel, A., Al-Shourbaji, I.: Design and evaluation of a hybrid system for detection and prediction of faults in electrical transformers. Int. J. Electri. Power Energy Syst. **67**, 324–335 (2015). https://doi.org/10.1016/j.ijepes.2014.12.005
3. Li, X., Peng, L., Hu, Y., Shao, J., Chi, T.: Deep learning architecture for Air quality predictions. Environ. Pollut. **231**, 997–1004 (2017). https://doi.org/10.1007/s11356-016-7812-9
4. Li, X., Peng, L., Yao, X., Cui, S., Hu, Y., You, C., Chi, T.: Long short-term memory neural network for air pollutant concentration predictions: Method development and evaluation. Environ. Pollut. 997–1004 (2017). https://doi.org/10.1016/j.envpol.2017.08.114
5. Ghoneim, O.A., Manjunatha, B.R.: Forecasting of Ozone Concentration in Smart City using Deep Learning, pp. 1320–1326 (2017). https://doi.org/10.1109/ICACCI.2017.8126024
6. Lifeng, W., Li, N., Yang, Y.: Prediction of air quality indicators for the Beijing-Tianjin-Hebei region. Clean. Prod. **196**(2018), 682–687 (2018). https://doi.org/10.1016/j.jclepro.2018.06.068
7. Popoola, O.A.M., Carruthers, D., Lad, C., Bright, V.B., Mead, M.I., Stettler, M.E.J., Saffell, J.R., Jones, R.L.: Use of networks of low-cost air quality sensors to quantify air quality in urban settings. Atmos. Environ. **194**, 58–70 (2018)
8. Wen, C., Liu, S., Yao, X., Peng, L., Li, X., Hu, Y., Chi, T.: A novel spatiotemporal convolutional long short-term neural network for air pollution prediction. Sci. Total Environ. **654**, 1091–1099 (2019). https://doi.org/10.1016/j.scitotenv.2018.11.086
9. Shang, Z., Deng, T., He, J., Duan, X.: A novel model for hourly PM2.5 concentration prediction based on CART and EELM. Sci. Total Environ. **651**, 3043–3052 (2019). https://doi.org/10.1016/j.scitotenv.2018.10.193
10. Li, H., Wang, J., Li, R., Haiyan, L.: Novel analysis forecast system based on multi-objective optimization for air quality index. Clean. Prod. **208**, 1365–1383 (2019). https://doi.org/10.1016/j.jclepro.2018.10.129
11. Buyya, R., Calheiros, R.N., Astjerdi, A.V.: Big data: principles and paradigms. Big Data: Principles and Paradigms, pp. 1–468 (2016). https://doi.org/10.1016/c2015-0-04136-3
12. Oussous, A., Benjelloun, F.-Z., Lahcen, A.A., Belfkih, S.: Big data technologies. Adv. Parallel Comput. 28–55 (2019). https://doi.org/10.3233/978-1-61499-814-3-28
13. Liu, S., Wang, Y., Yang, X., Lei, B., Liu, L., Li, S.X., Ni, D., Wang, T.: Deep learning in medical ultrasound analysis: a review. Engineering (2019). https://doi.org/10.1016/j.eng.2018.11.020
14. Al-Janabi, S., Alkaim, A.F.: A nifty collaborative analysis to predicting a novel tool (DRFLLS) for missing values estimation. J. Soft Comput. (2019). https://doi.org/10.1007/s00500-019-03972-x. Springer
15. Aunan, K., Hansen, M., Liu, Z., Wang, S.: The hidden hazard of household air pollution in Rural China. Environ. Sci. Policy **93**, 27–33 (2019). https://doi.org/10.1016/j.envsci.2018.12.004
16. Inácio, F., Macharet, D., Chaimowicz, L.: PSO-based strategy for the segregation of heterogeneous ro Botic swarms. J. Comput. Sci. 86–94 (2019). https://doi.org/10.1016/j.jocs.2018.12.008

17. Matos, J., Faria, R., Nogueira, I., Loureiro, J., Ribeiro, A.: Optimization strategies for chiral separation by true moving bed chromatography using Particles Swarm Optimization (PSO) and new Parallel PSO variant. Comput. Chem. Eng. 344–356 (2019). https://doi.org/10.1016/jcompchemeng.2019.01.020
18. Hu, M., Wang, H., Wang, X., Yang, J., Wang, R.: Video facial emotion recognition based on local en Hanced motion history image and CNN-CTSLSTM networks. J. Vis. Commun. Image Represent. 176–185 (2019). https://doi.org/10.1016/j.jvcir.2018.12.039
19. Al_Janabi, S., Mahdi, M.A.: Evaluation prediction techniques to achievement an optimal biomedical analysis. Int. J. Grid Util. Comput. (2019)
20. Al-Janabi, S., Alwan, E.: Soft mathematical system to solve black box problem through development the FARB based on hyperbolic and polynomial functions. In: IEEE, 2017 10th International Conference on Developments in eSystems Engineering (DeSE), Paris, pp. 37–42 (2017). https://doi.org/10.1109/dese.2017.23
21. Al_Janabi, S., Al_Shourbaji, I., Salman, M.A.: Assessing the suitability of soft computing approaches for forest fires prediction. Appl. Comput. Inf. **14**(2), 214–224 (2018). ISSN 2210-8327, https://doi.org/10.1016/j.aci.2017.09.006

# Practical Investigation on Bearing Fault Diagnosis Using Massive Vibration Data and Artificial Neural Network

Tarek Khoualdia[1,2](✉), Abdelaziz Lakehal[1], and Zoubir Chelli[3]

[1] Department of Mechanical Engineering, Mohamed Chérif Messaadia University, P.O. Box 1553, 41000 Souk-Ahras, Algeria
khoualdiatarek@hotmail.fr, lakehal21@yahoo.fr
[2] LGMREIU Laboratory, Mohamed-Cherif Messaadia University, Souk Ahras, Algeria
[3] Department of Electrical Engineering, Mohamed Chérif Messaadia University, P.O. Box 1553, 41000 Souk-Ahras, Algeria

**Abstract.** Bearing faults are one of the most probable causes for machine vibration. Early detections on bearing faults can save invaluable time and cost. In practice, vibration analysis is one of the techniques of condition-based maintenance that allows early detection and reliable diagnosis of bearing faults. In this paper, a test rig is implemented to bring a reliable approach for monitoring and diagnosis. A full spectrum analysis, derived from the fast Fourier transformation, is presented for the vibration signal to reveal combined fault signature of bearing. However, a monitoring system based on an artificial neural network (ANN) model is used to diagnose combined faults in bearings. To train and test the ANN, coded faults are used as output data. To determine the best ANN, a learning algorithm has been chosen and maintained. The proposed method gives more reliable diagnosis in complex faults scenario. Also, the best model can be used by practitioners in the industry as a decision support tool to increase the availability of the machine.

**Keywords:** Combined faults · Bearing · Vibration analysis · ANN

## 1 Introduction

The dynamic forces to which the bearings will be subjected reduce their lifetime considerably. For this reason it is important to correct problems such as misalignments and unbalance, which lead to a very important increase in dynamic efforts. However, in an evolution of concepts, ensuring optimum availability of rotating machines is not enough to correct faults. Maintenance has moved from corrective maintenance to predictive maintenance. For bearings and other machine elements there are condition-based maintenance techniques that allow early detection of faults.

Vibration and acoustic measurement techniques for early detection and diagnosis of faults in bearings are widely discussed in the literature [1]. Besides these fundamental techniques, the artificial intelligence methods have given a strong contribution to the early detection and automatic diagnosis of fault in rotating machines in general and


Y. Farhaoui (Ed.): BDNT 2019, LNNS 81, pp. 110–116, 2020.
https://doi.org/10.1007/978-3-030-23672-4_9

specially for bearings and gears [2, 3]. Artificial neural networks (ANN) have given good results and they are widely used for diagnosing bearing faults. Several authors have developed ANN where they are used vibration signals as input data into the algorithm. Vibration signals are used directly as input data in a deep structure of convolutional neural network for automatically fault diagnosis and which does not require any feature extraction techniques and achieves very high accuracy and robustness under noisy environments [4]. In the same context and for more accuracy and robustness of the algorithm it necessary that datasets contain big data extracted from massive measured signals involving different health conditions under various operating conditions [5].

Huo et al. [6] used an approach based on fault diagnosis of bearing signals vibration in different speed, by using an auto adapted wavelet transform components. In the same scopes Attoui et al. [7] Combined two techniques, the first one is a new data extraction and the second is classical called wavelet packet decomposition energy distribution, based on the selection of frequency bands with the greatest impulse, incorporating a time–frequency procedure. As well, Khoualdia et al. [8] developed an optimized intelligent system, based on an artificial neural network, in order to detect and diagnose faults of bearings, gears and their combined faults.

The proposed approach in this paper consists in early detection of bearing deterioration using the vibratory signal. The faults of the different parts of the bearing (outer ring, inner ring, rolling element and a combined state), are studied. We made several measurements to have a very rich database, in order to have good and precision experimental results.

## 2 Vibration Analysis

Rotating machines are never perfect and deteriorate with time in the form of several faults (unbalance, wear of bearings, gear wear, misalignment, lubrication failure, loosening, play, cracks, etc.). The techniques and methods used for monitoring and diagnosing bearings in a rotating machine are numerous and very diverse. Each of these methods corresponds to different levels of knowledge of the phenomenon and to the use of analysis materials. These techniques can be grouped into two large families (Fig. 1).

In this work, we will use the signal processing methods associated with appropriate methods, to highlight the monitoring with the goal of early failure bearing detection, using one of different indicators of temporal statistical methods. It should be noted that the methods and results in the field of vibration analysis are of direct relevance to industrial production.

The graphical representation of the vibration signal as a function of time remains rather "illegible". It does not allow analysis because all terms are superimposed. We need an additional mathematical tool. The Fast Fourier Transform (FFT), when applied to a function of time such as acceleration, velocity, or displacement, gives the result of another function whose variable is the frequency. Faults interaction in rolling element bearings generates a pulse signal. These pulses excite the natural frequencies of bearing elements resulting in an increase in the vibration energy at these high frequencies.

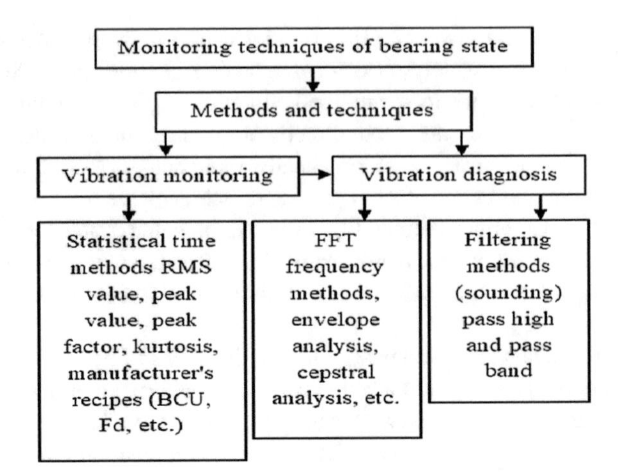


**Fig. 1.** Techniques of bearing faults monitoring and diagnostic

FFT analysis of a rolling element bearing faults produces peaks in high frequency in a spectrum. With the zoom function and envelope analysis the diagnostics is easier.

# 3 Experimental Study

## 3.1 Description of the Experimental Platform

For diagnosing rotating machinery, Spectra Quest's system is used to collect a large database of vibration signals of different faulty elements of the bearing. Different states of measures are presented in Table 1. U205 is the reference of the used roller bearing.

**Table 1.** Different faults

| States | Description | Fault location |
|---|---|---|
| 1 | Healthy bearing | |
| 2 | Defective bearing | Outer race |
| 3 | Defective bearing | Inner race |
| 4 | Defective bearing | Ball |
| 5 | Defective bearing | Combined |

Measurements were taken by piezoelectric sensor (ICP) with respectively the frequency and measurement ranges 0. 3 at 10000 Hz and ± 50 g, its sensitivity is 100 mV/g and it is mounted on the bearing by stud. The directions of the sensors are illustrated in Fig. 2; the measurements are taken simultaneously with three sensors at the same time.

The parameters for the Machine Fault Simulator (MFS) bearings and the calculated frequencies of the bearing: the Fundamental Train Frequency (FTF), The ball Pass

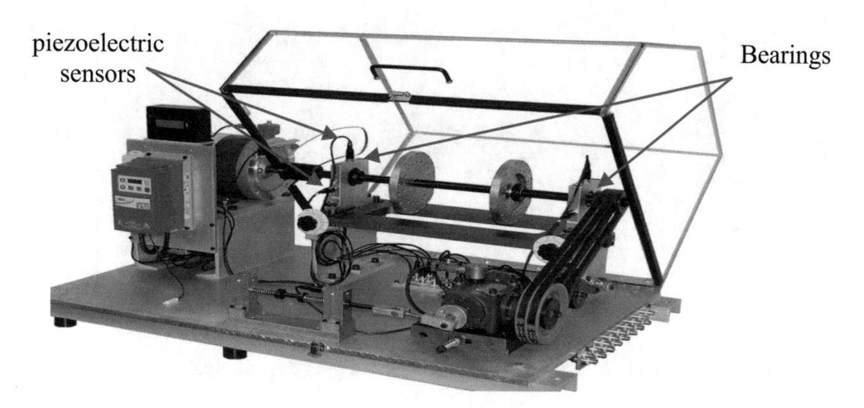


**Fig. 2.** Fault simulator platform.

**Table 2.** Ball bearing U205 parameters

| Parameters | Values |
|---|---|
| Number of rolling elements | 9 |
| Pitch circle diameter (mm) | 1.516 |
| Roller element diameter (mm) | 0.3125 |
| FTF | 0.402 |
| BPFO | 3.572 |
| BPFI | 5.43 |
| BSF | 2.322 |

Frequency Outer (BPFO), The ball Pass Frequency Inner (BPFI) and The ball Spin Frequency (BSF), (see Table 2):

### 3.2 Feature Extraction

In order to illustrate the signals measurements, an example plots are shown in this section with 15 Hz of rotating speed. A comparison of the healthy signal and the combined faults state signal is illustrate in Fig. 3. Another comparison is shown in Fig. 4 with the same bearing fault but in two different directions (Axial and radial).

The signals in time domain has an elevated amplitude in the cases of faults compared with the normal state, the elevation of the amplitude is explained by the flaking of the bearing, in the combined case we can see more pulses. And in the second comparison, the pulses of the radial direction are more important compared to axial direction.

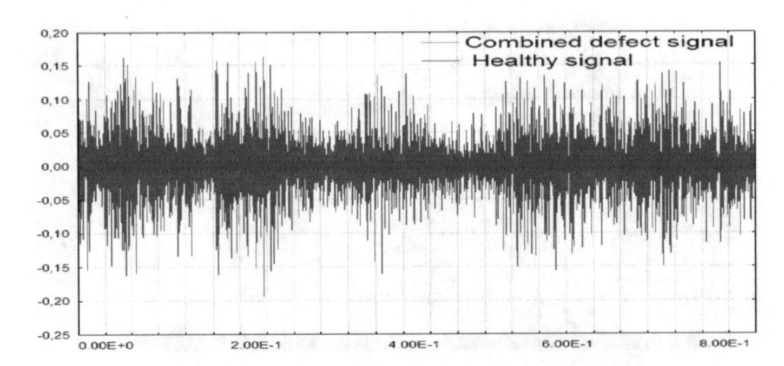


**Fig. 3.** Comparison of healthy signal and faulty.

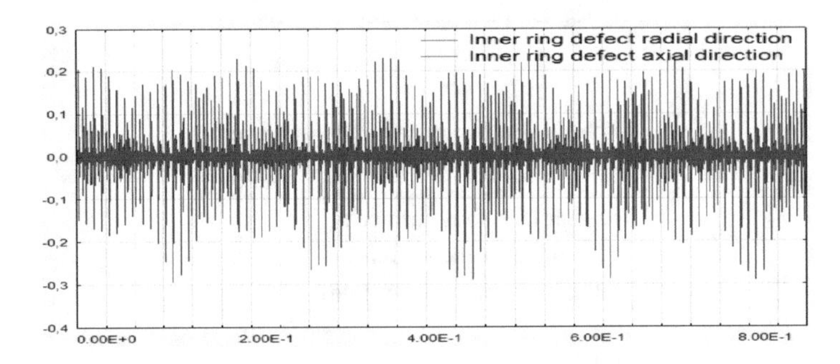


**Fig. 4.** Signals comparison in two different directions.

## 4 The Artificial Neural Network Models

Inspired and based on the structure and operation of a biological neural network, an ANN is modeled. A multi-layer perceptron (MLP) network training with back-propagation algorithm is used. The developed MLP neural network with LM algorithm consists of three layers, input layer, output layer (which are controlled, because they have a fixed number of neurons) and a hidden layer that is uncontrolled, because we must varied and train it to obtain the best neural model. The input layer is composed of the main indicators used in vibratory analysis, given in Table 3, and our output layers are composed of a fixed number of neurons represented by a codification of the different conditions of operation, bearing states, for three directions, axial, vertical and horizontal.

**Table 3.** Time domain features

| Feature | Equation |
|---|---|
| Maximum value | $\max\{\lvert x_i \rvert\}$ |
| RMS value | $\left(\frac{1}{N}\sum_{i=1}^{N} \mathrm{x}_i^2\right)^{\frac{1}{2}}$ |
| Mean square value | $\mathrm{MSV} = \frac{1}{N}\sum_{i=1}^{N} x_i^2$ |

*(continued)*

**Table 3.** (*continued*)

| Feature | Equation |
|---|---|
| Variance | $\frac{1}{N-1}\sum_{i=1}^{N}(\mathrm{x_i}-\mathrm{MSV})^2$ |
| Kurtosis factor | $\left(N\sum_{i-1}^{N}(x_i-\bar{x})^4\right)/\left(\sum_{i=1}^{N}(x_i-\bar{x})^2\right)$ |
| Crest factor | $\frac{\max\lvert x_i\rvert}{\mathrm{RMS}}$ |
| Clearance factor or margin factor | $(\max\lvert x_i\rvert)/\left(\frac{1}{N}\sum_{i=1}^{N}\lvert x_i\rvert^{\frac{1}{2}}\right)^2$ |

# 5 Results and Discussion

Compared to other learning methods, the total training time is very short because it does not exceed a few seconds. The Levenberg-Marquardt learning algorithm allowed our model to converge rapidly to the desired solution, according to the learning curve of Fig. 5, only 20 epochs were enough to reach this convergence.

According to the final results of our ANN model, the value of the global correlation coefficient is close to one hundred percent, because the value of 0.97344 is very encouraging for the chosen model. This model has good correlation coefficient values with error values close to zero, which gives our neural networks a great deal of power in diagnosing different failures.

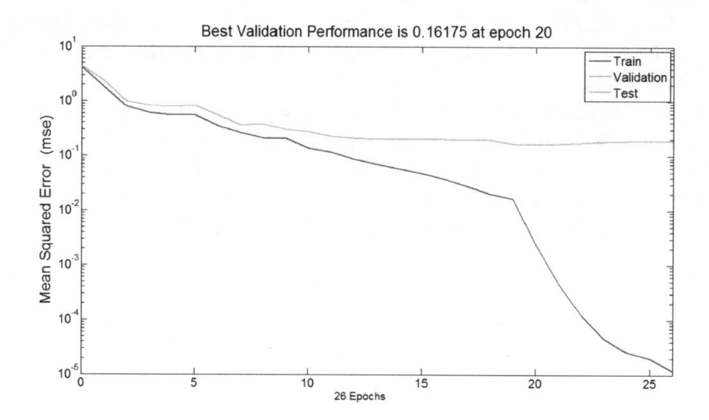


**Fig. 5.** Training of MLP ANN with LM algorithm.

# 6 Conclusion

In this work, experiences are made in order to bring a reliable approach for monitoring and diagnosis of bearings faults. The developed MLP neural network with the learning algorithm of Levenberg-Marquardt (LM) gives us a very good monitoring system to diagnose the different faults of bearings.

The developed ANN model with the Levenberg-Marquardt algorithm, reached the desired convergence rapidly, compared to other methods of learning algorithm, the training time is very short, because it does not exceed a few seconds.

The values of our correlation coefficient are very close to the one hundred percent, which gives the chosen neural network the diagnosis power of the different bearing faults. Therefore, the used method is powerful to study other cases of machine faults.

**Acknowledgment.** The authors like to thank the Algerian general direction of research (DGRSDT) for their financial support.

## References

1. Tandon, N., Choudhury, A.: A review of vibration and acoustic measurement methods for the detection of defects in rolling element bearings. Tribol. Int. **32**, 469–480 (1999)
2. Liu, R., Yang, B., Zio, E., Chen, X.: Artificial intelligence for fault diagnosis of rotating machinery: a review. Mech. Syst. Signal Process. **108**, 33–47 (2018)
3. Kumar, S., Goyal, D.K., Dang, R.S., Dhami, S., Pablab, B.S.: Condition based maintenance of bearings and gears for fault detection – a review. Mater. Today Proc. **5**(2), 6128–6137 (2018)
4. Duy-Tang, H., Hee-Jun, K.: Rolling element bearing fault diagnosis using convolutional neural network and vibration image. Cogn. Syst. Res. **53**, 42–50 (2019)
5. Jia, F., Lei, Y., Lin, J., Zhou, X., Lu, N.: Deep neural networks: a promising tool for fault characteristic mining and intelligent diagnosis of rotating machinery with massive data. Mech. Syst. Signal Process. **72–73**, 303–315 (2016)
6. Huo, Z., Zhang, Y., Francq, P., Shu, L., Huang, J.: Incipient fault diagnosis of roller bearing using optimized wavelet transform based multi-speed vibration signatures. IEEE Access **5**, 19442–19456 (2017)
7. Attoui, I., Fergani, N., Boutasseta, N., Oudjani, B., Deliou, A.: A new time–frequency method for identification and classification of ball bearing faults. J. Sound Vib. **397**, 241–265 (2017)
8. Khoualdia, T., Hadjadj, E.A., Bouacha, K., Abdeslam, D.O.: Multi-objective optimization of ANN fault diagnosis model for rotating machinery using grey rational analysis in Taguchi method. Int. J. Adv. Manuf. Technol. **89**(9–12), 3009–3020 (2017)

# Multi Objectives Optimization to Gas Flaring Reduction from Oil Production

Ayad F. Alkaim[1] and Samaher Al_Janabi[2(✉)]

[1] Department of Chemistry Science, Faculty of Science for Women (SCIW), University of Babylon, Babylon, Iraq
alkaim@iftc.uni-hannover.de

[2] Department of Computer Science, Faculty of Science for Women (SCIW), University of Babylon, Babylon, Iraq
samaher@itnet.uobabylon.edu.iq

**Abstract.** In recent years the emission results from the gas flaring become cause big problem of emission the environment. Therefore, this paper focuses on present a novel prediction tool based on developing MARS data mining technique through replace it kernel by multi objective optimization function. The objectives of this paper are: (i) Give a specific definition of Gas Flaring Reduction (GFR) from oil production with determined the main limitations and hypotheses of that problem. (ii) Study the nature of datasets related to oil extraction, pre-processing that datasets, then determined the main features for each dataset. (iii) Set the constrictions for each sub-optimization problem. (iv) Design a novel predictor based on develop MARS data mining technique through replace their kernel by multi objective optimization function based on their construction. (v) Calculate the rate of Gas flaring based on three main gases (CO2, CH4, N2O) in parallel at the same time that increase the performance and reduce the time used to find an optimal result. (vi) Compute the gas to oil rate (GOR) for traditional predictor "MARS" and Develop Predictor". Finally, evaluate and analyze the results of a proposed methodology.

**Keywords:** Big data analysis · Multi objectives optimization · Gas flaring · Linear combination · MARS · DOM · GOR · GHG

## 1 Introduction

Billions of cubic meters of gaseous petrol are flared yearly at oil production locales around the world. Flaring gas wastes a profitable vitality asset that could be utilized to help monetary development and advance. It additionally adds to environmental change by discharging a large number of huge amounts of $CO_2$ to the air. Amid oil generation, the related flammable gas is flared when hindrances to the improvement of gas markets and gas framework keep it from being utilized. The most recent satellite information shows that Iraq consumed more than 16 billion cubic meters of gas related with oil in 2015 just, making it the second biggest gas burner on the planet. As the Iraqi government has as of late received the activity to stop the programmed ignition of gas by 2030, which implies the state's dedication not to naturally consume the related gas in


Y. Farhaoui (Ed.): BDNT 2019, LNNS 81, pp. 117–139, 2020.
https://doi.org/10.1007/978-3-030-23672-4_10

any field of oil and to end this consuming in the current oil fields as quickly as time permits and no later than in 2030. Therefore, we will have designed optimization model to reduce the amount of gas flaring. To achieve this purpose, we will deal with very large databases (oil, gas, wells and cost). Given the difficulties related with gas flaring, its discharges, the effect on nature and the trouble in giving more exact and single worldwide assessments or information by different bodies concerned, care ought to be taken in choosing techniques being utilized. Wrong gauges or information of gas flared and the subsequent emanations as now and again announced, do not give the genuine effect of the operations of the oil and gas industry on the earth. So, there are four basic parameters (the lowest percentage of natural gas burning associated with oil extraction, the highest quantity of oil extracted at the lowest cost and the highest benefit) of the objective function each of them has several determinates. As we seek speed in dealing with databases, we will deal with a new technology that speeds up the implementation process by dividing the large task into several separate tasks and processing them separately and then assembling the final solution which is a major success point for the oil and gas companies namely GPU. Optimization problem is the problem of finding the best solution from all feasible solutions. Optimization is not new science. Optimization plays an important role in the process of designing a system. Successful application of optimization techniques requires no less than three conditions. These requirements are the ability to make mathematical models of problems encountered, knowledge of optimization techniques and knowledge of computer programs. The idea of optimization can be explained as a set of mathematical formulas and numerical methods for finding and identifying the best candidates from an arrangement of choices without having to unequivocally process and assess every single conceivable choice. Optimization is the process of maximizing or minimizing a function of purpose by remembering the existing constraints. Optimization in data mining is the side of our work. In this project, there is a need for some processes (normalization, balance data

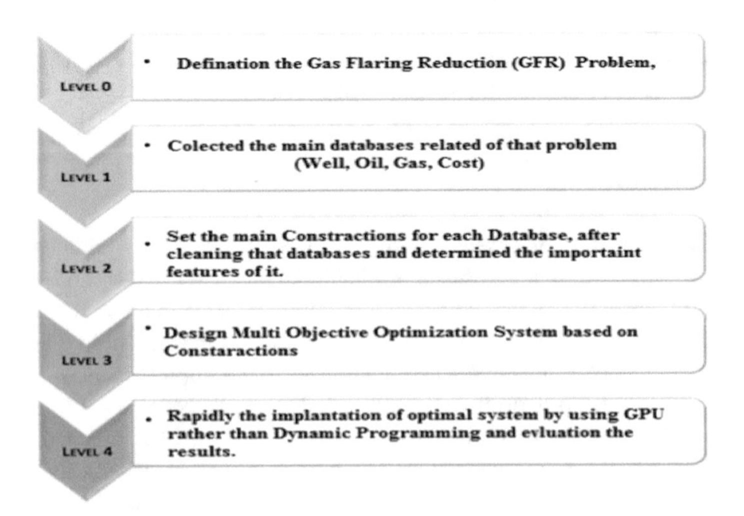


**Fig. 1.** Levels of proposed methodology

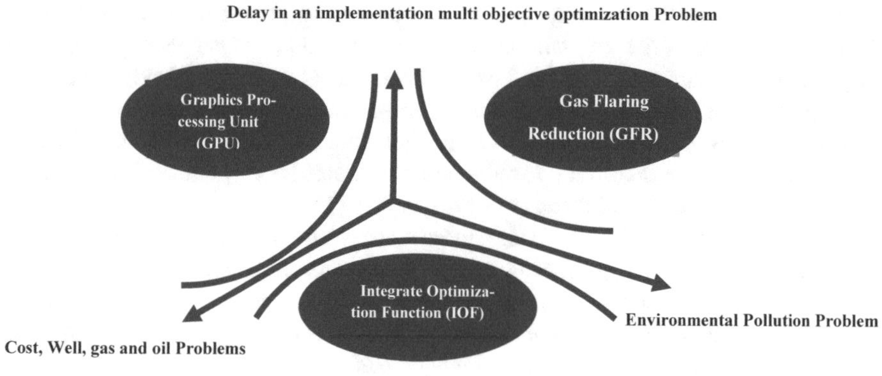

**Fig. 2.** Relationships among the main challenges

and feature selection) as preprocessing. The main level of the proposed optimization model will explain in Fig. 1, while the main challenges and their proposed solution will explain in Fig. 2.

## 2 Related Work

Talavera et al. (2010). This work used neural networks with reinforcement learning method for controlling oil production. The model displayed great comes about in controlling the yield of oil generation in the maker well indeed utilizing set focuses that were not portion of the preparing, too tried against startling irritations (issues) in layers of the maker well. Its emissions, the impact on the environment and the difficulty in providing more accurate and single global estimates or data by various bodies concerned, care should be taken in selecting methods being employed. Inaccurate estimates or data of gas flared and the resulting emissions as sometimes reported, do not give the true impact of the operations of the oil and gas industry on the environment [11].

Rahimpour et al. (2011). This work intends to limit ecological and conservative hindrances of consuming flare gas. It utilized three rationalities: (1) Gas-to-Fluid (GTL) creation, (2) control age with a gas turbine and, (3) weight and implantation into the refinery pipelines. All of these strategies require higher taken a toll. This work same of this paper in utilizing oil database and in the point of decreasing gas flaring, but it contrasts with it in strategies utilized it [8].

Ozgen Karacan et al. (2011). This work surveys the specialized viewpoints of CMM capture in and from coal mines, the fundamental variables influencing CMM aggregations in underground coal mines, strategies for capturing methane utilizing boreholes. It similar to this work in using oil database and differs in all another part [12].

Porzio et al. (2014). These authors work on an industrial database and we need first to apply discretization for data as preprocessing, after that the prepare gasses network of a genuine coordinates steelwork is analyzed and modeled to the point of defining an optimization issue considering two clashing targets. The multi objective optimization issue is firstly confronted through linear programming (LP) and illuminated by abusing the limitation strategy. That come about tall execution and the realized framework has illustrated its capabilities of optimizing the dissemination of prepare gasses inside coordinates press and steel plant and can constitute a substantial choice bolster device for prepare administrators. Previous study parallel with this paper at two points: gas reduction, and uses of multi objective function, but it different with that paper in database used it and it focus on co2 reduction. It was handling of finding the conditions that grant the greatest or least esteem of a function [1].

Saheed and Ezaina (2014). This study considered different reaction types and operating conditions for gas flaring and it predicts is of paramount importance to governments, environmental agencies and the oil and gas industrial. This work reduces the type and quantity of chemical species in flaring emissions that can degrade our environment and impact negatively on human health. But it did not apply done on the flare system by oil and gas companies to ensure efficiencies [5].

Ezaina and Saheed (2015). This work using mass balance concept (reaction type) to predict gas emission quality. This model predicted gaseous emissions based on the possible individual combustion types and conditions anticipated in gas flaring operation. It will assist in the effort by environmental agencies and all concerned to track and measure the extent of environmental pollution caused by gas flaring operations in the oil and gas industry.

This work similar to this paper in the aim for reducing gas flaring and in challenges: Its emissions, the impact on the environment and the difficulty in providing more accurate and single global estimates or data by various bodies concerned, care should be taken in selecting methods being employed. Inaccurate estimates or data of gas flared and the resulting emissions as sometimes reported, do not give the true impact of the operations of the oil and gas industry on the environment [7].

Li and Long (2017). This work solved oil/water separation problem by using a two-way separation T-tube device was designed by integrating a pair of meshes with opposite wettability, i.e., underwater super oleophobic and super hydrophobic/super oleophilic properties. Such integrated system can continuously separate both oil and water phase from the oil/water mixtures simultaneously through one-step procedure with high flux. It needs extra cost for designing the hardware [3].

Lakehal (2017). In this study, a Bayesian network was developed to predict failures of electrical transformers and using Duval triangle method as a development to the Bayesian model. One of advantages was can use the same model for predicting as for diagnosing transformer faults. Probability of fault, and conversely, any information on the nature of the fault can modify the knowledge that we have on the released gas probability [13].

Sotirvo et al. (2018). This study is used artificial neural networks for learning and to estimations between these two concepts we use intuitionistic fuzzy sets for recognizing the type of crude oil. This thesis achieve to its aim is of recognize the type of the crude oil based on 6 of their properties, but it used neural network in learning although the outputs of the neural network are not so correct if we use the data from the regular measurement process with standard hardware and standard errors [4].

The following table explain compared among the previous works from multi points (i.e., techniques, database, advantages, disadvantages and evaluation measures) (Table 1).

**Table 1.** Compare among previous works

| Researcher name(s) | Techniques | Data base | Preprocessing | Advantages | Disadvantage | Evaluated measure |
|---|---|---|---|---|---|---|
| Talavera et al. 2010 [11] | Neural networks with reinforcement learning | Oil data base | Non | In controlling the output of oil production, the MPC-RL model presented good results in the producer well even using set points that were not belong to the training set, also tested against unexpected perturbations (faults) in layers of the producer well. | Did not take into account the cost calculation | MAPE, RMSE, U-Theil, Linear R |
| Rahimpour et al. 2011 [8] | Gas-to-Liquid (GTL) production, electricity generation with a gas turbine and, compression and injection into the refinery pipelines | Oil database | Non | The aim of these methods are minimization of environmental and economical disadvantages of gas flaring | All mothed need high cost | Rate of Return for Capacity Increment (ROR). |
| Ozgen Karacan et al. 2011 [12] | Coal mine methane (CMM) and association with coal mining activities | oil database | Non | Generally low Destruction efficiencies between 98% and 99% cost destruction option | For the most part requires a methane concentration of 30% Concerns over security of flares at mining locations | |
| Giacomo Filippo Porzio et al. 2014 [1] | LP formulation and a Multi-Objective Optimization Evolutionary Algorithm | Industrial data (dedicated Data base) | Discretization | The exhibitions accomplished are much higher (50% lower $CO_2$ emanations and 15–20% higher profit). The realized framework has illustrated its capabilities of optimizing the dispersion of prepare gasses inside an coordinates press and steel plant and can constitute a substantial choice back instrument for handle operators | The issue definition is restricted by the utilize of a framework of linear conditions: the modeling of a complex framework can become challenging, particularly in the event that incremental changes have to be made | Evolutionary algorithm |
| Saheed Ismail and Ezaina Umukoro 2014 [5] | Mass balance concept 2(reaction type) | Oil data base | Non | Reduce the type and quantity of chemical species in flaring emissions that can degrade our environment and impact negatively on human health | Should be done on the flare system by oil and gas companies to ensure efficiencies as suggested in this work and many other literatures are adhered to by oil corporations | Mass |

(*continued*)

**Table 1.** *(continued)*

| Researcher name(s) | Techniques | Data base | Preprocessing | Advantages | Disadvantage | Evaluated measure |
|---|---|---|---|---|---|---|
| Ezaina Umukoro and Saheed Ismail 2015 [7] | Mass balance concept(reaction type) | Oil database | Non | Assist in the effort by environmental agencies and all concerned to track and measure the extent of environmental pollution caused by gas flaring operations in the oil and gas industry | Its emissions, the impact on the environment and the difficulty in providing more accurate and single global estimates or data by various bodies concerned, care should be taken in selecting methods being employed. Inaccurate estimates or data of gas flared and the resulting emissions as sometimes reported, do not give the true impact of the operations of the oil and gas industry on the environment | mmt |
| Li and Long 2017 [3] | Tow-way separation T-tube device | Oil database | Non | High-flux and efficient separation of both oil (light and heavy) and water from the mixture at the same time | Increasing in cost | Separation efficiency |
| Lakehal 2017 [13] | Bayesian Duval Triangle Method | Oil database | Non | We can use the same model for predicting as for diagnosing transformer faults. That is to say, any new information on any gas can change the knowledge that we have about the nature and the occurrence probability of fault, and conversely, any information on the nature of the fault can modify the knowledge that we have on the released gas probability. The method presented in this paper has made the Duval triangle a good tool for prediction and assessment of failures beside its traditional role as a diagnostic tool | | |
| Sotirvo et al. 2018 [4] | Neural naetworks and intuitionistic fuzzy set | Oil database | Non | Assess the quality of the designed neural network. This estimates how the real data corresponds to the predicted values | The outputs of the neural network are not so correct if we use the data from the regular measurement process with standard hardware and standard errors | MSE |

# 3 Main Topics & Materials

There are many concepts is related to this work, we will explain it briefly.

## 3.1 Big Data

**Data Science** "is term refer to combination three domains include computer science, mathematical and knowledge to extract insights from data and transform it into actions that have an impact in the particular domain of application" (Fig. 3).

**Fig. 3.** Main domains of data science [9]

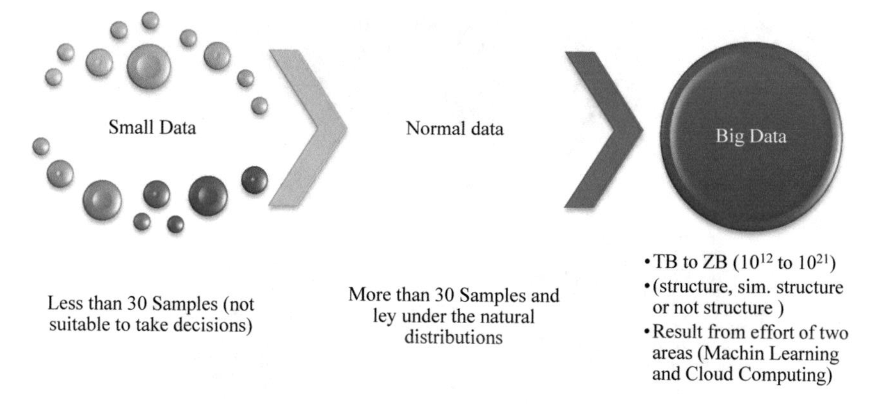


**Fig. 4.** Data types [9]

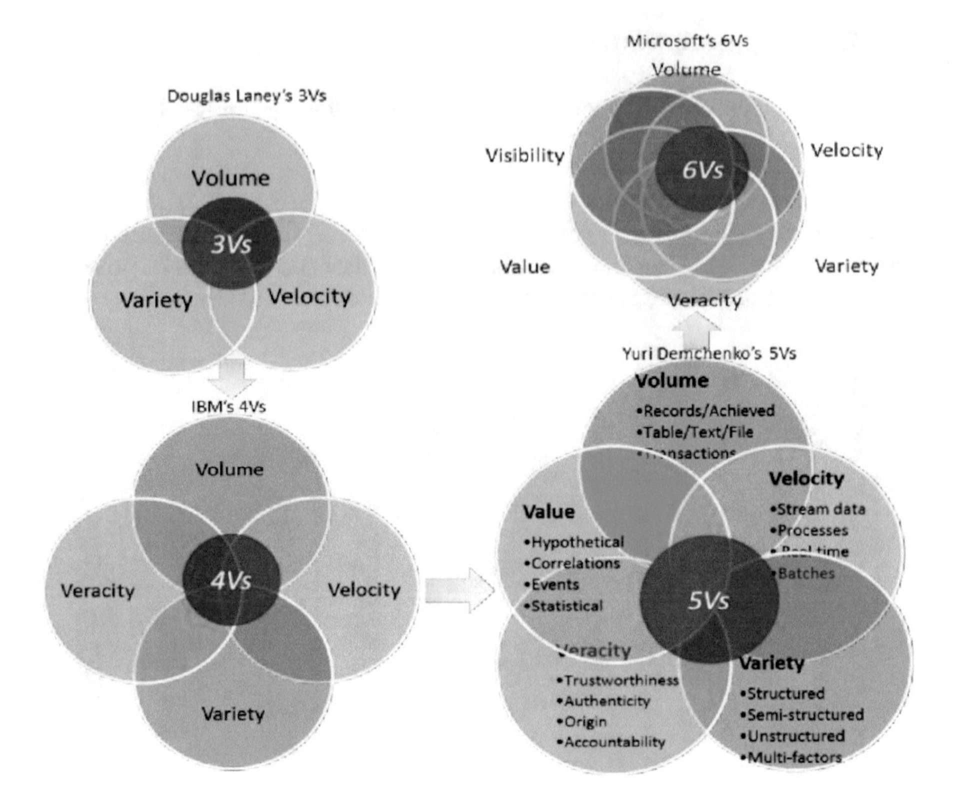


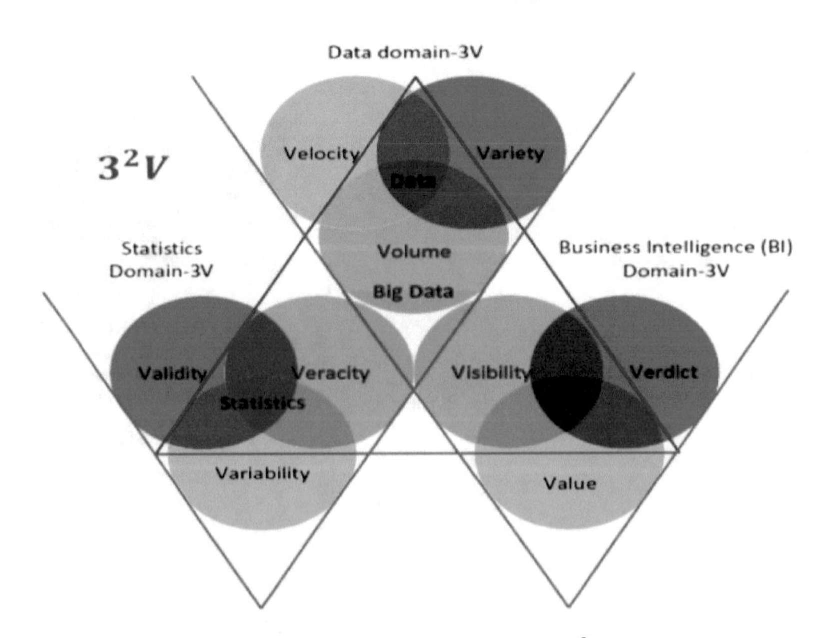


**Fig. 5.** Exploitering big data from 3Vs to $3^2$Vs [9, 10]

In general, data can be divided into the three types (i.e., "small data, normal data, and big data") as explained in Fig. 4. While Big data is informational indexes that are so voluminous and complex that conventional information preparing the application, programming are lacking to manage them. Big data challenges incorporate catching information, information stockpiling, information investigation, look, sharing, exchange, perception, questioning, refreshing and data protection. There are three measurements to enormous information known as Volume, Variety and Velocity. [1] but this definition is developed to covert Big data from 3Vs to 9 Vs as explained in Fig. 5.

### 3.2 Intelligent Data Analysis (IDA)

"Intelligent Data Analysis is considered one of the key regions in real applications and computer science. The concept of cleverly data analysis to create other ways of demonstrating during the discovery or recovery pattern planning us learn the instruments to discover patterns of information based on artificial intelligence. when we make any type of data analysis and handling to utilize the comes about in the application. It requires the presence of a real issue predefined and concrete to be solved, and the presence of factual data to be analyzed and at that point to clarify the reason best technology to construct a model to analyze the information after it is pre-determined, what is the objective of the analysis is the reason of building rules or troubleshooting optimization or Resultified data or forecast of values or make a summarized in keen and reasonable way" [10].

### 3.3 Knowledge Discovery Database (KDD)

Knowledge discovery database is the mechanized or helpful extraction of designs speaking to information certainly put away or captured in expansive databases, information stockrooms, the Internet, other gigantic data storehouses, or information streams. A knowledge discovery process incorporates data cleaning, data integration, data selection, data transformation, data mining, pattern evaluation, and knowledge presentation. [2]. In general, **Data mining** is a core step in KDD that blends data analysis methods with sophisticated algorithms for processing large volumes of data. It has also opened up exciting opportunities for exploring and analyzing old types of data in new ways [3].

### 3.4 Optimization

Optimization is the act of getting the finest result beneath given circumstances. In plan, development, and support of any building framework, engineers have to take numerous innovative and administrative choices at a few stages. The extreme goal of all such choices is either to play down the exertion required or to maximize the craved benefit. Since the exertion required or the benefit craved in any viable circumstance can be communicated as a function of certain choice factors, optimization can be defined as the process of finding the conditions that give the maximum or minimum value of a function [7].

### 3.5 Oil and Gas

The amount of oil and gas production is increase by 80% from last 28 year ago and 7 next years as explain in next figures. Figures 6 and 7.

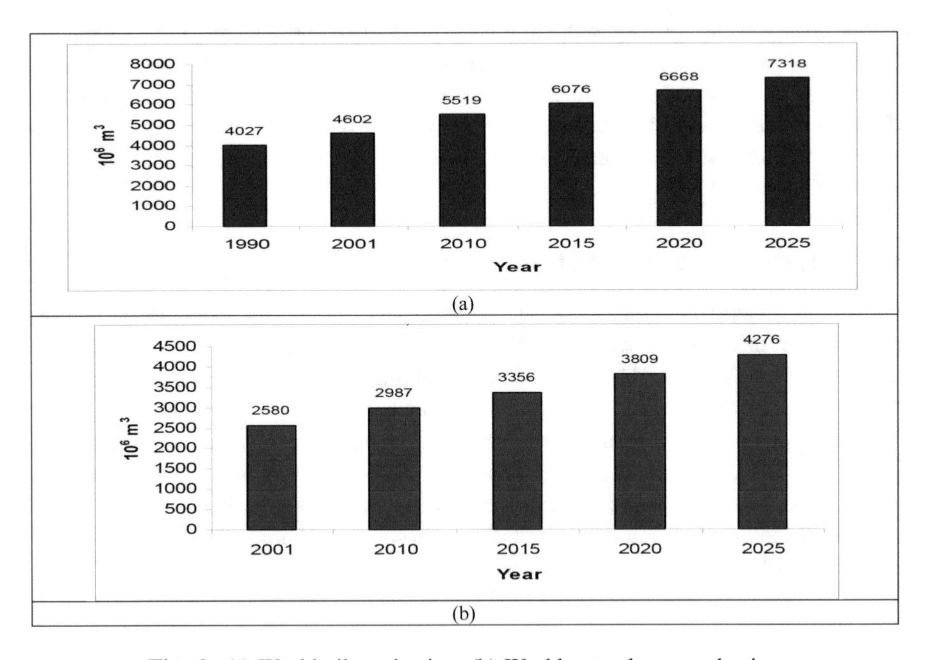


**Fig. 6.** (a) World oil production, (b) World natural gas production

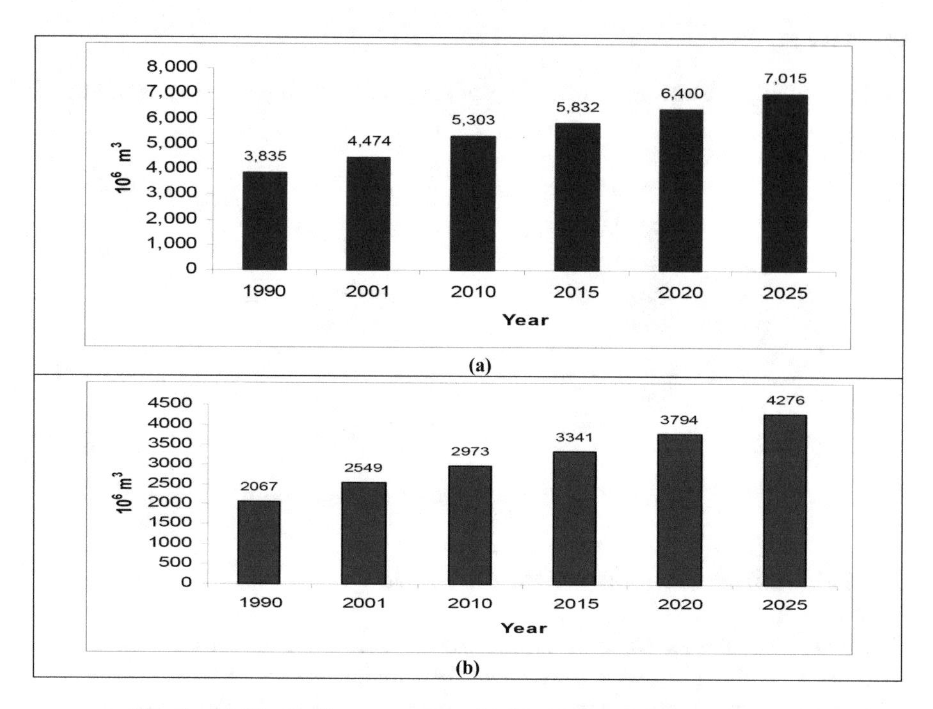


**Fig. 7.** (a) World oil consumption, (b) World natural gas consumption

## 3.6 Natural Gas

The main components [4–6] of natural gas shown in Fig. 8. In general, we remove the Methane and non-hydrocarbons ($H_2O$, $CO_2$, and $H_2S$) and remined only natural gas liquid.

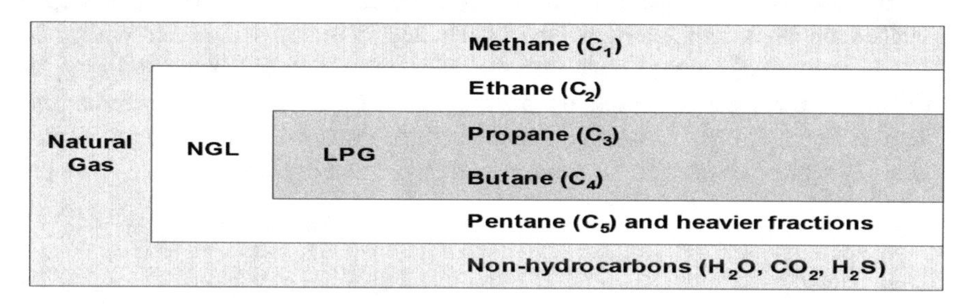


**Fig. 8.** Component of natural gas

### 3.7 Gas Flaring

It refers to the gas association with oil production, Fig. 9 show the trends of both.

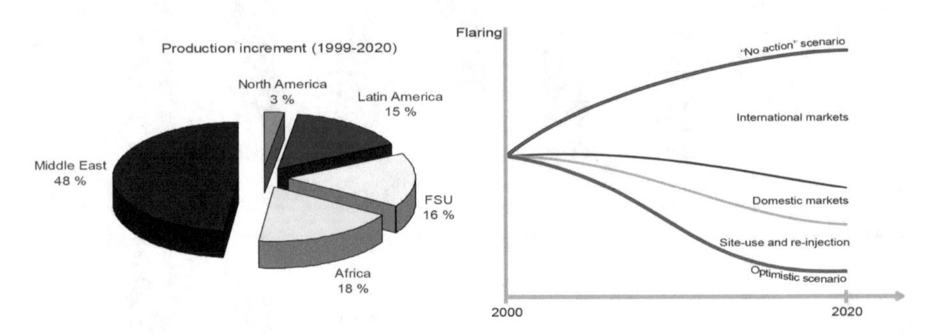


**Fig. 9.** Future oil production and flaring trends

## 4 Multi Objectives Optimization Model

The main goal of this paper is to develop an accurate and efficient optimization technique to building software implementation rapidly. Figure 10 show that model

We can summarization the fundamental research questions by the following points:

- Is the proposed optimization scheme suitable for the Gas Flaring problem compared to other such comparable techniques?
- Does it satisfy the main challenges of optimization (the max rate of oil extraction, high precision of design well, min rate of gas flaring and less cost)?
- Is the proposed technique suitable enough to lead to an optimal solution to satisfy the requirements?
- Are Graphics processing units (GPUs) reduce the computation times, after known GPS are a normal outgrowth of the computational requests put by today's applications on such stringently power-capped and bounded wire-delay equipment. GPUs accomplish higher computational efficiency than CPUs by utilizing less difficult centres and covering up memory idleness by exchanging absent from slowed down strings. Generally, the GPU throughput-oriented execution demonstrate is a great coordinate for numerous data-parallel applications.

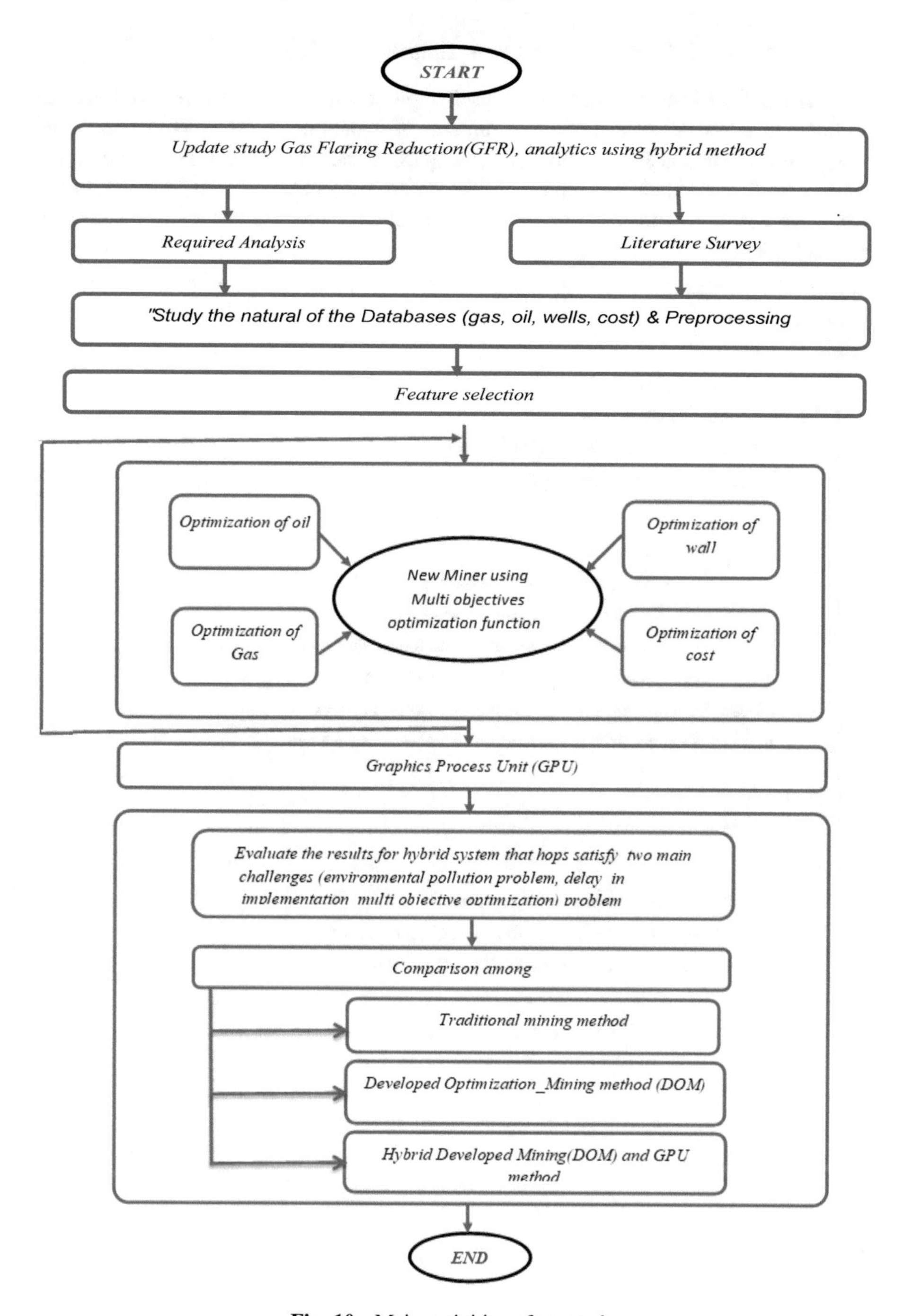


**Fig. 10.** Main activities of research

### 4.1 Multivariate Adaptive Regression Splines (MARS)

The MARS [10] is a data driven regression procedure that built a model based on "divide and conquer" concept from a number of equations (basis function) and coefficients, each equation for a region in input space. It can handle high dimensional data (from 2 to 20 variables) that represents the main problem in other techniques. In this paper, we will dial with multi optimization function inside the MARS to achieve the required goals.

**Pseudo Code of Traditional data mining MARS**

**Input:** Datasets. M,

**Output:** Optimal Values.

Specify Target variable Y.

Building Model by following steps:

While complexity of Model < M, do following steps:

For each variable x, do following

For each Knot of variable x, Test each Knot according to equation:

$$(x-t)_+ = \begin{cases} x-t & x > t \\ 0 & otherwise \end{cases}$$

Choose Knot for variable x, which decrease prediction error.

Add new basis function from variable x with knot to the Model.

IF complexity of Model >= M, then Stop building Model.

For each basis function in the Model make pruning as following steps:

Calculate Generalized Cross Validation error according to equation:

$$GCV = \frac{\sum_{i=1}^{N} (y_i - f(x_i))^2}{(1 - \frac{C}{N})^2}$$

Where N: number of samples, C = 1+ cd

Remove function with high Generalized Cross Validation error.

After Model is completed, predict value of Y by using this equation:

$$y = f(x) = \beta_0 + \sum_{m=1}^{M} \beta_m H_{km}(x_{v(km)})$$

End.

## 4.2 Develop Optimization Mining (DOM)

**Pseudo Code of Develop Optimization Mining DOM**

**Input:** Datasets. M,
**Output:** optimal Values.

Specify Target variable Y.

Determined the general equation for estimations emissions from flares as follow

$$CO_2\ Emissions = Volume\ Flared \times Molar\ volume \times \sum\left(\frac{mole\ Hydrocarbon}{mole\ gas} \times \frac{X\ mole\ C}{mole\ Hydrocarbon}\right) \times Combustion\ efficiency\left(\frac{0.98\ mole\ CO_2\ formed}{mole\ C\ combusted}\right) \times MW\ CO_2$$

$$CH_4\ Emissions = Volume\ Flared \times CH_4\ Mole\ fraction \times \%\ residual\ CH_4 \times Molar\ volume \times MW\ CH_4$$

$$N_2O\ Emissions = Volume\ Flared \times N_2O\ emission\ factor$$

Building Model by following steps:

While complexity of Model < M, do following steps:

For each variable x, do following

For each Knot of variable x, Test each Knot according to equation:

$$(x-t)_+ = \begin{cases} x-t & x > t \\ 0 & otherwise \end{cases}$$

Choose Knot for variable x, which decrease prediction error.

Add new basis function from variable x with knot to the Model.

IF complexity of Model >= M, then Stop building Model.

For each basis function in the Model make pruning as following steps:

Calculate Generalized Cross Validation error according to equation:

$$GCV = \frac{\sum_{i=1}^{N}(y_i - f(x_i))^2}{(1-\frac{C}{N})^2}$$

Where N: number of samples, C = 1+ cd

Remove function with high Generalized Cross Validation error.

After Model is completed, predict value of Y by using this equation:

$$y = f(x) = \beta_0 + \sum_{m=1}^{M} \beta_m H_{km}(x_{v(km)})$$

**End.**

# 5 Results

These are six kinds of gas classify as greenhouse gases “carbon dioxide (CO2), methane (CH4), nitrous oxide (N2O), sulphur hexafluoride (SF6), hydrofluorocarbons (HFCs) and perfluorocarbons (PFCs) families” [8] (Table 2).

The general equations for estimating emissions from flares are:

$$\mathrm{CO_2\ Emissions} = \mathrm{Volume\ Flared} \times \mathrm{Molar\ volume} \times \sum\left(\frac{\mathrm{mole\ Hydrocarbon}}{\mathrm{mole\ gas}} \times \frac{\mathrm{X\ mole\ C}}{\mathrm{mole\ Hydrocarbon}}\right) \times \mathrm{Combustion\ efficiency}\left(\frac{0.98\ \mathrm{mole\ CO_2 formed}}{\mathrm{mole\ C\ combusted}}\right) \times \mathrm{MW\ CO_2}$$

$$\mathrm{CH_4\ Emissions} = \mathrm{Volume\ Flared} \times \mathrm{CH_4\ Mole\ fraction} \times \%\ \mathrm{residual\ CH_4} \times \mathrm{Molar\ Volume} \times \mathrm{MW\ CH_4}$$

$$\mathrm{N_2O\ Emissions} = \mathrm{Volume\ Flared} \times \mathrm{N_2O\ emission\ factor}$$

**Table 2.** Greenhouse gases (GHG) emission factors for gas flaring

| *Original units (tons/106 m$^3$ or tons/1000 m$^3$)* | | | | |
|---|---|---|---|---|
| Flare source | Emission factors[a] | | | Units |
| | $CO_2$ | $CH_4$ | $N_2O$ | |
| Flaring - gas production | 1.8 | 1.1E−02 | 2.1E−05 | tons/10$^6$ m$^3$ gas production |
| Flaring - conventional oil production | 67.0 | 5.0E−03 - 2.7E−01 | 6.4E−04 | tons/1000 m$^3$ conventional oil production |
| *Units Converted to tons/10$^6$ scf or tons/1000 bbl* | | | | |
| Flare source | Emission factors[a] | | | Units |
| | $CO_2$ | $CH_4$ | $N_2O$ | |
| Flaring - gas production | 5.1E−02 | 3.1E−04 | 5.9E−07 | tons/10$^6$ scf gas production |
| Flaring - conventional oil production | 10.7 | 7.9E−04- 4.3E−02 | 1.0E−04 | tons/1000 bbl conventional oil production |

[a] While the presented emission factors may all vary appreciably between countries, the greatest differences are expected to occur with respect to venting and flaring, particularly for oil production due to the potential for significant differences in the amount of gas conservation and utilisation practiced.

## 5.1 Formal Calculation According to Gas Flaring

$$CO_2\ Emissions = Volume\ Flared \times Molar\ volume \times \sum\left(\frac{mole\ Hydrocarbon}{mole\ gas} \times \frac{X\ mole\ C}{mole\ Hydrocarbon}\right) \times Combustion\ efficiency\left(\frac{0.98\ mole\ CO_2\ formed}{mole\ C\ combusted}\right) \times MW\ CO_2$$

$$CH_4 = volume\ flared \times CH_4\ mole\ fraction \times \%\ residual\ CH_4 \times molar\ volume \times MW\ CH_4$$

$$N_2O = volume\ gas\ production \times 2.1 \times 10^{-5}$$

Let: A production facility produces 84,950 m$^3$/day of natural gas. In a given year 566,337 m$^3$ of field gas are flared at the facility. The flare gas composition is unknown.

Assumptions: Since test results or vender data are not available, emissions will be calculated based on 98% combustion efficiency for CO2 emissions and 2% uncombusted CH4. This is consistent with published flare emission factors, fuel carbon combustion efficiencies, control device performance, and results from the more recent flare studies.

Calculations:

$$CO_2: \frac{566,337\text{m}^3\text{ gas}}{\text{yr}} \times \frac{\text{lbmole gas}}{10.74\text{m}^3\text{ gas}} \times \left( \begin{array}{l} \frac{0.80\text{ lbmole CH}_4}{\text{lbmole gas}} \times \frac{\text{lbmole C}}{\text{lbmole CH}_4} \\ + \frac{0.15\text{ lbmole C}_2\text{H}_6}{\text{lbmole gas}} \times \frac{2\text{ lbmole C}}{\text{lbmole C}_2\text{H}_6} \\ + \frac{0.05\text{ lbmole C}_3\text{H}_8}{\text{lbmole gas}} \times \frac{3\text{ lbmole C}}{\text{lbmole C}_3\text{H}_8} \end{array} \right)$$

$$\times \frac{0.98\text{ lbmole CO}_2\text{formed}}{\text{lbmole C combusted}} \times \frac{44\text{ lb CO}_2}{\text{lbmole CO}_2} \times \frac{\text{ton}}{2204.62\text{ lb}} = 1,289\text{ tons CO}_2/\text{yr}$$

$$CH_4: \frac{566,337\text{m}^3\text{ gas}}{\text{yr}} \times \frac{0.80\text{ scf CH}_4}{0.02831685\text{ m}^3\text{ gas}} \times \frac{0.02\text{ scf noncombusted CH}_4}{\text{scf CH}_4\text{ total}} \times \frac{\text{lbmole CH}_4}{379.3\text{ scf CH}_4} \times \frac{16\text{ lb CH}_4}{\text{lbmole CH}_4}$$

$$\times \frac{\text{ton}}{2204.62\text{ lb}} = 6.1\text{ tons CH}_4/\text{yr}$$

$$N_2O: \frac{84,950\text{ m}^3}{\text{day}} \times \frac{365\text{ days}}{\text{yr}} \times \frac{2.1 \times 10^{-5}\text{tons N}_2\text{O}}{10^6\text{m}^3\text{ gas}} = 6.51 \times 10^{-4}\text{tons N}_2\text{O/yr}$$

**Total GHG Emission**

$$\begin{aligned} &= (1 \times 1.289) + (21 \times 6.1) + (310 \times 6.51 \times 10 - 4) \\ &= 1417.302\text{ tons CO2 equivalent} \end{aligned}$$

## 5.2 Costs

There are four scenarios for determined the costs of gas flaring reduction Scenario 0: all associated gas is flared

- Scenario 1: switch power plant's fuel from oil to gas
- Scenario 2: scenario 1 + transfer associated gas to a brick factory
- Scenario 3: scenario 2 + Liquefied Petroleum Gas (LPG)

Depending on the possibilities for gas use, the percentages of associated gas which can be used will vary.

For each scenario, there are two production cases:

- Case #1 no gas is flared and the oil production is reduced to the level required to achieve this goal of no flaring.
- Case #2 the refinery is assumed to have a throughput reflecting the demand for petroleum products in Chad even though this entails some flaring.

| Investment Cost | Case 1 minimized flaring | Case 2 reflecting demand |
|---|---|---|
| Scenario 0 | 78.1 | 98.0 |
| Scenario 1 | 114.1 | 134.0 |
| Scenario 2 | 114.1 | 134.0 |
| Scenario 3 | 114.1 | 134.0 |

Gas-to-oil ratio (GOR) is used to calculate how much volume of gas at atmospheric pressure is produced per unit of oil produced in cubic feet/barrel. GOR is calculated using known gas and oil volumes at surface conditions.

In this paper, the calculations of GOR will be done simply by dividing the amount of associated gas by the amount of oil production.

$$GOR = \frac{amount\ of\ associated\ gas,\ in\ \mathrm{m}^3}{oil\ production,\ in\ \mathrm{m}^3}$$

For dataset that contain the missing values handle it by mean and median measures. The mean (or average) is the most popular and well-known measure of central tendency. It can be used with both discrete and continuous data, although its use is most often with continuous data. So, if we have n values in a data set and they have values $x_1, x_2, \ldots, x_n$, the sample mean, usually denoted by $\bar{x}$ (pronounced x bar), is:

$$\bar{x} = \frac{(x_1 + x_2 + \ldots + x_n)}{n}$$

The median is the middle score for a set of data that has been arranged in order of magnitude. The median is less affected by outliers and skewed data.

### 5.3 Analysis and Compare from Both Sides the Performance and Cost of (TM, DOM, Hybrid) Methods

A. S(1) = 1.962963 + 1.037037 * CO2 [0.6,1]
   Result = −8.326673e − 016 + 1 * S(1)
   ***Function used y = p1 + p2 * x1***

B. S(3) = 1.867384 + 0.917563 * CO2 [0.6,1] + 0.430108 * CH4 [0,0.5]
   S(1) = 0.080197 − 0.176564 * N2O [0.6,1] + 0.981529 * S(3)
   S(8) = 2 + 1 * CO2 [0.6,1] − 0.142857 * N2O [0.6,1]
   S(6) = −0.146513 + 0.429503 * CH4 [0.6,1] + 1.050176 * S(8)
   Result = −0.220885 + 0.734999 * S(1) + 0.373571 * S(6)
   ***Function used y = p1 + p2 * x1 + p3 * x2***

C. S(3) = 1.956989 + 0.430108 * CH4 [0.6,1] + 1.043011 * N2O [0.6,1] − 0.16129* CO2 [0.6,1]
S(6) = 1.968379 + 0.632411* CH4 [0.6,1] + 1.031621* N2O [0.6,1] − 0.201581 * CO2 [0.6,1]
S(2) = −1.749966−0.924511 * N2O[0.6,1] + 0.93643 * S(3) + 0.955062 * S(6)
S(11) = 1.910864 + 0.866295 * N2O [0.6,1] − 0.181058 * CO2 [0.6,1] + 0.445682 * CH4 [0,0.5]
S(8) = −0.231453 − 0.144285 * CH4 [0.6,1] + 1.221628 * CO2 [0.26,1] + 1.127296 * S(11)
Result = −1.472006 − 0.862308 * N2O [0.6,1] + 0.781598 * S(2) + 0.971161 * S(8)
***Function used y = p1 + p2 * x1 + p3 * x2 + p4 * x3***

D. Result = 2−9.742932e + 017 * CO2 [0.6,1] + 9.742932e + 017 * CO2 [0.6,1]^2
***Function used y = p1 + p2 * x1 + p3 * x1^2***

E. Result = 2−2.85384e + 017 * CO2[0.6,1] + 2.640157e + 017 * CO2[0.6,1] ^2 + 2.0534e + 033 * CH4{0_5} − 2.301949e + 049 * CH4{0_5}^2 + 2.301949e + 049 * CO2 [0.6,1] * CH4{0_5}
***Function used y = p1 + p2 * x1 + p3 * x1^2 + p4 * x2 + p5 * x2^2 + p6 * x1 * x2***

F. Result = 2 − 1.724097e + 016 * CH4 [0.6,1] + 1.583826e + 033 * CH4 [0.6,1] ^2 − 1.583826e + 033 * CH4 [0.6,1]^3
***Function used y = p1 + p2 * x1 + p3 * x1^2 + p4 * x1^3***

G. S(1) = 2.086957 − 0.253623 * CH4 [0.6,1] * CO2 [0.6,1]
S(4) = 1.962963 + 1.037037 * N2O [0.6,1] * CO2 [0.6,1]
Result = 0.199661 + 0.442886 * S(1) * S(4)
***Function used y = p1 + p2 * x1 * x2***

H. Result = 2.787881 + 0.168732/(CO2 [0.6,1] − 0.204544)
***Function used y = p1 + p2/(x1 + p3)***

I. S(1) = 3.012749 − 2.067351/(1 + exp(5.113522 * (CO2 [0.6,1] − 0.006095)))
Result = 3.131673 − 17.47943/(1 + exp(2.164804*(S(1) − 0.745341)))
***Function used y = p1 + p2/(1 + exp(−p3 * (x1 − p4)))***

J. Result = 0.386292 + 3.017885 * exp (−((CO2 [0.6,1] − 0.67977) ^2)/0.710772)
***Function used y = p1 + p2 * exp (−((x1 − p3) ^2)/p4)***

K. Result = 2.112856 + 1.385081 * log(CH4 [0.6,1] + 0.897432)
***Function used y = p1 + p2 * log(x1 + p3)***

L. S (1) = 2.840061 − 0.877098 * exp(-5.67461 * (CH4 [0.6,1] + 0.1))
Result = 5.689994 − 7.553622 * exp(−0.351476 * (S(1) − 1.862963))
***Function used y = p1 + p2 * exp(p3 * (x1 + p4))***

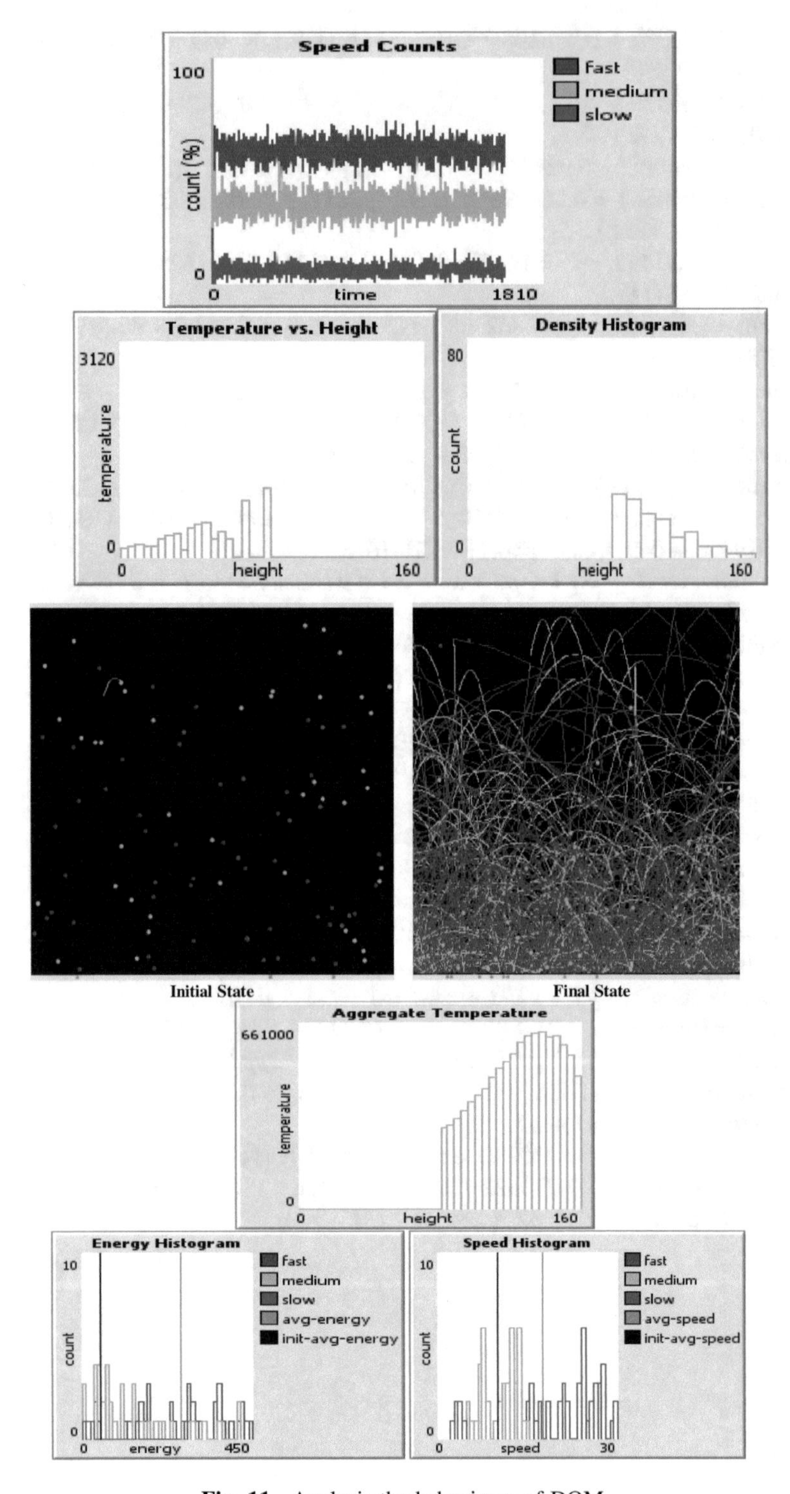


**Fig. 11.** Analysis the behaviours of DOM

**Table 3.** Compare the performance of three mining method

| Hybrid (DOM & GPU) | Develop optimization mining (DOM) | (TM) Traditional mining |
|---|---|---|
| 0.952 | 0.712 | 0.547 |
| 0.955 | 0.345 | 0.549 |
| 0.95 | 0.138 | 0.546 |
| 0.962 | 0.089 | 0.553 |
| 0.952 | 0.831 | 0.547 |

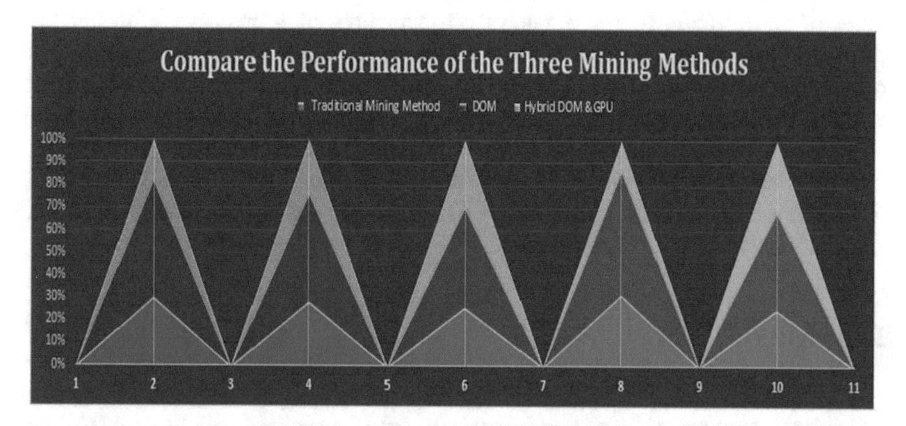


**Fig. 12.** Compare the performance among the three mining methods (i.e., traditional, optimal and hybrid)

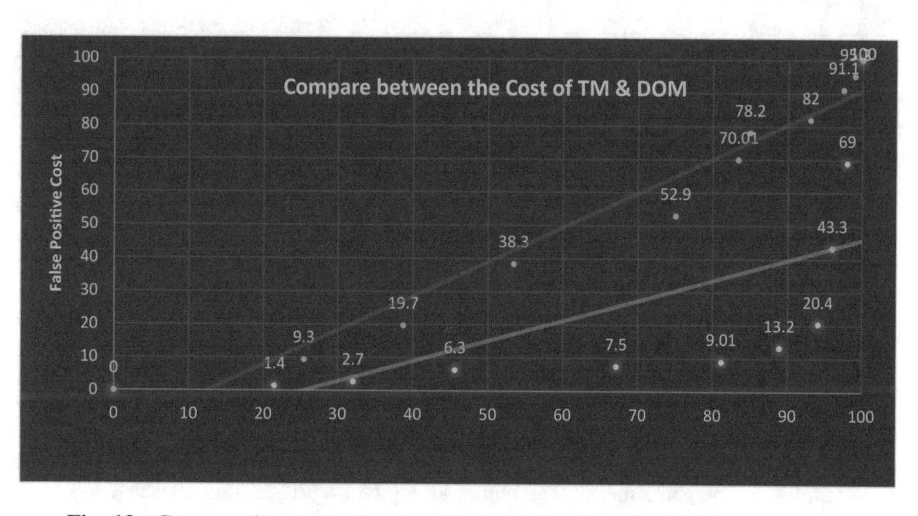


**Fig. 13.** Compare the cost between traditional & develop optimal mining methods

## 6 Conclusion

As a results shown in Table 3 with Figs. 11, 12 and 13, predication techniques are powerful tools for solving Multi objectives optimization problems. From experiments analysis of these techniques; we find DOM give very useful result based on multi objectives function and by replace the traditional question in MARS by optimization equations. In addition, it considers more powerful and faster. The proposed optimization scheme (DOM) suitable for the Gas Flaring problem compared to other traditional techniques; in this study attempt to satisfy balance among the main challenges of optimization (the max rate of oil extraction, high precision of design well, min rate of gas flaring and less cost), really we deal with three challenges only "gas, cost and oil" while the study not handle the challenges related to design well and left it to future work. The DOM lead to an optimal solution that satisfy the requirements determined in problem formal. To ensure that the DOM can handle such data and proof results, we used multi verification measures from such as GOR, cost, etc., We can say DOM give optimal solution because it utilized features of mathematics such as multi optimization functions and integration.

## References

1. Porzio, G.F., Nastasi, G., Colla, V., Vannucci, M., Branca, T.A.: Comparison of multi-objective optimization techniques applied to off-gas management within an integrated steelwork. Appl. Energy **136**(C), 1085–1097 (2014)
2. Sarbazi-azad, H.: Advances In GPU Research And Practice, 1st edn. Morgan Kaufmann, Elsevier (2017)
3. Li, J., Long, Y., Xu, C., Tian, H., Wu, Y., Zha, F.: Continuous high-flux and efficient oil/water separation assisted by an integrated system with opposite wettability. Appl. Surf. Sci. **433**(2018), 374–380 (2017)
4. Sotirov, S., Sotirova, E., Stratiev, D., Stratiev, D., Sotirov, N.: An application of neural network to heavy oil distillation with recognitions with intuitionistic fuzzy estimation. Springer (2018)
5. Ezaina Umukor, G., Saheed Ismail, O.: Modelling emissions from natural gas flaring. J. King Saud Univ. Eng. Sci. **28**(2016), 130–140 (2015)
6. Garcia, S., Luengo, J., Herrera, F.: Data Preprocessingin Data Mining. Springer, London (2015)
7. Saheed Ismail, O., Ezaina Umukor, G.: Modelling combustion reactions for gas flaring and its resulting emissions. J. King Saud Univ. Eng. Sci. **29**(2017), 178–182 (2014)
8. Rahimpour, M.R., Jamshidnejad, Z., Jokar, S.M., Karimi, G., Ghorbani, A., Mohammadi, A. H.: A comparative study of three different methods for flare gas recovery of Asalooye gas refinery. J. Nat. Gas Sci. Eng. **4**(2012), 17–28 (2011)
9. Al_Janabi, S.: Smart system to create optimal higher education environment using IDA and IOTs. Int. J. Comput. Appl. (2018). https://doi.org/10.1080/1206212x.2018.1512460
10. Al_Janabi, S., Mahdi, M.A.: Evaluation prediction techniques to achievement an optimal biomedical analysis. Int. J. Grid Util. Comput. (2019)
11. Talavera, A.L., Vellasco, M.M.B.R.: Controlling oil production in smart wells by MPC strategy with reinforcement learning. In: SPE International (2010)

12. Ozgen karacan, C., Ruiz, F.A., Cote, M., Phipps, S.: Coal mine methane: a review of capture and utilization practices with benefits to mining safety and to greenhouse gas reduction. Int. J. Coal Geol. **86**, 121–156 (2011)
13. Lakehal, A.: Bayesian duval triangle method for fault prediction and assessment of oil immersed transformers. Meas. Control. **50**(4), 103–109 (2017)

# Nesterov Step Reduced Gradient Algorithm for Convex Programming Problems

Abdelkrim El Mouatasim[1(✉)] and Yousef Farhaoui[2]

[1] Faculty of Polydisciplinary Ouarzazate (FPO), Ibn Zohr University, B.P. 284, 45800 Ouarzazate, Morocco
a.elmouatasim@uiz.ac.ma

[2] Faculty of Sciences and Techniques, Department of Computer Science, Moulay Ismail University, IDMS Team, Errachidia, Morocco
y.farhaoui@fste.umi.ac.ma

**Abstract.** In this paper, we proposed an implementation of method of speed reduced gradient algorithm for optimizing a convex differentiable function subject to linear equality constraints and nonnegativity bounds on the variables. In particular, at each iteration, we compute a search direction by reduced gradient, and line search by bisection algorithm or Armijo rule. Under some assumption, the convergence rate of speed reduced gradient (SRG) algorithm is proven to be significantly better, both theoretically and practically. The algorithm of SRG are programmed by Matlab, and comparing by Frank-Wolfe algorithm some problems, the numerical results which show the efficient of our approach, we give also an application to ODE, optimal control, image and video co-localization and learning machine.

**Keywords:** Matlab code · Linear constraints · Convex optimization · Reduced gradient algorithm · Bisection method · Nesterov step · Armijo rule

## 1 Introduction

Optimization problems with linear equality constraints contain the implicit difficulty arising from the need to solve a system of equations while optimizing a cost function.

Many algorithms have proposed for solving linearly-constrained optimization problems, such as the reduced gradient see, for instance [3], proceed by solving a sequence of subproblems in which the number of variables has been implicitly reduced. These reduced problems are obtained by using the linear constraints to express certain variables, designated as 'basic', in terms of other variables.

Wolfe [17] introduced the reduced gradient method, which can be considered as extensions of the classical simplex method. It is well known that the convex simplex method fails by converging to a nonoptimal point.


Y. Farhaoui (Ed.): BDNT 2019, LNNS 81, pp. 140–148, 2020.
https://doi.org/10.1007/978-3-030-23672-4_11

There exist several reduced gradient algorithm application areas like Engineering problems, Water distribution [4], Power flow [15], Machine learning [13] and Optimal control [5].

The main techniques that have been proposed for solving constrained optimization problems in this work is the reduced gradient method.

In this paper, we are interested in design and test a Matlab code of SRG algorithm and comparing with reduced gradient algorithm (RG) and Frank-Wolfe algorithm (FW).

The organization of paper is as follows: Principe of the reduced gradient method is recalled in Sect. 2, the algorithm of SRG are given in Sect. 3, while its convergence is proved in Sect. 4. The results of some numerical experiments of linear constraints optimization are given in Sect. 5.

# 2 Reduced Gradient Method

## 2.1 Reduced Problem

From now on, we consider a nonlinear programming problem with linear equality constraints of the form

$$\begin{cases} \text{minimize} & f(x) \\ \text{subject to} & Ax = b \\ & 0 \leq x \end{cases} \tag{1}$$

where $f : I\!R^n \longrightarrow I\!R$ is convex twice continuously differentiable function, $A$ is $m \times n$ matrix with $m \leq n$ and $b$ is a vector in $I\!R^m$.

The reduced gradient method begins with a basis $B$ and a feasible solution $\mathbf{x}^k = (\mathbf{x}_B^k, \mathbf{x}_N^k)$ such that $\mathbf{x}_B^k > 0$. The solution $\mathbf{x}$ is not necessarily a basic solution, i.e. $\mathbf{x}_N$ do not has to be identically zero. Such a solution can be obtained e.g. by the usual first phase procedure of linear optimization. Using the basis $B$ form $B\mathbf{x}_B + N\mathbf{x}_N = b$, we have

$$x_B = B^{-1}b - B^{-1}Nx_N,$$

hence the basic variables $x_B$ can be eliminated from the problem (1)

$$\begin{cases} \text{minimize :} & F(x_N) \\ \text{subject to :} & 0 \leq x_N, \end{cases} \tag{2}$$

where $F(x_N) = f(B^{-1}b - B^{-1}Nx_N, x_N)$. Using the notation

$$\nabla f(\mathbf{x})^t = \left[\nabla_B f(\mathbf{x})^t, \nabla_N f(\mathbf{x})^t\right],$$

the gradient of $F$ which is the so-called *reduced gradient*, can be expressed as

$$r = \nabla F(x)^t = -\left(\nabla_B f(x)^t B^{-1} N\right) + \left(\nabla_N f(x)^t\right).$$

We determine the optimal step as the value $\eta_k^*$ such that

$$f(x^k + \eta_k^* d^k) = \min_{0 \leq \eta_k \leq \eta_{\max}} \{f(x^k + \eta_k d^k)\}. \tag{3}$$

In this paper we used the bisection algorithm for solving the unconstrained optimization problem (3).
Let bis(g,a,b,$\epsilon$) denote the recursive bisection procedure. The inputs for this procedure are: the procedure for calculation values of $g$, the segment $[a, b]$ and the accuracy $\epsilon$. The outputs are the estimation $x_m$ for the minimizer $x^*$ and $g_m$ for the value of the minimum of the function $g$ over the segment [a,b].
The iteration of the recursive procedure includes the following steps.

Step 0 If $b - a \geq \epsilon$, go to step 1, otherwise stop.
Step 1 Compute

$$c = \frac{a+b}{2}, a' = \frac{a+c}{2}, b' = \frac{c+b}{2}, g(c), g(a'), g(b')$$

Step 2 If $g(a') \leq g(c) \leq g(b')$, set $b = b'$.
If $g(a') \geq g(c) \geq g(b')$, set $a = a'$.
If $g(c) \leq \min\{g(a'), g(b')\}$, set $a = a', b = b'$
Step 3 Execute bis(g,a,b,$\epsilon$) with new inputs.

## 3 Algorithm of Reduced Gradient

### 3.1 Algorithm of RG

We are now ready to present the details of our RG algorithm.

Step 0 (Initialization).
Choose a feasible point $\mathbf{x}^0 \in I\!R^n$, and $I_B^0$, $I_N^0$ such that $B^0$ is nonsingular. Set the iteration counter $k = 0$.
Step 1 (Independent variables choice).
If $k \neq 0$ choose the sets $I_B^k$, $I_N^k$;
Step 2 (Search direction computation).
1. Let $(d_N)_j = \begin{cases} -r_j & \text{if } r_j < 0 \quad \text{or } (x_N)_j > 0, \\ 0 & \text{otherwise} \end{cases}$
2. If $d_N$ is zero, stop; the current point is a solution. Otherwise, fined $d_B = -B^{-1} N d_N$.

Step 3 (Optimal line search computation: Bisection algorithm).
Fined $\eta_{\max}, \eta_k$ achieving respectively,

$$\min_{1 \leq j \leq n} \{\frac{x_j^k}{-d_j^k} : d_j^k < 0\}$$

$$\min_{0 \leq \eta \leq \eta_{\max}} f(x^k + \eta d^k) \tag{4}$$

Step 4 (Next point computation). Put $x^{k+1} = x^k + \eta_k d^k$;
Step 5 If $\eta_k < \eta_{\max}$, return to Step 2. Otherwise, declare the vanishing variable in the dependent set independent and declare a strictly positive variable in the independent set dependent. Update $B$ and $N$
Step 6 (Basic variables choice). Choose $I_B^{k+1}, I_N^{k+1}$. Let $k = k + 1$ and go to Step 1.

### 3.2 Speed Reduced Gradient Algorithm

Reduced gradient and Nesterov step

1. Select a point $\mathbf{y}_0 \in \mathbf{C}$. Put
$$k = 0, \quad \mathbf{b}_0 = 1, \quad \mathbf{x}^{-1} = \mathbf{y}_0.$$
2. kth iteration.
   (a) Compute direction of reduced gradient $\mathbf{d_k}$.
   (b) Put
$$\begin{aligned} \mathbf{x}^k &= \mathbf{y}_k + \eta_k \mathbf{d}_k, \\ \mathbf{b}_{k+1} &= 0.5(1 + \sqrt{4\mathbf{b}_k^2 + 1}\ ), \\ \mathbf{y}_{k+1} &= \mathbf{x}^k + (\tfrac{\mathbf{b}_k - 1}{\mathbf{b}_{k+1}})(\mathbf{x}^k - \mathbf{x}^{k-1}), \end{aligned} \tag{5}$$
   see for instance [14].

The recalculation of the point $\mathbf{y}_k$ in (5) is done using a "ravine" step, and $\eta_k$ is the optimal step (4).

## 4 Convergence Analysis

The reduced gradient method stops when the direction vector with respect to the non basic variables $d_N^k = 0$. The justification of this stopping criterion is presented in [6].

### 4.1 Assumptions

Here we give a proof of some typical convergence results for the SRG algorithm.
We assume that there is a minimizer of $\mathbf{f}$, say $\mathbf{x}^*$.
We choose $\eta_k$ by an Armijo rule see for instance [2], then

$$f(y_k) - f(y_k + \eta_k d_k) \geq 0.5\eta_k \|\mathbf{d}_{k+1}\|^2. \tag{6}$$

**Theorem 1.** *If the sequence $\{\mathbf{x}_k\}_{k\geq 0}$ is constructed by SRG algorithm, then there exist a constants* $\mathbf{C}$ *such that* $\forall k \geq 0$

$$\mathbf{f}(\mathbf{x}_k) - \mathbf{f}(\mathbf{x}^*) \leq \frac{\mathbf{C}}{(k+2)^2}. \tag{7}$$

**Proof.** Let $\mathbf{p}_k = (\mathbf{b}_k - 1)(\mathbf{x}^{k-1} - \mathbf{x}^k)$, then

$$\mathbf{p}_{k+1} - \mathbf{x}^{k+1} = \mathbf{p}_k - \mathbf{x}_k - \mathbf{b}_{k+1}\eta_{k+1}\mathbf{d}_{k+1}.$$

Consequently,

$$\begin{aligned} \|\mathbf{p}_{k+1} - \mathbf{x}^{k+1} + \mathbf{x}^*\|^2 = \|\mathbf{p}_k &- \mathbf{x}^k + \mathbf{x}^*\|^2 \\ &+ 2(\mathbf{b}_{k+1} - 1)\eta_{k+1}\mathbf{d}_{k+1}^t \mathbf{p}_k \\ &+ 2\mathbf{b}_{k+1}\eta_{k+1}\mathbf{d}_{k+1}^t(\mathbf{x}^* - \mathbf{y}_{k+1}) \\ &+ \mathbf{b}_{k+1}^2 \eta_{k+1}^2 \|\mathbf{d}_{k+1}\|^2. \end{aligned} \tag{8}$$

Using inequality (6) and the convexity of $f(x)$, we obtain

$$\begin{aligned}
\mathbf{d}_{k+1}^t(\mathbf{y}_{k+1} - \mathbf{x}^*) &\geq f(\mathbf{x}^{k+1}) - f^* + 0.5\eta_{k+1}\|\mathbf{d}_{k+1}\|^2, \\
0.5\eta_{k+1}\|\mathbf{d}_{k+1}\|^2 \leq & \ f(\mathbf{y}_{k+1}) - f(\mathbf{x}_{k+1}) \leq f(\mathbf{x}_k) - f(\mathbf{x}_{k+1}) \\
& -\mathbf{b}_{k+1}^{-1}\mathbf{d}_{k+1}^t\mathbf{p}_k.
\end{aligned}$$

We substitute these two inequalities into (8):

$$\begin{aligned}
&\|\mathbf{p}_{k+1} - \mathbf{x}^{k+1} + \mathbf{x}^*\|^2 - \|\mathbf{p}_k - \mathbf{x}^k + \mathbf{x}^*\|^2 \\
&\leq 2(\mathbf{b}_{k+1} - 1)\eta_{k+1}\mathbf{d}_{k+1}^t\mathbf{p}_k \\
&-2\mathbf{b}_{k+1}\eta_{k+1}(\mathbf{f}(\mathbf{x}^{k+1}) - \mathbf{f}(\mathbf{x}^*)) + (\mathbf{b}_{k+1}^2 - \mathbf{b}_{k+1})\eta_{k+1}^2\|\mathbf{d}_{k+1}\|^2 \\
&\leq -2\mathbf{b}_{k+1}\eta_{k+1}(\mathbf{f}(\mathbf{x}^{k+1}) - \mathbf{f}(\mathbf{x}^*)) \\
&+2(\mathbf{b}_{k+1}^2 - \mathbf{b}_{k+1})\eta_{k+1}(\mathbf{f}(\mathbf{x}^k) - \mathbf{f}(\mathbf{x}^{k+1})) \\
&= 2\mathbf{b}_k^2\eta_{k+1}(\mathbf{f}(\mathbf{x}^k) - \mathbf{f}(\mathbf{x}^*)) - 2\mathbf{b}_{k+1}^2\eta_{k+1}(\mathbf{f}(\mathbf{x}^{k+1}) - \mathbf{f}(\mathbf{x}^*)) \\
&\leq 2\mathbf{b}_k^2\eta_k(\mathbf{f}(\mathbf{x}^k) - \mathbf{f}(\mathbf{x}^*)) - 2\mathbf{b}_{k+1}^2\eta_{k+1}(\mathbf{f}(\mathbf{x}^{k+1}) - \mathbf{f}(\mathbf{x}^*)).
\end{aligned}$$

Thus

$$\begin{aligned}
&2\mathbf{b}_{k+1}^2\eta_{k+1}(\mathbf{f}(\mathbf{x}^{k+1}) - \mathbf{f}(\mathbf{x}^*)) \\
&\leq 2\mathbf{b}_{k+1}^2\eta_{k+1}(\mathbf{f}(\mathbf{x}^{k+1}) - \mathbf{f}(\mathbf{x}^*)) + \|\mathbf{p}_{k+1} - \mathbf{x}^{k+1} + \mathbf{x}^*\|^2 \\
&\leq 2\mathbf{b}_k^2\eta_k(\mathbf{f}(\mathbf{x}^k) - \mathbf{f}(\mathbf{x}^*)) + \|\mathbf{p}_k - \mathbf{x}^k + \mathbf{x}^*\|^2 \\
&\leq 2\mathbf{b}_0^2\eta_0(\mathbf{f}(\mathbf{x}^0) - \mathbf{f}(\mathbf{x}^*)) + \|\mathbf{p}_0 - \mathbf{x}^0 + \mathbf{x}^*\|^2 \\
&\leq \|\mathbf{y}_0 + \mathbf{x}^*\|^2
\end{aligned}$$

It remains to observe that

$$\mathbf{b}_{k+1} \geq \mathbf{b}_k + 0.5 \geq 1 + 0.5(k+1).$$

And there exist a real $\beta > 0$ such that $\eta_k \geq \beta \quad \forall k \geq 0$, see for instance [2]. Let

$$\mathbf{C} = \|\mathbf{y}_0 - \mathbf{x}^*\|^2/\beta,$$

then theorem hold.

## 5 Numerical Experiments

The code of proposed algorithm SRG, RG, FW is written by using Matlab programming language. We test SRG method and compare it with RG and FW algorithm.

The optimal line search process of RG, FW and SRG fined by using bisection algorithm, we set $\epsilon = 10^{-4}$.

We stop the iteration if the KKT condition $\|d_N\| \leq 10^{-5}$ is satisfied or its equal maximal iteration.

The algorithms are run on a workstation HP Intel(R) Celeron(R) M processor 1.30 GHz., 224 Mo RAM. The row cpu gives the mean CPU time in seconds for one run.

**Table 1.** Comparing results of test problems between FW, RG and SRG algorithms.

| Problem | | | Algorithm | | | | | | | | |
|---|---|---|---|---|---|---|---|---|---|---|---|
| | | | FW | | | RG | | | SRG | | |
| # | $n$ | $n_c$ | $CPU$ | $f^*$ | $Iter$ | $CPU$ | $f^*$ | $Iter$ | $CPU$ | $f^*$ | $Iter$ |
| HS44 | 4 | 6 | 0.16 | −15 | 4 | 0.06 | −15 | 5 | 0.05 | −15 | 5 |
| HS48 | 5 | 2 | 1.25 | 6.23e−9 | 49 | 0.09 | 1.81e−9 | 47 | 0.06 | 1.99e−9 | 41 |
| HS49 | 5 | 2 | 4.22 | 1.9e−5 | 219 | 0.48 | 1.9e−5 | 198 | 0.10 | 7.9e−6 | 32 |
| HS50 | 5 | 3 | 0.11 | 5.6303 | 10 | 0.02 | 2.9e−4 | 10 | 0.02 | 2.9e−4 | 10 |
| HS51 | 5 | 3 | 0.19 | 8.6e−8 | 14 | 0.03 | 1.6e−8 | 11 | 0.03 | 1.6e−8 | 11 |
| HS53 | 5 | 8 | 0.08 | 4.093 | 6 | 0.05 | 4.093 | 6 | 0.02 | 4.093 | 6 |
| HS55 | 6 | 8 | 0.05 | 6.3333 | 3 | 0.02 | 6.3333 | 2 | 0.02 | 6.3333 | 2 |
| HS76 | 4 | 3 | 0.94 | −4.6708 | 50 | 0.06 | −4.6818 | 9 | 0.02 | −4.6818 | 9 |
| HS86 | 5 | 10 | 4.47 | −32.3242 | 250 | 0.03 | −32.3487 | 8 | 0.02 | −32.3487 | 8 |
| HS110 | 10 | 20 | 0.04 | −45.7785 | 2 | 0.04 | −45.7785 | 3 | 0.03 | −45.7785 | 3 |
| HS118 | 15 | 47 | 0.63 | 664.8205 | 8 | 0.81 | 664.8205 | 39 | 0.40 | 664.8205 | 19 |
| HS119 | 16 | 8 | 41.50 | 244.9144 | 2000 | 2.50 | 244.8997 | 398 | 0.41 | 244.8997 | 221 |
| F7 | 20 | 10 | 2.31 | −5099 | 100 | 0.44 | −7251.7 | 96 | 0.39 | −7251.7 | 96 |
| F8 | 24 | 10 | 3.20 | 18423 | 150 | 0.84 | 3258500 | 150 | 0.43 | 15990 | 137 |
| K6 | 30 | 60 | 5.20 | 2.41e−13 | 100 | 6.42 | 6.64 | 100 | 3.17 | 2.48e−14 | 45 |

### 5.1 Test Problems

This algorithms has been tested on some problems from Hock and Schittkowski (HS#) [8], Floudas (F#) [7] and Kiwiel (K6) [11] where linear constraints are present. We test the performance of FW, RG and SRG methods on the following test problems with given initial feasible points $x^0$. The results are listed in Table 1, $n$ stands for the dimension of tested problem and $n_c$ stands for the number of constraints. We will report the following results: the CPU time, the optimal value $f^*$, the number of iteration $Iter$.

From Table 1 above, we can see that our algorithm SRG can find their solutions with a small number of iterations or CPU time, and the computation results illustrate that our algorithm SRG executes well for those problems.

In contrast to the numerical results of Frank-Wolfe algorithm and reduced gradient algorithm, the results in Table 1 show that the numbers of iterations for some problems are larger, while this may not very important, since the computational complexity in a single iteration and the CPU time are the main effort of this work for applications.

### 5.2 Applications

**Application 1 (Ordinary differential equation (ODE)).** *Given a positive integer k, the problem is defined as follows :*

$$\begin{array}{ll} \textit{minimize:} & \frac{1}{2}\sum_{i=1}^{k-2}(x_{k+i+1}-x_{k+i})^2, \\ \textit{subject to:} & x_{k+i}-x_{i+1}+x_i=0, \quad i=1,\ldots,k-1, \\ & \alpha_i \le x_i \le \alpha_{i+1}, \quad i=1,\ldots,k, \\ 0.4(\alpha_{i+2}-\alpha_i) \le & x_{k+1} \le 0.6(\alpha_{i+2}-\alpha_i) \quad i=1,\ldots,k-1, \end{array}$$

*where the constants $\alpha_i$ are defined by*

$$\alpha_i = 1.0 + (1.01)^{i-1}.$$

*These problems arise in the optimal placement of nodes in a scheme for solving ordinary differential equations with given boundary values [12]. We solve these problems for different values of $k = 500$.*

**Application 2 (Optimal Control).** *Consider the optimal control of a servo-motor [5]*

$$\begin{cases} \textit{minimize } J(\mathbf{x}_0,\mathbf{u}) = \int_0^\infty [\mathbf{x^T(t)Qx(t)} + \mathbf{ru^2(t)}]\mathbf{dt} \\ \qquad \textit{subject to } \frac{\partial \mathbf{x}}{\partial \mathbf{t}} = \begin{bmatrix} 0 & 1 \\ 0 & -1 \end{bmatrix} \mathbf{x} + \begin{bmatrix} 0 \\ 1 \end{bmatrix} \mathbf{u} \end{cases} \tag{9}$$

*where $Q = \begin{bmatrix} 1 & 0 \\ 0 & 0 \end{bmatrix}$, $r > 0$.*

*The state and control variables are parametrized by the proposed approximation method in [16].*

$$\begin{cases} \textit{minimize} & \frac{h}{2}(\sum_1^m \mathbf{y}_i^2 + r\sum_1^m \mathbf{u}_i^2) \\ \textit{subject to :} & y_1 = y_0 + \frac{h}{2}z_1 \\ & y_i = y_{i-1} + \frac{h}{2}(z_i + z_{i-1}), i = 2,\ldots,m \\ & z_1 = z_0 + \frac{h}{2}(-z_1 + u_i) \\ & z_i = z_{i-1} + \frac{h}{2}(-z_i + u_i - z_{i+1} + u_{i-1}), i = 2,\ldots,m \end{cases} \tag{10}$$

*where $m = 250$, $\mathbf{x} = (\mathbf{y},\mathbf{z})$; the time interval $h = 2$.*

*We choose $r = 0.5$; $\mathbf{u}_0 = 1000 * ones(1,m)$ and $\mathbf{x}_0 = [3.25; 4.95]$.*

**Application 3 (Image and video co-localization).** *The application comes from image and video co-localization. The approach used by [9] is formulated as a quadratic program (QP) over a flow polytope, the convex hull of paths in a network. In this application, the linear minimization oracle is equivalent to finding a shortest path in the network, which can be done easily by dynamic programming [10].*

*For comparing with Frank-Wolfe algorithm, we re-use the code provided by [9] and their included aeroplane dataset resulting in a QP over 660 variables (Table 2).*

**Table 2.** Comparing results between FW, RG and SRG algorithms for applications.

| Application | | | Algorithm | | | | | | | | |
|---|---|---|---|---|---|---|---|---|---|---|---|
| | | | FW | | | RG | | | SRG | | |
| # | $n$ | $n_c$ | *CPU* | $f^*$ | *Iter* | *CPU* | $f^*$ | *Iter* | *CPU* | $f^*$ | *Iter* |
| 1 | 999 | 499 | 12.73 | 0.0047 | 150 | 6.18 | 0.0049 | 50 | 3.66 | 0.0047 | 30 |
| 2 | 750 | 500 | 1.0469 | 0 | 4 | 1.6875 | 0 | 4 | 1.0938 | 0 | 4 |
| 3 | 660 | 33 | 5.97 | 0.098437 | 2001 | 3.10 | 0.098419 | 673 | 2.83 | 0.098419 | 664 |
| 4 | 6251 | 1 | 174.22 | −0.9581 | 1000 | 19.32 | −0.9561 | 1000 | 14.44 | −0.9597 | 585 |

**Application 4 (Support vector machines (SVM)).** *The Lagrange dual of standard convex optimization setup for structural SVMs [10] has n variables. Writing $\alpha_i$ for the dual variable associated with the training example i, The dual problem is given by*

$$\begin{cases} \text{minimize} & \frac{\lambda}{2}\|A\alpha\| - b^T\alpha \\ \text{subject to :} & \sum_i \alpha_i = 1 \end{cases} \tag{11}$$

*where the matrix $A \in \Re^{d\times m}$, and the vector $b \in \Re^m$.*

*We evaluate our approach on the OCR datasets ($n = 6251$, $d = 4028$) for handwritten character recognition [10].*

## 6 Conclusion

We have proposed a new reduced gradient method, for linearly constrained smooth optimization. The global convergence of SRG algorithm is guaranteed for convex function. The implementation and test of SRG algorithm proposed show that this approach is very faster than reduced gradient algorithm in CPU time and SRG algorithm converge for large problem.

## References


1. Baushev, A.N., Morozova, E.Y.: A multidimensional bisection method for minimizing function over simplex. In: Lectures Notes in Engineering and Computer Science, vol. 2, pp. 801–803 (2007)
2. Bertsekas, D.P.: Nonlinear programming, 2nd edn. Athena Scientific, Belmont (1999)
3. Dembo, R.S.: Dealing with degeneracy in reduced gradient algorithms. Math. Program. **31**, 363–375 (1985)
4. El Mouatasim, A., Ellaia, R., Al-Hossain, A.: A continuous approach to combinatorial optimization: application to water system pump operations. Optim. Lett. J. **6**(1), 177–198 (2012)
5. El Mouatasim, A., Ellaia, R., Souza de Cursi, J.E.: Stochastic perturbation of reduced gradient & GRG methods for nonconvex programming problems. J. Appl. Math. Comput. **226**, 198–211 (2014)

6. El Mouatasim, A.: Implementation of reduced gradient with bisection algorithms for non-convex optimization problem via stochastic perturbation. J. Numer. Algorithms **78**(1), 41–62 (2018)
7. Floudas, C.A., Pardalos, P.M.: A collection of test problems for constrained global optimization algorithms. In: Lecture Notes in Computer Science, vol. 455. Springer-Verlag, Berlin (1990)
8. Hock, W., Schittkowski, K.: Test examples for nonlinear programming codes. In: Lecture Notes in Economics and Mathematical Systems, vol. 187. Springer (1981)
9. Joulin, A., Tang, K., Fei-Fei, L.: Efficient image and video co-localization with Frank-Wolfe algorithm. In: European Conference on Computer Vision (ECCV) (2014)
10. Lacoste-Julien, S., Jaggi, M., Schmidt, M., Pletscher, P.: Block-coordinate Frank-Wolfe optimization for structural SVMs (2013)
11. Kiwiel, K.C.: Proximity control in bundle methods for convex nondifferentiable minimization. Math. Program. **46**, 105–122 (1990)
12. Martinez, J.M., Pilotta, E.A., Raydan, M.: Spectral gradient methods for linearly constrained optimization. J. Optim. Theory Appl. **125**(3), 629–651 (2005)
13. Nanuclef, R., Frandi, E., Sartori, C., Allende, H.: A novel Frank-Wolfe algorithm. Analysis and applications to large-scale SVM training. Inf. Sci. **285**, 66–99 (2014)
14. Nesterov, Y.E.: A method for solving the convex programming problem with convergence rate $O(1/k^2)$. Dokl. Akad. Nauk SSSR **269**, 543–547 (1983)
15. Penêdo de Carvalho, E., Júnior, A., Fu Ma, T.: Reduced gradient method combined with augmented Lagrangian and barrier for the optimal power flow problem. Appl. Math. Comput. **200**, 529–536 (2008)
16. Wang, F., Su, C., Liu, Y.: Computation of optimal feedforward and feedback control by a modified reduced gradient method. Appl. Math. Comput. **104**, 85–100 (1999)
17. Wolfe, P.: The Reduced Gradient Method. Rand Document, Santa Monica (1962)

# Recommendation System of Big Data Based on PageRank Clustering Algorithm

Samaher Al_Janabi(✉) and Noora Kadiam

Department of Computer Science, Faculty of Science for Women (SCIW), University of Babylon, Babylon, Iraq
samaher@itnet.uobabylon.edu.iq

**Abstract.** Big data is an area focused on storing, processing and visualizing huge amount of data. Today data is growing faster than ever before. We need to find the right tools and applications and build an environment that can help us to obtain valuable insights from the data. We use clustering algorithms to discover communities (clusters) in the data. The algorithms finding graph clusters using the personalized PageRank vectors to determine a set of clusters and optimizing the jumping parameter $\alpha$ subject to several cluster variance measures in order to capture the graph structure according to PageRank clustering algorithms and then use the clusters for building a recommendation system that can recommend products to customers based on their buying behavior. Recommendation system plays important role in Internet world and used in many applications. It has created the collection of many application, created global village and growth for numerous information. This paper generated in recommendation system by PageRank clustering algorithm.

**Keywords:** Big data analysis · PageRank-clustering · gSpan · Recommendation system · PRS-CPR · Forward and backward rules

## 1 Introduction

Data can be defined any object have set of features, or object recognize by signal specific feature or collection of objects with their feature. Data supplies the overall standards that the user can get benefit of various kinds of data such as financial, medical and the like. There are different forms of data (i.e., numbers, text and fact). In addition, Data can be classified into three types, small data is data not enough to take any decision because number of sample less than 30 sample not lay to the normal distribution, normal data is data can used to take different type of decision and it's structure in standard format such as table content of rows and columns, while the last of type is big data can be define as structure, unstructured, semi structure, this data recognized by size lay in rang (1 TB to 1 ZB) also can take decision by it through combination between machine learning and cloud computing. Nowadays data is increasing quicker than in the past. It is necessary for users to find the proper applications and instruments to create an environment of great help for them to gain worthy insights and perceptions out of the given data. One of the domains in which large amount of business data can


Y. Farhaoui (Ed.): BDNT 2019, LNNS 81, pp. 149–171, 2020.
https://doi.org/10.1007/978-3-030-23672-4_12

be gathered every day is retail. Thus, retailers require comprehending the purchasing patterns and conduct of their customers for typical business decisions [1]. It is worthy to mention another type of data that is called data science which "is a discipline that merges concepts from computer science (algorithms, programming, machine learning, and data mining), mathematics (statistics and optimization), and domain knowledge (business, applications, and visualization)" to extract insights from data and transform it into actions that have an impact in the particular domain of application [2].

Data analysis is meaner to extraction useful knowledge from any data science. It has six stages, that stages integration and cooperative with other to achieve the purpose of analysis. The first stage is the understanding of the problem which means a comprehensive understanding of business domain like market basket, products, and orders. The second stage is the understanding of data which means comprehending features of data set within data base. The third stage is preprocessing of data that may contain some problems such as random error, duplicate, outlier etc. The fourth stage is the building of various types of models such as classification, clustering, association rules etc. The fifth stage is the evaluation of the built model to insure the results. The last stage is the deployment which is responsible for the model marketing [3, 16].

Recommender systems are defined as recommendation inputs given by the people, which the system then aggregates and directs to appropriate recipients. It can be further defined as a system that produces individualized recommendations as output or has the effect of guiding the user in a personalized way to interesting objects in a larger space of possible options. There are majorly six types of recommender systems which work primarily in the Media and Entertainment industry: Collaborative Recommender system, Content-based recommender system, Demographic based recommender system, Utility based recommender system, Knowledge based recommender system and Hybrid recommender system [4, 15]. The remind of this paper is organization follow: section two show Literature Survey, section three explains the main concept used to handle this problem, section four show the prototype of recommendation system, section five show the results and finally the section six description the conclusion and feature work of the problem statement.

## 2 Related Works

Many researches work to solve the increase growth of data and given recommendation of customer to select the suitable products. Also many techniques present to deal with this problem as explained below:

Sabastian and Ivan 2014 [9] discuss novel approach based on advanced retail. Indeed affiliation rules deliver comes about that are troublesome to interpret. These realities propel us to utilize a novel approach that produces communities of items. One of the most preferences of these communities is that are significant and easily interpretable by retail investigators. This approach permits the handling of billions of exchange records inside a sensible time, agreeing to desires of companies. This work differs with our work from a several of aspects, first the data is used transactional records from a retail chain in Chile. In terms of volume, we have around half billion records gathered within a period of twenty months, approximately 2, 200, 000 customers and over 42, 000 SKUs1 globally, second the pre-processing of data is used three hierarchical level structures. Each level belongs to its predecessor based on an ad–hoc developed taxonomy by the retailer, third the technique is used network & graph, retail analysis, fourth the evaluation measure is used top three heavy edges threshold (TTHET). our work similar it in the design for the same purpose represent by recommendation system.

Jevin, Ian and Carl 2015 [10] present Eigenfactor Recommends for improving scholarly navigation by used the hierarchical structure of scientific knowledge, making possible multiple scales of relevance for different users. This work differs with our work from a number of aspects, first the data is used 35 million articles from various bibliographic databases including the AMiner dataset, second the pre-processing of data is used citation graph, third the evaluation measure is used using an A-B testing environment developed by the Social Science Research Network SSRN. This work is similar with our work from point the type technique used in clustering and the goal.

Ukrit et al. 2016 [11] discuss methods including the visual-clustering recommendation (VCR) method, the hybrid between the VCR and user-based methods, and the hybrid between the VCR and item-based methods. The user-item clustering is based on the genetic algorithm (GA). This work differs with our work from a number of aspects, first type of the data set used, second, pre-processing of dataset, third, techniques used in clustering reprehend by GA, finally, by the evaluation measure include precision, recall, and F1 score. While it similar our work by design system of recommendation.

Krishna and Suresh 2017 [12] discuss the system intended to develop stores the user's purchase and rating data in a backend database. It also includes a smart beacon that connects to the offline shopper's mobile phone and gains data about the particular customer from backend database. Where, users are clustered based on their demographic data. Recommender system employed analyses each cluster and finds out top rated items in each cluster. It then sends out customised offers for a particular customer. This work differs with our work from a number of aspects, first the data used (i.e., Book Rating Data Set), second the preprocessing techniques used to handle data, third the technique used for clustering users based on location and age and plotting them on a scatter plot, fourth the evaluation measure is used collaborative filtering and demographics. Our work similar with it in the goal represent by design recommendation system.

Harno [7] present the kind of information mining strategies for the retail segment to utilize, and how can these advances be utilized when arranging focused on showcasing. This prototype shows how the information approximately a retailer's clients is collected, what kind of information is collected and how the circumstance is in Finnish basic supply stores. After that, diverse kind of information mining devices will be presented beginning from the advertise wicker container investigation, taking after the essentials of proposal frameworks and diverse information mining apparatuses utilized for proposal frameworks. This work differs with our work from a number of aspects, first the data used the biggest Finnish grocery store chains, second the preprocessing of data is used K-Group, third the technique is used is association rule, fourth the evaluation measure is used collaborative filtering based on the Jaccard distance. While similar with it in the goal.

Priya [5] presented mine transaction data by modelling the data as graphs. He used clustering algorithms to discover communities (clusters) in the data and then use the clusters for building a recommendation system to recommend products of customers based on their buying behaviour. This work differs with our work it used online retail service that delivers groceries. While, it similar with our work, from multi points include, missing values and random error handled as prepressing stage. Clustering techniques from type (Louvain algorithm) rather than PageRank clustering also our work have the same goal.

Table 1 explained compare among the previous works based on the seven points determine by our, these points are: names of anthers with year, dataset used, and type of preprocessing, techniques used to solve problem, advantages of this methodology, disadvantage of this methodology, type of evaluation measures used to prove the results. As a result, we can say our work near from researcher [5] present by Rashmi priya 2018. From the type of the data used and preprocessing stage but our work will used intelligent computation represent by (PageRank clustering and gSpan).

**Table 1** Compare among the previous works

| Name of author | Database & dataset | Pre-processing | Techniques | Models | Evaluations Measured |
|---|---|---|---|---|---|
| Sebastian and Ivan 2014 [9] | Transaction records from a retail chain in Chile | Ad-hoc developed taxonomy by the retailer | Network & Graph, retail analysis | Recommendation system | Top three heavy edges threshold (tthet) |
| Jevin, Ian and Carl 2015 [2] | Dataset related to 35 million articles from various bibliographic | AMiner Citation graph | Hierarchical clustering | Recommendation system | Using an A-B testing environment developed by SSRN |
| Ukrit, Nipon and Sansanee 2016 [10] | KDD-CUP2000 database related to the transaction of purchasing at Thaiherbs-Thaimassage shop (TTS) | Ten-Fold cross-validation | Clustering method by changing the fitness function in the genetic algorithm | Recommendation system | Precision, recall, and F1 score |
| Krishna and Suresh 2017 [11] | Book Rating Data Set: http://www2.informatik.uni-freiburg.de/~cziegler/BX | • Split the dataset in to three main tables<br>• Determined the main information for each article (name of author (s), name of journal/book, etc.) | Clustering Users based on location and age and plotting them on a scatter plot | Recommendation system | Collaborative Filtering and Demographics |
| Harno 2017 [7] | The biggest Finnish grocery store chains | K-Group | Association rule (the Apriori algorithm) | Recommendation system | Collaborative filtering based on the Jaccard Distance |
| Priya 2018 [5] | Online retail service that delivers groceries: https://tech.instacart.com/3-million-instacart-orders-open-sourcedd40d29ead6f2/ | Inspected the data for missing value and random error | Clustering (Louvain algorithm) | Recommendation system | Caveman graph |

# 3 Main Concept

In this section, we will show the main concept used to definition and solve problem.

## 3.1 Big Data (BD)

(BD) refers to huge amount of structured, semi structured and unstructured data. This data is also called the three V's that refers to "Volume, Variety and Velocity". *Volume* represents the amount of data that is generated and collected and it ranges from terabytes to petabytes of data. *Variety* represents the wide range of data gathered from different sources in various formats such as tweets, blogs, videos, music, catalogs, photos etc. *Velocity* represents the speed with which the data is grown everyday such as the emails exchanged, social media posts, retail transactions descriptions and the speed with which it is analyzed, whether in real time or near real time [5].

## 3.2 Intelligent Data Analysis

One of the new field of science that concord with the design novel and pragmatic method to solve the real problem that have specific definition. In addition, it is analysis the results and put in the form suitable of the user. In big data environment, applications of IDA begin to enlarge their exposure of different researches types such as climate, graphic processing, bioscience, geography, atomic physics etc. [6]. IDA is new trend of science to solve the problem of knowledge discovery in database and data mining. In general, IDA can be define from three sides.

A. **Real problem:** find problem have specific definition with known parameters.
B. **Model:** design a novel or pragmatic model to salsify to goal of that problem based on parameter.

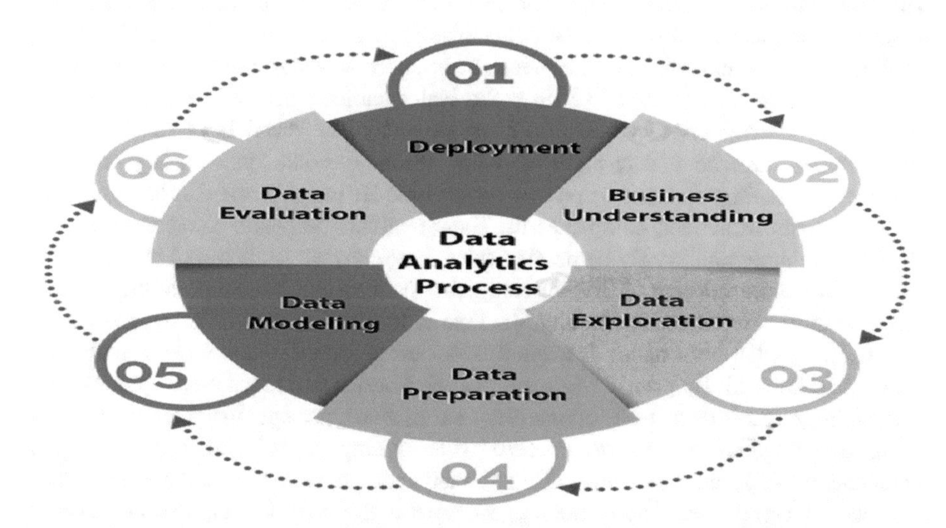


**Fig. 1.** Big Data Analysis Stages

C. **Analysis:** generated deep analysis of result, to convert it into understandable form by any user such as histogram, report, rules or mathematical form. Figure 1 explain Big Data Analysis Stages.

### 3.3 Clustering

Clustering is one of the unsupervised machine learning techniques that connect set of objects have the same attributes together. In general, the set of objects is considered cluster if satisfy the condition minimization the distance between objects inside of cluster and the maximization the distance a many different clusters. As a data mining technique, clustering refers to a large number of methods, for example k-means and self-organizing map, which are some of the most popular clustering methods. However, most of the clustering algorithms can be divided into two categories; partitional or hierarchical. Partitional clustering implies that every object can be a member of only one cluster, so that the clusters do not intersection with each other. Partitional clustering means that each object can belong to only one cluster, so that the clusters do not overlap with each other. Hierarchical algorithms on the other hand divide data objects into nested clusters. Hierarchical algorithms can be further divided into two categories: agglomerative or divisive. In agglomerative algorithms, all the objects have their “own clusters” at the beginning from where they are merged into bigger clusters until all the objects are in one big cluster. For the divisive clustering algorithms, the idea is very similar, however, now the objects are first in one big cluster, which is then iteratively divided into smaller clusters in a way that each cluster is always split into two new clusters [7].

### 3.4 Retail Data Analytics [13]

Retail analytics focuses on providing insights related to sales, inventory, customers, and other important aspects crucial for merchants’ decision-making process. The discipline encompasses several granular fields to create a broad picture of a retail business’ health, and sales alongside overall areas for improvement and reinforcement. Essentially, retail analytics is used to help make better choices, run businesses more efficiently, and deliver improved customer service analytics. The field of retail analysis goes beyond superficial data analysis, using techniques like data mining and data discovery to sanitize datasets to produce actionable BI insights that can be applied in the short-term. Moreover, companies use these analytics to create better snapshots of their target demographics. By harnessing sales data analysis, retailers can identify their ideal customers according to diverse categories such as age, preferences, buying patterns, location, and more. Essentially, the field is focused not just on parsing data, but also defining what information is needed, how best to gather it, and most importantly, how it will be used. By prioritizing retail analytics basics that focus on the process and not exclusively on data itself, companies can uncover stronger insights and be in a more advantageous position to succeed when attempting to predict business and consumer need. There are several excellent retail analytics examples that are relevant to a variety of companies. One of the biggest benefits the field delivers to companies is optimizing their inventory and procurement. Thanks to predictive tools, businesses can use historical data and trend analysis to determine which products they should order, and in what quantities instead of relying exclusively on past orders. In addition, they can optimize inventory management to emphasize products customers need, reducing wasted space and associated overhead costs. Apart from inventory activities, many retailers use analytics to identify customer trends and changing preferences by combining data from different areas. By merging sales data with a variety of factors, businesses can identify emerging trends and anticipate them better. This is closely tied to marketing functions, which also benefit from analytic [5]. Figure 2 explain Retail Data Analytics stages

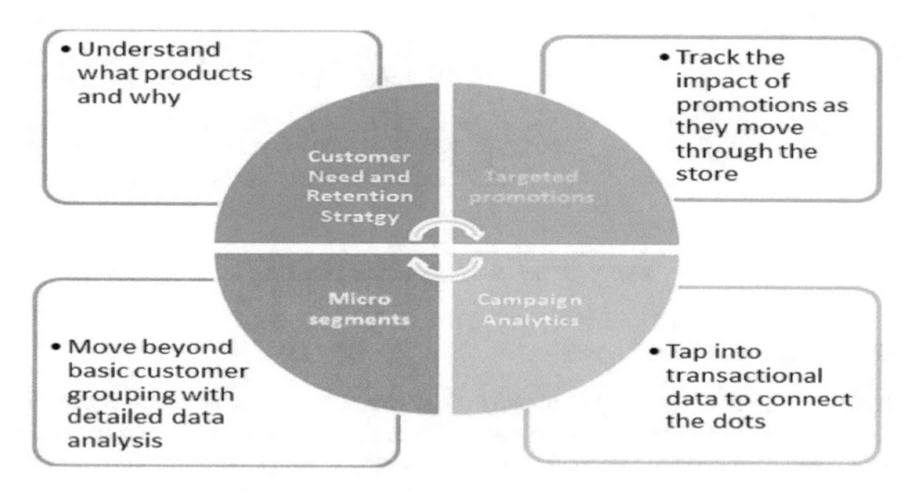


**Fig. 2.** Retail Data Analytics.

### 3.5 Graph Mining Techniques

Graph Mining (GM) is one of the new methods for mining the dataset denoted by graph building. Graph mining discoveries its applications in different issue spaces, counting, chemical reactions, bioinformatics, social networks, computer networks etc. different graph mining approaches have been suggested. Each of these methods is based on either clustering; decision trees or classification data mining methods. GM methods have been characterized by following. ***Graph clustering*** is the assignment of gathering the vertices of the graph into clusters taking into thought the edge structure of the graph. This a way that there should be many edges within each cluster and relatively few between the clusters? Graph clustering in the sense of grouping the vertices of a given input graph into clusters, graph clustering is based on unsupervised learning technique in which the classes are not known in prior to clustering. The graph clusters are formed based on some similarities in the underlying graph structured data graph. ***Graph Classification***; in graph classification the main task is to classify separate, individual graphs in a graph database into two or more categories/classes. Classification is based on supervised/semi supervised learning technique in which the classes of the data are defined in prior. ***Sub graph mining*** is a graph whose vertices and edges are subsets to other graph. The frequent sub graph mining problem is to produce the set of sub graphs occurring in at least some given threshold of the given n input example graphs [8]. Figure 3 explain Graph Mining Techniques.

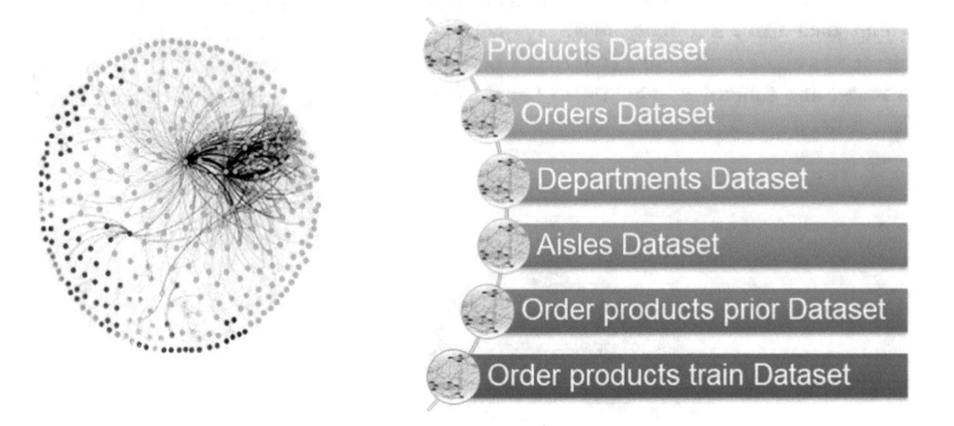


**Fig. 3.** Types of dataset in this Database

### 3.6 Recommendation System

Recommendation System is part of daily life where persons depend on knowledge for making decision of their individual interest. It is subclass of data filtering to predict preferences to the objects utilized by or for customers, which personalize suggestion and deals with data overload. Recommendation system could be a sharp system that gives idea about products to customers that might interest them various samples are movies in movie lens, music by last.fm, amazon.com. A variety of methods has been utilized to offer suggestion like content based, collaborative filtering and hybrid approach. Various Algorithms and methods are there to offer suggestion that may utilize rating or content information. In content-based method, similar objects to the ones the customer preferred in past will be suggested to the customer whereas in collaborative filtering, objects that similar group persons with similar tastes and preferences like will be suggested. In order to overcome the limitations of both methods hybrid systems are suggested that combines both methods in some way [4]. Figure 4 explain types of recommendation system.

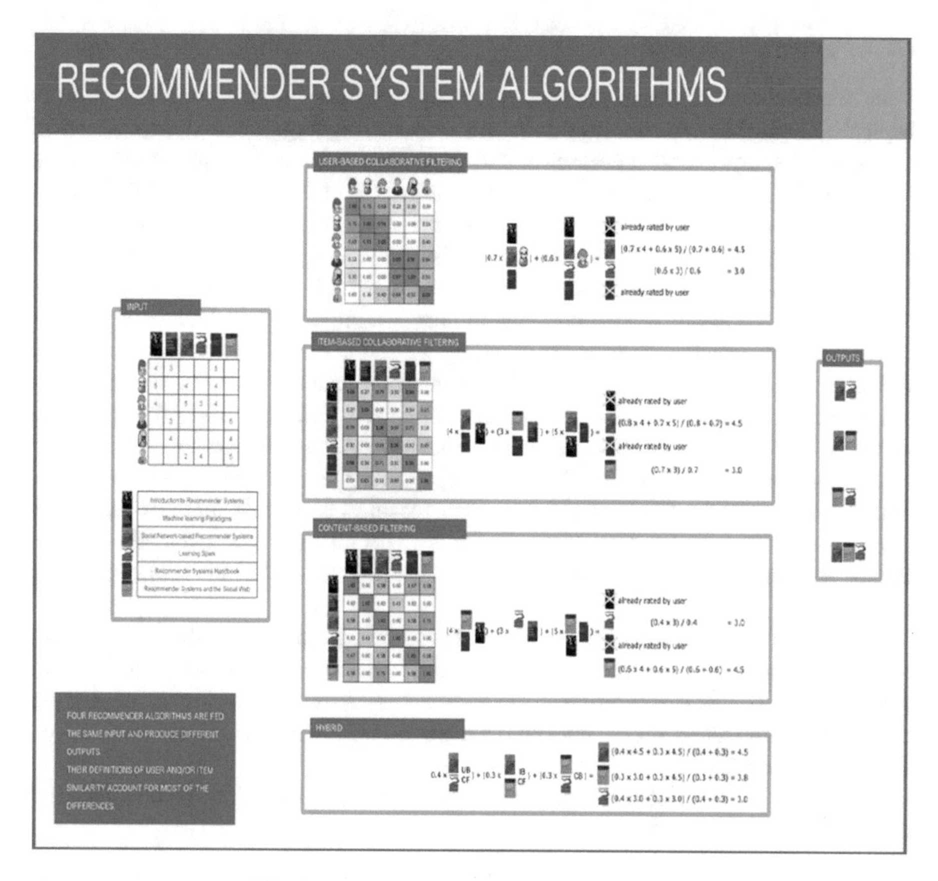


**Fig. 4.** types of recommendation system [14]

## 4 A Prototype of Recommendation System (PRS-CPR)

In this paper we will build a recommendations system to help customers to choose products easily and quickly, where we will use page rank clustering algorithm (PRS-CPR). To generate clusters where we go through several stages first generate weights for each node after the stage of finding weights, we will find the distance between the nodes through which to determine the groupings and if the distance is close, they will be within the same assembly. If the distances are spaced generated another clustering. For further clarification the first step is to get the database and display the statistics and the value of processing the database by using the average and average equation after the data was processed, we will convert the data into indirect graph then generate the weight for each node using personal PageRank algorithm. The second step is to use a distance PageRank algorithm to find the distance between the nodes. The third step is to use the PageRank clustering algorithm to generated clustering.

## 4.1 Problem Statement

Data is considered as one of the most precious resources in the world. The vast and huge development of technology provides us with big/huge amount of data which needs to be analysis and processed accurately. This data, in turn, represents a great challenge for intelligent computation's researcher [6]. Moreover, the availability of such massive data especially in the social media complicated things for the clients to selected goods that meet their needs [5]. This study will present an intelligent computation method to eliminate the abovementioned two above challenges by merging two essential concepts related to data mining techniques. The first concept used derivative from graph mining represented by ***PageRank clustering*** algorithm which will divide the network into several cluster then verification from the accuracy of it using two measures (i.e., ***Modularity and Silhouette Validity***). Then the extraction useful patterns for each subgraph using ***gSpan algorithm*** that based on depth first search and verification from results based on applying ***forward and backward rules***. Finally, these patterns (i.e., ***rules***) arrange decreasing based their ***forward and backward rules*** to choose the five optimal of it and display as recommendation for the clients. As an effect, we will simplify the selection process of the clients and save their time.

To handle this problem, we used two algorithms; the first algorithm is PageRank clustering to generate communities. When we use these packages, we encountered a problem. They take a great time to execute. The second algorithm is gSpan, the algorithm to generate a set of association rules through which to get the best suggestions for the user.

To satisfy the goal of this research, we will attempt to building recommendation consist of the following Stages:

- It deals with database having five parts (products, order, contains the products purchased in each order, aisles and departments dataset).
- The pre-processing stage involves handling missing values for each datasets and covert datasets into graph based on three method (i.e., random, direct, undirected).
- It builds the new model using PageRank clustering algorithm to find the optimal number of products in each cluster and using gSpan algorithm for each subgraph to get the best patterns.
- It computes the evaluation measure of cluster by using two measures (i.e., Modularity and Silhouette Validity) and verification from results based on applying forward/backward edges.

The main steps of PRS-CPR system show in Algorithm 1

**Algorthim#1: PRS-CPR**

***Input:*** Database called delivers groceries have five parts (F1, F2, F3, F4 and F5)

***Output:*** List of recommendation.

***// Pre-Processing Stage***

**1**: F6= call Joiner (F1, F2)

**2**: Ds'= F6

**3**: HDs' = Call missing values (Ds')

**4**: Gi=Call convert F6 to graph

***// Build NRGM Model***

**5**: Split datasets into training &testing datasets

***// PageRank Clustering***

**6**: **For** each training dataset

**7**: Call PPRV // Personal PageRank Vector

**8**: Call DPR // Distance PageRank

**9**: Call PRC //PageRank Clustering

**10: End** for

**// Validation of Clustering**

**11**: Call *Evaluation* base on Modularity

**12**: Call *Evaluation* base on Silhouette Validity

**// Generating Association Pattern by gSpan**

**13**: **For** optimal cluster (subgraph)

**14**: Call gSpan

**15: End** for

**16: For** each pattern generating by gSpan

**17:** Call *Evaluation* base on forward /backward edges

**18: End** for

**20: End** PRS-CPR

Where F1: dataset of orders, F2: dataset of products, F3: dataset of contains the products purchased in each order, F4: dataset of aisles, F5: dataset of departments (Fig. 5).

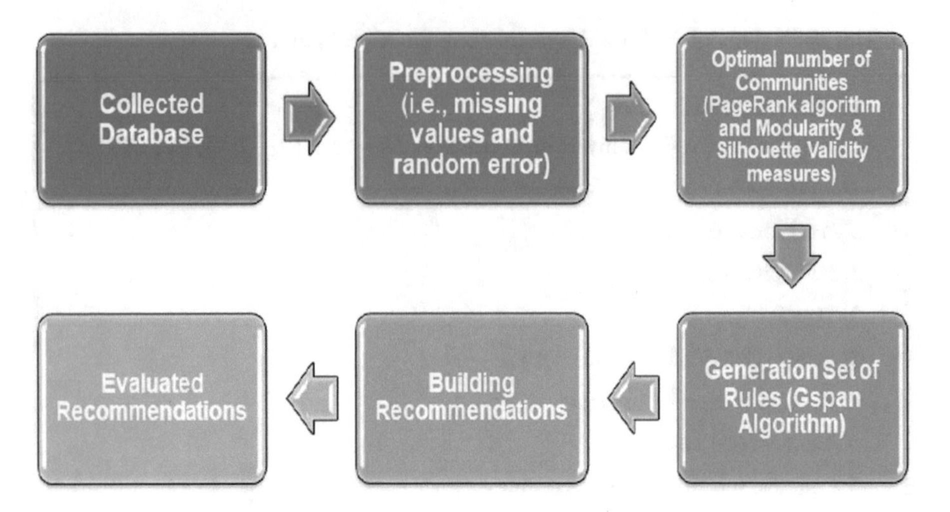


**Fig. 5.** Block diagram for generated recommendation based on PRS-CPR

# 5 Experiment

Using (PPRV) algorithm in Python to generate the weight of each node and use (DPR) algorithm to extract the distance between nodes, Then use the algorithm (PRC) to generated optimal number of clustering.

## 5.1 Collect Data

The database used in our project is an open-source database called “delivers groceries”. Shopping through internet becomes today one of the main service that present to customer from different web application. In general, the customers received their delivered in speed manner, that data base published in 2017 for research purposes. The database contains over 3 million grocery orders from more than 200,000 customers. The database have five dataset (products, order, contains the products purchased in each order, aisles and departments dataset) is anonymized and does not contain any customer information. It only contains a unique identifier (id) for each customer. The data is available in multiple comma-separated-value (csv) files. We saw the relational data model of the data in Fig. 6.

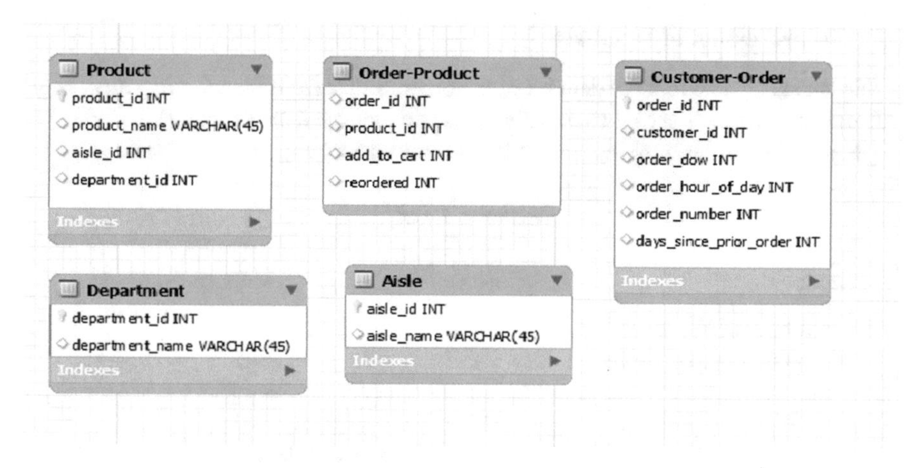


**Fig. 6.** The relational data model

In this database containing five datasets we will take the data set (i.e. order data set) as shown in the table (Table 2) below.

**Table 2.** Data set of orders

| | Order_Id | User_Id | Eval_Set | Order_Number | Order_Dow | Order_Hour_Of_Day | Days_Since_Prior_Order |
|---|---|---|---|---|---|---|---|
| 0 | 2539329 | 1 | Prior | 1 | 2 | 8 | nan |
| 1 | 2398795 | 1 | Prior | 2 | 3 | 7 | 15.0 |
| 2 | 473747 | 1 | Prior | 3 | 3 | 12 | 21.0 |
| .......<br>....... | | | | | | | |
| 1048570 | 1805240 | 63100 | Prior | 1 | 0 | 11 | nan |
| 1048571 | 2719402 | 63100 | Prior | 2 | 0 | 8 | 14.0 |
| 1048572 | 2855250 | 63100 | Prior | 3 | 2 | 13 | 16.0 |
| 1048573 | 2355832 | 63100 | Prior | 4 | 5 | 16 | 30.0 |
| 1048574 | 2362552 | 63100 | Train | 5 | 1 | 13 | 30.0 |

Statistical measures of this data set (i.e. orders data set) can be described as shown in the table (Table 3) below.

**Table 3.** Statically measures orders data set

| | Order_Id | User_Id | Eval_Set | Order_Number | Order_Dow | Order_Hour_Of_Day | Days_Since_Prior_Order |
|---|---|---|---|---|---|---|---|
| Count | 1.048575e +06 | 1.048575e +0 | 1.048575e +06 | 1.048575e +06 | 1.048575e +06 | 1.048575e+06 | 985475.000000 |
| Mean | 1.710968e +06 | 3.156404e +04 | 1.717418e +01 | 1.717418e +01 | 2.778637e +00 | 1.345229e+01 | 11.102506 |
| Std | 9.875774e +05 | 1.816695e +04 | 1.771094e +01 | 1.771094e +01 | 2.047751e +00 | 4.219144e+00 | 9.186663 |
| Min | 6.000000e +00 | 1.000000e +00 | 1.000000e +00 | 1.000000e +00 | 0.000000e +00 | 0.000000e+00 | 0.000000 |
| 25% | 8.554670e +05 | 1.589600e +04 | 5.000000e +00 | 5.000000e +00 | 1.000000e +00 | 1.000000e+01 | 4.000000 |
| 50% | 1.710657e +06 | 3.160800e +04 | 1.100000e +01 | 1.100000e +01 | 3.000000e +00 | 1.300000e+01 | 7.000000 |
| 75% | 2.566970e +06 | 4.729700e +04 | 2.300000e +01 | 2.300000e +01 | 5.000000e +00 | 1.600000e+01 | 15.000000 |
| Max | 3.421083e +06 | 6.310000e +04 | 1.000000e +02 | 1.000000e +02 | 6.000000e +00 | 2.300000e+01 | 30.000000 |

### 5.2 Preprocessing of Database

In this database, note that the data set of the orders suffer from the missing value, so we will process this data set by two methods (mean and median) (Table 4).

***Step1:*** Handel the missing values based on mean as explain in Table 6.

$$\text{Mean} = \sum xi/\text{n}$$

Where,

xi: is a sum of values
n: is the number of values

**Table 4.** Results of handle missing values by mean

| | Order_Id | User_Id | Eval_Set | Order_Number | Order_Dow | Order_Hour_Of_Day | Days_Since_Prior_Order |
|---|---|---|---|---|---|---|---|
| 0 | 2539329 | 1 | prior | 1 | 2 | 8 | 11.102506 |
| 1 | 2398795 | 1 | prior | 2 | 3 | 7 | 15.0000000 |
| 2 | 473747 | 1 | prior | 3 | 3 | 12 | 21.000000 |
| ....... | | | | | | | |
| ....... | | | | | | | |
| 1048570 | 1805240 | 63100 | prior | 1 | 0 | 11 | 11.102506 |
| 1048571 | 2719402 | 63100 | prior | 2 | 0 | 8 | 14.000000 |
| 1048572 | 2855250 | 63100 | prior | 3 | 2 | 13 | 16.000000 |
| 1048573 | 2355832 | 63100 | prior | 4 | 5 | 16 | 30.000000 |
| 1048574 | 2362552 | 63100 | train | 5 | 1 | 13 | 30.00000 |

***Step2:*** Handel the missing values based on median as explain in Table 5.

The first step is to arrange the items either ascending or descending. Second, choose the middle element (Table 5).

**Table 5.** Results of handle missing value median

| | Order_Id | User_Id | Eval_Set | Order_ Number | Order_ Dow | Order_Hour_ Of_Day | Days_Since_ Prior_Order |
|---|---|---|---|---|---|---|---|
| 0 | 2539329 | 1 | prior | 1 | 2 | 8 | 7.0 |
| 1 | 2398795 | 1 | prior | 2 | 3 | 7 | 15.0 |
| 2 | 473747 | 1 | prior | 3 | 3 | 12 | 21.0 |
| ....... | | | | | | | |
| ....... | | | | | | | |
| 1048570 | 1805240 | 63100 | prior | 1 | 0 | 11 | 7.0 |
| 1048571 | 2719402 | 63100 | prior | 2 | 0 | 8 | 14.0 |
| 1048572 | 2855250 | 63100 | prior | 3 | 2 | 13 | 16.0 |
| 1048573 | 2355832 | 63100 | prior | 4 | 5 | 16 | 30.0 |
| 1048574 | 2362552 | 63100 | train | 5 | 1 | 13 | 30.0 |

## 5.3 Convert Data Set into Graph

In this step convert order data set into three types of graph (direct graph, indirect graph, random graph).

- The first type direct graph for two orders as shown in the figure (Fig. 7) below

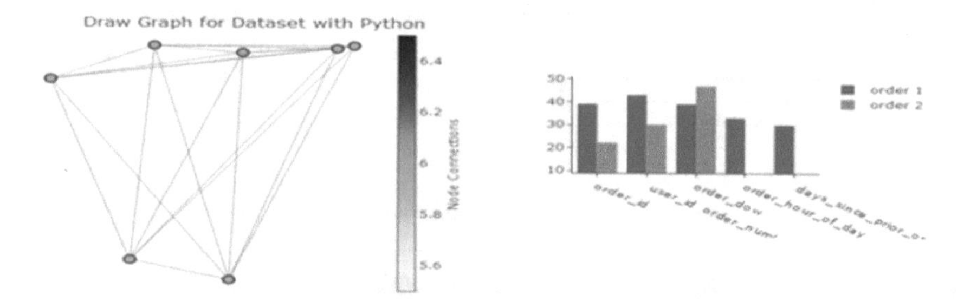


**Fig. 7.** Show direct graph for orders data set

- The second type indirect graph for two orders, the results of this method are used to generate communities as shown in the figure (Fig. 8) below

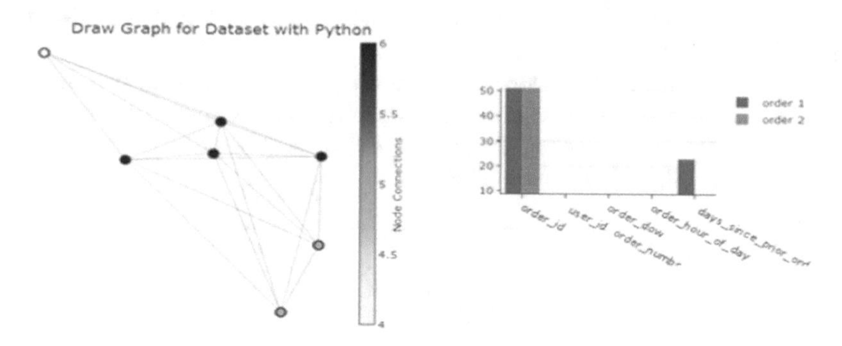


**Fig. 8.** Show indirect graph for orders data set

- The third type random graph for two orders as shown in the figure (Fig. 9) below

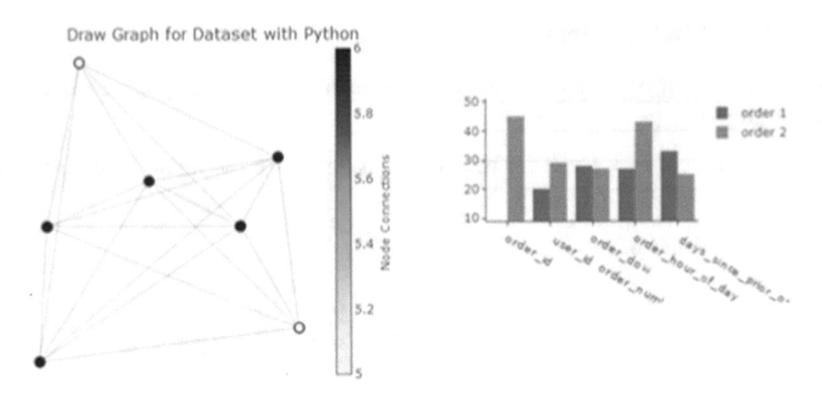


**Fig. 9.** Show random graph for orders data set

## 5.4 Generated Optimal Number of Communities

Using PageRank clustering algorithms to discover optimal number of communities (clusters). As shown in the Fig. 10 below.

***Step 1:*** Generate the weight for each node in graph based on Personalized PageRank vector (PPRV). In this algorithm used undirected graphs, let dv denote the degree of v which is the number of neighbors of v, the volume of T is defined to be vol (T) = $\sum$v$\epsilon$T dv. Let D denote the diagonal degree matrix and A the adjacency matrix of G, with the transition probability matrix defined by P = D¯ 1*A and we denote the lazy walk by W = (I + P)/2.

Personalized PageRank vectors are based on random walks with two governing parameters: a seed vector s, representing a probability distribution over V, and a jumping constant jc, controlling the rate of diffusion. The personalized PageRank vector pr (jc; s). Algorithm 2 show PPRV.

**Algorithm#2: PPRV**
***Input***: jc: jumping constant, sv: seed vector
***Output***: weight for each node in G
**1: For** each nodes in G Do
**2:** Compute adjacency matrix of G according to equation
A i, j= 1 if connected
0 otherwise
**3:** Compute diagonal degree matrix D of G according to equation
D= {deg (v) | v ∈ V (G)}.
**4:** Compute transition propriety matrix P of G according to equation
P = D^-1 *A
**5:** Compute identity matrix I of In=daig {1, 1, 1,…., 1}
**6:** Compute W= (I+P)/2
**7:** Compute ppr according to equation
Pr (jc, s) = jc*s + (1 -jc) ppr (jc,s)W
**8: End** PPRV

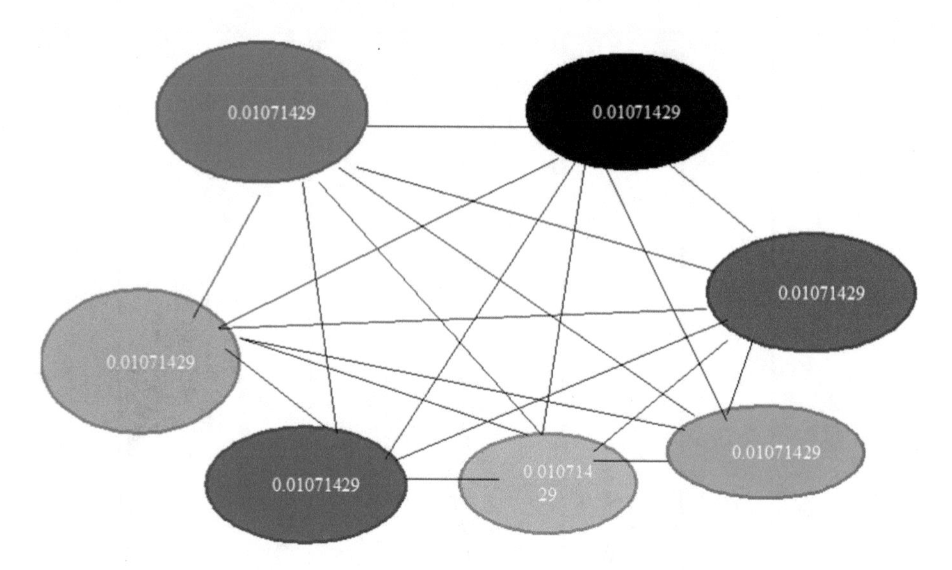


**Fig. 10.** Personal PageRank

***Step 2:*** Compute the distance among sample using distance PageRank (DPR). Algorithm 3 explain how can achieved this step. In this step we extracted the distance between order 1 and order 2, he result was (0.139806)

> **Algorithm#3: (DPR)**
> ***Input:*** set of samples (i.e., u&v € G)
> ***Output***: distance among sample
> **1:** Call of PPRV algorithm for (u) represent by pr (jc, s)
> **2:** Call of PPRV algorithm for (v) represent by pr (jc, s)
> **3:** Distance= ||PPRV (jc; u) D^1/2 – PPRV (jc, v)D^1/2||
> **4: END DPR**

***Step 3:*** Find the optimal number of communities (clusters) through applying PageRank clustering (PRC). The base point of it explained in Algorithm 4. We will use products data set to generate special clusters of this data, where we note that the products contain a (49688) product which is a large number so it is difficult to give recommendations to the consumer so we will conduct the process of clustering using PageRank clustering (PRC) algorithm e, when using this algorithm note the generation of 21 clusters as shown in Table 6.

**Table 6.** Show clustering of product data set

| #Group | #term in group | #name of term |
|---|---|---|
| G1 | 4007 | [Chocolate Sandwich Cookies, All-Seasons Salt,..............................., Hazelnut Spread With Skim Milk & Cocoa, Adhesive Bandages Active Flex Regular Advanced Healing] |
| G2 | 547 | [Fat Free Seasoned Croutons Restaurant Style, Organic Sparkling Apple Peach Juice,..............................................., Garbage Disposal Freshener And Cleaner Tangerine w/Lemon Grass, Non Drowsy 24 Hour Allergy Relief] |
| ...................... | | |
| G11 | 6561 | [Instant Gluten Free Blueberries Strawberries & Brown Sugar Oatmeal, Peach Yoghurt,——, Sports Drink, Coconut Water, Orange-Mango, Balsamic Vinaigrette Dressing] |
| —— | | |
| G21 | 1255 | [Single Serve Chunk Light Tuna In Sunflower Oil, Country Inn Chicken and Wild Rice,....................................................., Artisan Baguette, Smartblend Healthy Metabolism Dry Cat Food] |

**Algorithm 4: Page Rank Clustering (PRC)**
**Input:** G: graph
jc: jumping constant € : constant
**Output**: Optimal Number of Clustering
**1: For** all v ∈ G
**2:** Compute PPRV (jc,s )
**3: End** for
**4**: Find the roots of ϕ (jc) (There can be more than one root if G has a layered clustering structure.)
**5: For** all roots jc do
6: Compute ϕ (jc) according to equation

ϕ (jc) =dv||PPRV (jc, s) D^-1/2-PPRV(jc, PPRV(jc,s))D^-1/2||^2
**7:** **IF** ϕ (jc) < = € then

**8:** Compute ψ (jc) according to equation
ψ (jc) = dv|| PPRV (jc, PPRV(jc,v))D^-1/2-πD^-1/2||^2
**9:** **Else**
**10:** Go to the next jc
**11:** **End** if
**12:** **IF** k < ψ(jc)- 2 - € then
**13:** Go to the next jc
**14:** **Else**
**15:** Select c log n sets of k potential centers, randomly chosen according to π
**16:** **End** if
**17:** **For** all sets S = {v1; : : : ; vk} do
**18:** Let C be the set of centers of mass where ci = pr(_jc; vi).
**19:** Compute μ(C) and ψα(C) according to equation

$$\mu(C) = \sum_{v \in V} \left\| \frac{\mathrm{PPRV(jc,v)D^{-1}}}{2} - \frac{\mathrm{PPRV(jc,PPRV(jc,v))D^{1}}}{} 2 \right\|^2$$

ψα(C)= $\sum_{c \in C} vol(Rc)$ ||PPRV(jc, v)D^ − 1/2 − PPRV(jc, PPRV(jc, v))D^1/2||^
**20:** **IF** | μ(C) - ϕ(jc) |<= € and |ψjc(C)- ψ(jc) |<= € then
**21**: Determine the k Voronoi regions according to the DPR algorithm using C and turn them.
**22:** **End** if
**23:** **End** for
**24: End** for
**25: End** PRC

### 5.5 Generate Associations Pattern

We used the output of PageRank cluster as input to gSpan that is explained with details in chapter. The main purpose of this stage is building set of association patterns. This algorithm consists of four steps.

Step1: Traveling each graph through depth first search (DFS) to find the path from start point to goal point, to generated bank contain all path related trees. Algorithm 5 shows the details of applying this step.

**Algorithm#5: (DFS)**
***Input:*** G Graph object containing vertices and edges
v Root vertex of the graph G
***Output:*** preordering and post ordering of the vertices.
**1:** pd = 0; pt= 0
**2: For** each vertex v ∈V
**3:** Mark v unvisited
**4: End** for
**5: For** each vertex v ∈V calls Procedure DFS for each unvisited vertex in v
**6: If** v is marked unvisited then DFS (v)
7: **End** for
***Procedure DFS (v)***
**1:** Mark v visited
**2:** pd= pd+ 1
**3: For each edge (v, w) ∈E**
**4: If** w is marked unvisited then DFS (w)
**5: End** for
**6:** pt= pt+ 1
**7: End DFS**

Step 2: Applying Breath first search (BFS) to find the right most path on DFS path to return the shortest path from v to T (i.e., Target), we use BFS because it have linear complexity O(n + m), this step determined the optimal tree.

Step 3:

A. Extend the optimal tree through (nodes & edges), this lead to generated multi new trees.
B. Find sequence of edges for each new tree (forward/backward edges).
C. Remove the duplication tree based on the condition (i.e., kept the tree have less than number of edges).
D. Remove the error path (edges) based on the law of forward/backward edges.

Step 4: We using gSpan algorithm to generated associations pattern. Algorithm 6 explain how can achieved that.

***Algorithm 6: gSpan***

**Input**: D: a graph data set, s: result from Third step

min sup: the minimum support threshold.

**Output:** The frequent graph set S

**1:** S = ∅

**2:** Call gSpan(s, D, min sup, S)

***Procedure gSpan (s, D, min sup, S)***

**1: IF** s ≠DFS(s) then

**2: Return**

**3:** Insert s into S

**4:** Set C to ∅

**5:** Scan D once; find all the edges e such that s can be right-most extended insert right-most extended into C and count its frequency

**6:** Sort C in DFS order;

**7: For** each frequent s _r e in C do

**8:** gSpan(s _r e, D, min sup, S);

**9: Return**

**10: End** gSpan

We will illustrate the concept of association rules through a simplified example that contains a (3) consumer and their association with the products through consumer demand for products.

**Table 7.** Show relationship between customer and product

| | G1 | G2 | G3 | G4 | G5 | G6 | G7 | G8 | G9 | G10 | G11 | G12 | G13 | G14 | G15 | G16 | G17 | G18 | G19 | G20 | G21 |
|---|---|---|---|---|---|---|---|---|---|---|---|---|---|---|---|---|---|---|---|---|---|
| C1 | | 2 | | 4 | | | 7 | | | | 11 | | | | | | | | | | |
| C2 | 1 | | 3 | 4 | | 6 | | | 9 | | 11 | | 13 | 14 | 15 | | | | | | |
| C3 | 1 | 2 | | | 5 | | | 8 | | 10 | | 12 | | | | | | | | | |

## 6 Discussions and Conclusions

In this paper, we utilize graph mining for three types (direct, indirect, random) as presented in Figs. (8–10), but we use an indirect graph because PRC algorithm relies on it, then apply PRC algorithm to generate the optimal number of communities (clusters) we have applied this algorithm to product data set containing (49688) products and the production was 21 clustering, then we apply one of association rules algorithm (gSpan) on only 3 customer and their association with the products through consumer demand for products to get the best association rule for building a recommendation system that can recommend products to customers based on their buying behavior to take better decisions benefited for their business and fast at the same time. Lastly, to prove graphs are very intuitive and easy to work on for high-dimensional data. In general, we can summary the main questions that paper answer it as follow:

- **How can graph mining create useful knowledge help to building association patterns?**
  Through convert delivers groceries database into undirected graph then apply PageRank clustering to find set of subgraph (communities) and determined the best community based on modularity measure, PageRank clustering provides the new information represent by the communities that used as input of the gSpan(i.e., association pattern algorithm). This meaning graph mining provides multi separated communities.
- **How can PageRank clustering generated better communities?**
  PageRank clustering used PageRank vectors and validation measures (i.e., Modularity & Silhouette) to draw attention to local graph structure within a larger network to finding better communities and determined. PageRank provides essential structural relationships between nodes and is particularly well suited for clustering analysis. Furthermore, PageRank vectors can be computed more efficiently than performing a dimension reduction for a large graph.
- **How can gSpan algorithm find association patterns?**
  Optimal community result from PageRank clustering consider as input to gSpan that discovers all frequent subgraph without candidate generation and false positives pruning. It combines the growing and checking of frequent subgraph into one procedure thus reduces the mining process. gSpan apply four main steps (i.e., DFS, BFS, forward/backward Extended and remove duplication candidate) to determine all the association patterns from the optimal community.
- **How can determine the optimal recommendations by PRS-CPR?**
  The result of gSpan (i.e., set of association patterns) pass through multi conditions called forward and backward rules. These rules considers as a filters to determine the optimal patterns by two sides forward and backward. As a result these patterns will present to the customer as list of recommendation.

## References

1. Pawar, A.M.: Big data mining: challenges, technologies, tools and applications. Database Syst. J. **II**(2), 28–33 (2016)
2. Getting Started with Big Data Analytics in Retail, intel (2014)
3. Ridge, M., Johnston, K.A., O'Donovan, B.: The use of big data analytics in the retail industries in South Africa. Afr. J. Bus. Manag. **9**(19), 688–703 (2015)
4. Bhatt, B., Patel, P.J., Gaudani, H.: A review paper on machine learning based recommendation system. IJEDR **2**(4) (2014). ISSN 2321-9939
5. Priya, R.: Retail Data Analytics Using Graph Database, 30 May 2018
6. Kong, W.c., Wu, Q., Li, L., Qiao, F.: Intelligent data analysis and its challenges in big data environment. In: IEEE International Conference on System Science and Engineering (ICSSE), Shanghai, China, 11–13 July 2014
7. Harno, P.: Techniques for mining transactional data for personalized marketing actions. Business Technology, 23 May 2017. Bachelor's Thesis
8. Rehman, S.U., Khan, A.U., Fong, S.: Graph mining: a survey of graph mining techniques. 978-1-4673-2430-4112/$31.00 © IEEE.2012
9. Ríos, S.A., Videla-Cavieresb, I.F.: Generating groups of products using graph mining techniques. Procedia Comput. Sci. **35**, 730–738 (2014)
10. West, J.D., Wesley-Smith, I., Bergstrom, C.T.: A recommendation system based on hierarchical clustering of an article-level citation network. IEEE Trans. Big Data **2**, 113–123 (2016)
11. Marung, U., Theera-Umpon, N., Auephanwiriyakul, S.: Top-N recommender systems using genetic algorithm-based visual-clustering methods. Received: 6 April 2016, Accepted: 17 June 2016; Published 24 June 2016
12. Krishna Kishore, G., Suresh Babu, D.: Recommender system based on customer behaviour for retail stores. J. Comput. Eng. (IOSR-JCE) **19**(3), 06–17 (2017). e-ISSN 2278-0661, p-ISSN 2278-8727, Ver. I
13. Al-Janabi, S., Alkaim, A.F.: A nifty collaborative analysis to predicting a novel tool (DRFLLS) for missing values estimation. Soft Comput. (2019). https://doi.org/10.1007/s00500-019-03972-x
14. Mesas, R.M., Bellogín, A.: Exploiting recommendation confidence in decision-aware recommender systems. J. Intell. Inf. Syst. 1–34 (2018)
15. Al-Janabi, S., Salman, M.A., Fanfakh, A.: Recommendation system to improve time management for people in education environments. J. Eng. Appl. Sci. **13**, 10182–1019 (2018). https://doi.org/10.3923/jeasci.2018.10182.10193. http://medwelljournals.com/abstract/?doi=jeasci.2018.10182.10193
16. Al_Janabi, S.: Smart system to create optimal higher education environment using IDA and IOTs. Int. J. Comput. Appl. (2018). https://doi.org/10.1080/1206212x.2018.1512460

# Intelligent System for Monitoring and Detecting Water Quality

Jamal Mabrouki[1](✉), Mourade Azrour[2], Yousef Farhaoui[2], and Souad El Hajjaji[1]

[1] Laboratory of Spectroscopy, Molecular Modeling, Materials, Nanomaterial, Water and Environment, CERNE2D, Faculty of Science, Mohammed V University in Rabat, BP1014, Agdal, Rabat, Morocco
jamalmabrouki@gmail.com

[2] Faculty of Sciences and Techniques, Department of Computer Science, IDMS Team, Moulay Ismail University, Errachidia, Morocco

**Abstract.** Testing water quality has a significant role in environment controlling. Whenever, the water quality is bad it can affect the aquatic life and surrounding environment. Due to the importance of some parameters to show the quality of water, we have designed an intelligent system that can measures remotely five parameters of water. The captured values are sent to the database which is connected to the platform. The platform can process the received values. The user can connect to the application via Internet Protocol for monitoring the measured parameters. The outcomes demonstrate that with fitting alignment, a dependable observing framework can be built up. This will enable catchment administrators to consistently observing the nature of the water at higher spatial goals than has recently been doable, and to keep up this reconnaissance over an all-inclusive timeframe. Moreover, it comprehends the conduct of sea-going creatures in respect to water contamination utilizing information investigation.

**Keywords:** Water quality · Monitoring · Intelligent system · IoT · Arduino · Technology

## 1 Introduction

Water is a natural resource that is indispensable and obligatory to the life of all living beings. So, life will not be possible in places where there is no water. In addition, it will not be normal if the water is polluted. Today, water quality is affected by the industries and people who produce daily tons of waste thrown directly or indirectly into the water. As a result, the checking and monitoring water quality before its consumption is indispensable.

The basic parameters of the water quality fluctuate dependent on the utilization of water. For instance, for aquariums, it is important to keep up the temperature, pH level, broke down oxygen level, turbidity, and the dimension of the water in a specific ordinary range so as to guarantee the wellbeing of the fish inside the aquarium. For the mechanical and family unit applications, in any case, a few parameters of the water are


Y. Farhaoui (Ed.): BDNT 2019, LNNS 81, pp. 172–182, 2020.
https://doi.org/10.1007/978-3-030-23672-4_13

more fundamental to be checked as often more possible than the others, contingent upon the utilization of the water [1, 2].

The customary strategy for observing of the water quality is with the end goal that the water test is taken and sent to the research center to be tried physically by systematic techniques. Despite the fact that by this strategy the compound, physical, and natural specialists of the water can be examined, it has a few disadvantages. Right off the bat, it is tedious and work concentrated [3].

Conventionally, the water analyzer must move to the source of water that he want analyze then take the sample to the laboratories where the analysis is done. However, this traditional method has various problems that are detailed in the following: During the water analyzing some values must be measured in the field immediately such as temperature. Furthermore, other values my change if the conditions of transporting the simple are not good, for example pH value can change if the bottle is not very washed or it is washed with water of the tap [4]. The parameters values of water are not stable, but they are changing continuously from hour to hour. We can also add some problem attached to the availability of information in every time and in everywhere for everyone. In order to overcome these problems, we have designed a new monitoring system. The development of this new intelligent system that aims to control and measure water characteristics remotely, without having to move and take action on the field, will facilitate the task of detecting and monitoring water quality anywhere.

In this paper, we present our proposed system that can measure the water parameters continuously then sends the measured values to distance application which stores the collected data in an SQL database. So, the user can connect to the platform and consult the data remotely.

The rest of this paper is organized as following. The section two delivers the related works to our designed system. In the section three, we detailed our proposed system. The analysis tests and obtained results are presented in the section four. The last section concludes our work.

## 2 Related Work

Before the invention of new technologies, to test water quality one have to collect the sample of water and bring it to the laboratories where he can do the experimental analysis [5]. Nevertheless, in our times, the new technologies have growing very rapidly. So, they offer the new methods and ways to survey remotely the quality of water without need to move to the fields and take water sample. In recent years, various systems have been proposed to analyze water quality. In this section, we cite some recent propositions related to our work.

In 2012, Deqing et al. [2] proposed an automatic measurement and reporting system of water quality based on GSM that can captures the values of pH, conductivity, dissolved oxygen, and turbidity. The proposed system is created based on the microcontroller that process the data, and GSM module for forecasting the collected and processed data. However, this system is expensive and it cannot be afforded by common people.

In 2013, Rao et al. [6] designed a low cost and autonomous water quality monitoring system based on Arduino Mega 2560. The system can measure the value of temperature, pH, conductivity, dissolved oxygen, light, and oxidation reduction potential. Then, the captured values are transferred to a local computer in which they are saved.

In 2014, Cloete et al. [7] designed a water quality monitoring system that offers to user the information in real-time. The system measures temperature, pH, conductivity and the oxidation reduction potential. It consists on microcontroller-based measuring node that process the data and a ZigBee which transfers the processed information.

In 2015, Awsale et al. [8] suggested water environment monitoring system based on WSN. The proposed system use Arduino Atmega328, 8 bit micro-controller and two sensors (pH and temperature). It consists of three parties: data monitoring node, wireless sensing node and remote monitoring station. The system allows sending the captured data via wireless network. However, it measures only two values pH and temperature which are not sufficient to notice that the water have a good or bad quality.

In December 2015, Hongpin et al. [9] designed a new Real-time remote monitoring system for aquaculture water quality. The proposed system can measure the values of temperature, pH, and dissolved oxygen level and ammonia nitrogen. For processing data the system use STM32F103 chip, while GPRS and ZigBee are responsible to transmit the collected values to the storage and monitoring parties.

In 2017, Ranjbar and Abdalla [10] proposed low-cost, real-time, autonomous water quality testing and notification system. The proposed architecture consists of four sensors (pH, temperature, turbidity and Ultrasonic) that are connected to a microcontroller. The monitor LC and four Light Emitting Diode (LED) are used to communicate the captured values to local user. For distant user, he receives an SMS via GSM shield.

In 2018, Samsudin [11] presented a smart monitoring of a water quality detector system that can measures the pH and turbidity values. The collected data are processed by Arduino then are sent to the database through Wi-Fi module.

## 3 Our Developed System

### 3.1 Over All Structure of Our Proposed System

In order to provide the water data parameters to the water analyzers in permanent way and simplifying the controlling water task, we have designed a new smart system for monitoring and detection water quality. As illustrated in Fig. 1, our system consists of five modules: perception module, communication module, processing module, storage module, and application module. The perception module has as role sensing environment and measuring water characteristics. The communication module allows the transferring the captured values to the applications and database; While, the storage module is a database that can store the received values and information system. The processing module has as charge to process and trait the received data. The last module is called application module, it exploits the treatment results in order to help the administrators to make a good decisions. Generally, this module has a mission to send the alerts to the system administrators and to send the commands to the actuators.

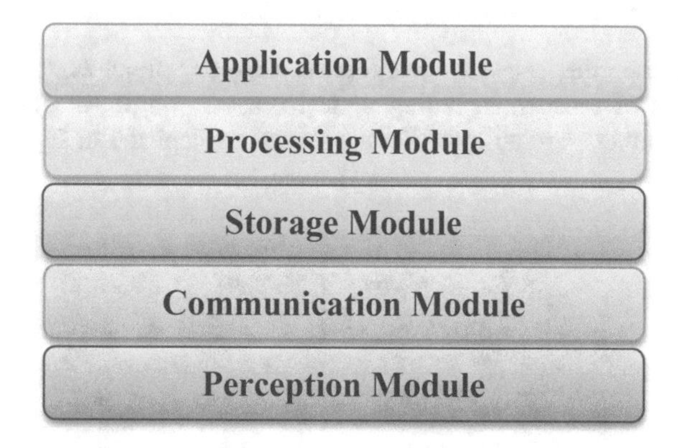


**Fig. 1.** The five modules of developed system

## 3.2 System Hardware Architecture

As illustrated in Fig. 2, the designed system is composed on sensors, Arduino, Wi-Fi module, Bluetooth module, gateway, and computer where the database and platform are installed. Each component is detailed in the following.

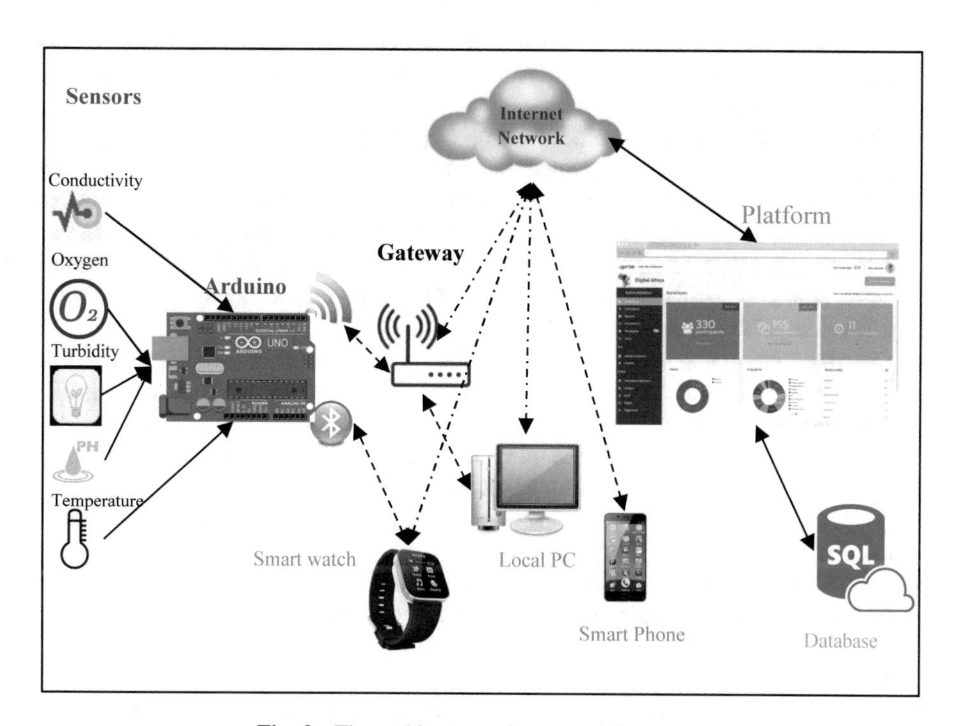


**Fig. 2.** The architecture of proposed de system

**A. Sensors**
In this subsection, the sensors incorporated into our framework are introduced. Distinctive sorts of sensors, including optical sensors, warm sensors, and attractive sensors, among others, are utilized. Some sensors are depicted in Fig. 3.

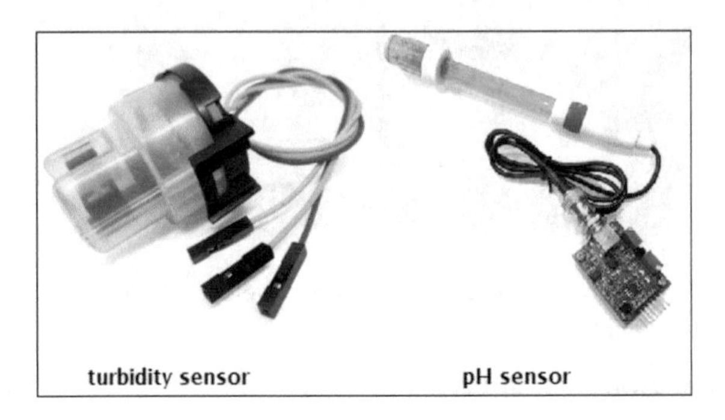


**Fig. 3.** Employed components for pH and turbidity

To begin with, the sensor: for checking water quality are exhibited:

- **pH sensor**: for measuring the pH value, the system need pH meter(SKU: SEN0161) [12]. It measures the all pH values between 0 and 14. Its measuring temperature is ranging between 0 and 60 °C.
- **Turbidity sensor:** The turbidity sensor is portrayed underneath. There are two fundamental strategies for estimating turbidity, the acoustic technique and the optical strategy. In our situation, the optical technique is chosen. The sensor utilized is portrayed in [13]. It comprises of an infrared (IR) light producing diode and an infrared photodetector. The IR LED utilized is the Vishay TSHG6200, the most extreme wavelength of this LED is 850 nm. Furthermore, the IR photodetector utilized is BPW83 from Vishay [14].
- **Dissolved oxygen sensor**: In order to reflect water quality we have used *Analog Dissolved Oxygen Sensor*. This sensor can measure the value of dissolved oxygen in the water.
- **Temperature sensor**: For determining the value of the temperature we need sensor DS18B20 for Arduino. It is a very small digital thermometer which can be linked to Arduino with no trouble through any digital input. It does not require very hard effort to be installed.
- **Conductivity sensor**: To capture water conductivity we *use Analog Electrical Conductivity Sensor,* which can be connected to Arduino card.

**B. Arduino**
In this subsection, the hub utilized and their arrangements are point by point. Gather information from all sensors, we have chosen a module of the Arduino Uno board which is a microcontroller board worked around the ATmega328 (information sheet). It has 14 computerized information/yield pins (6 of which can be utilized as PWM yields,

PWM), 6 simple sources of info, a 16 MHz gem oscillator, a USB connector, a power jack, an ICSP base, and a reset catch. The Uno card contains all that is essential for the task of the microcontroller. To utilize it, essentially associate it to a PC with a USB link, or feed it with an outer power supply or batteries. This remote interface type card to incorporate it into a WSN and capacity the information limit of the sensors is constrained. The Arduino UNO hardware is pictured in Fig. 4.

**Fig. 4.** Adruino UNO hardware

**C. Network Tools**

In our designed architecture various tools are used in order to connection our system, to the IP network. Firstly, for transmitting the captured values in local space, we utilize an ESP8266 Wi-Fi module [15]; which is a tool that allows to the Arduino card to communicate the captured values to the Access point. By using the Bluetooth module HM-10 [16] for Arduino, the card also can send the measured parameters to smart material equipped by the Bluetooth such as smart watch. Secondly, for transferring the obtained values, the access point is linked to the internet. The last one receives the data coming from the local area then transfers them to local computer and to the distance database.

### 3.3 Utility of Measured Parameters

In order to check the quality of controlled water our developed system measures five parameters. The utility and importance of each parameter is detailed in the following:

- ***pH***: it is a measurement that determine the acidity or basicity of the solution(ex. Water) [17]. It takes values between 0 and 14 according to the Arrhenius [18] an acid will release the ion of hydrogen (H+) when it dissolves in the water, will the basic will release ion of hydroxyl (OH-) in the water. Due to the importance of value of pH to determine the acidity or basicity of water, we have used pH meter to detect the pH value of controlled water.
- ***Turbidity***: it is the qualitative characteristic that determines the relative clarity of the water. It means existing of strong objects and solids in the water. These objects and solids have negative effects on the life under water in the streams, lakes, seas… it block the sun light to reach the submerged aquatic plants. Therefore, the photosynthesis operation will stop, and then the dissolved oxygen will be reduced.

- ***Dissolved oxygen***: the free oxygen present in the water is called the dissolved oxygen. Due to its importance for any living beings. The oxygen is a significant parameter for verifying the water quality. Its value can help us to determine if the water have a good or a bad quality.
- ***Temperature***: Water temperature is a physical property that specifies the thermal energy of water it is a considerable parameter to check water quality. The changing of water temperature has several effects on the aquatic life. Furthermore, it has an influence on the other parameters.
- ***Conductivity Electrical***: conductivity is the inverse of resistivity. The conductivity of a homogeneous material is equivalent to the conductance of a tube shaped conductor made of this material, partitioned by its area, and duplicated by its length.

# 4 Results of Analysis and Discussion

## 4.1 Results and Discussion

In this subsection, the consequences of the adjustment of the utilized sensors for water quality observing are appeared. Sometimes, as turbidity or temperature sensors, alignments have just been completed that relate the natural parameter and the opposition of the electronic part.

By the by, for activity in the framework, it is important to acquire the information as the yield voltage (Vout) that lands at the hub. In addition, it might be expected to incorporate a voltage divisor to guarantee that present in the hub isn't excessively high. In this manner, it will be important to change the information and to get another adjustment condition for the temperature and turbidity sensor. On account of the conductivity sensor, a similar alignment and conditions that has been utilized in [19] are used.

Firstly, for the temperature sensor, in light of the information offered by the producer [19] it is conceivable to extricate the normal protections at various temperatures. A voltage divider must be utilized to amplify the Vout distinction between the base and most extreme estimations of the NTC opposition.

A R2 of 12 KW must be utilized and the NTC is utilized as R1. We can compute the Vout of the temperature sensor at various temperatures. The information and the numerical model that pursues this information are introduced in Fig. 5. The connection among Vout and Temperature can be found in our condition. The relationship coefficient of our condition is 0.9. While the goals of the hub is 10 bits and the limit of Vout that can be recorded is 6 V, the base contrast of Vout that the hub can recognize is 3 mV. Along these lines, the base temperature variety recognizable by our framework is 1 °C.

The goals of the temperature sensor are adequate for observing aquaculture. In Fig. 5(a), we can see the code utilized in the hub to peruse the estimation of the temperature sensor. The recipe that relates the temperature and the voltTemp esteem is taken.

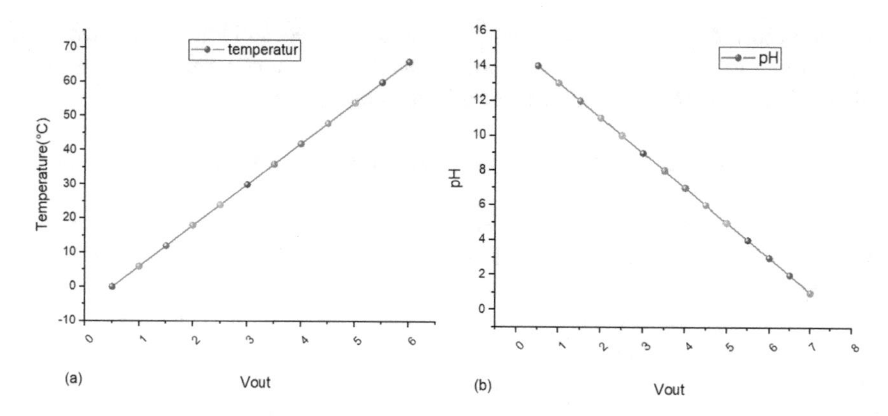


**Fig. 5.** (**a**) Data of the temperature sensor (**b**) Data of the pH sensor

$$\text{Temperature}(^\circ\text{C}) = 0.03 \times \text{V}_{\text{out}}(\text{V}) + 1.1 \tag{1}$$

$$\text{pH} = -0.03 \times \text{V}_{\text{out}} + 1.1 \tag{2}$$

Secondly, for the turbidity sensor we utilize the information appeared [27], where an adjustment has just been finished. All things considered, the information acquired among this adjustment must be changed from the opposition of the IR photodetector to Vout. By and by, a voltage divider must be utilized. Utilizing the 3 V wine, the IR photodetector like R1 and a 6 MW R2, we augment the distinction between the most extreme and the base of Vout. Information after utilization of voltage divider and numerical model that fit these information are exhibited in Fig. 5(a) and (b). The connection coefficient of the numerical model, is 0.9. Given the goals of the hub, the base turbidity variety that can be recognized by the turbidity sensor increments from 2 NTU under low turbidity conditions to 3 NTU under high turbidity conditions. In Fig. 6, the code brought into the hub for perusing the Vout from the turbidity sensor and changing this voltage into turbidity esteem is appeared. In addition to the method

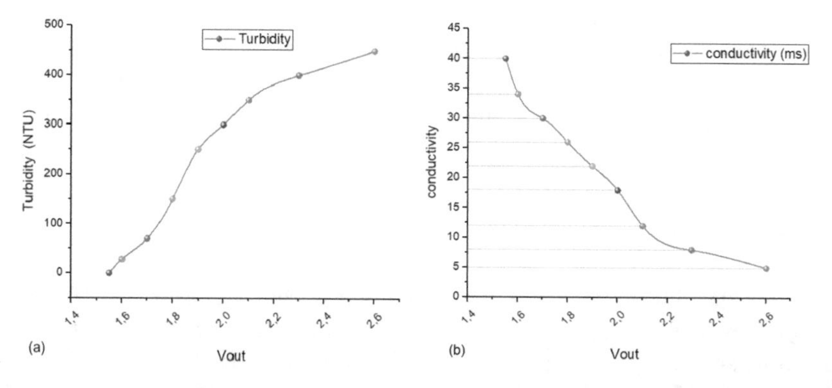


**Fig. 6.** (a) Data of the turbidity sensor (**b**) Data of the conductivity sensor

concerning turbidity, the system esteem is acquired by consolidating the information appeared in Fig. 6, voltage as the autonomous variable, and turbidity as the needy variable. The relationship coefficient of this equation is 0.9.

$$\text{Turbidity(NTU)} = 1620 + 1032 \times V_{out}^2 - 2740 \times V_{out} \tag{3}$$

Chart book Scientific Platinum Conductivity Sensor is utilized as an electrical conductivity sensor. A conductivity sensor K = 1 was utilized which estimates the conductivity in the range 1.3 to 40 ms.

Now, we will make an examination between our created arrangement and the present frameworks depicted in the comparing take a shot at supplies and lakes. We outline the qualities of the present frameworks and our proposition in Fig. 6(b). The primary contrasts are that different frameworks control between multiple times, while our frameworks can screen multiple times before alignment. What's more, different recommendations utilize business tests and our framework depends without anyone else sensors. This makes the cost of the framework decline. In our proposition, we likewise consider conservation of the environment by observing of water quality. At long last, in our answer, we showed the sensor area in the tank. Most of papers don't focus on the area of the sensors.

### 4.2 Comparison with Other Systems

In our proposed technique, a large portion of the above issues have been considered. In light of past work, it was felt that water parameters ought to be observed in practically all applications. In this way, the proposed plan has been created to quantify water parameters, for example, temperature, pH level, turbidity (darkness) and conductivity (amount) of water and oxygen beneath, as demonstrated in the outcomes above. The magnificence of our proposed framework lies in the way that it is a minimal independent gadget that can show water settings on a little computer screen or your smartphone, while additionally having the capacity to send information remotely to the users. Also, its minimal effort has created it and can furnish water quality estimations with worthy precision. It tends to be additionally improved by including more sensors and utilizing water-safe segments later on.

The galvanic broke up oxygen sensor is like the EC test and measures the disintegrated oxygen content in a scope of 0 to 20 mg/L. It works at a most extreme temperature of 60 °C. The disintegrated oxygen yield is directed to an OD circuit that gives the outcomes in RS232 position. This is interfaced to the second Ardunio UNO R3 sequential port.

## 5 Conclusion

We have planned a practical and dependable water parameter estimation framework that is not difficult to utilize and can be actualized in any geographic area with a straightforward change. As it is a remote framework, it tends to be utilized as a parameter estimating gadget by Wi-Fi. Our planned framework is valuable in numerous spots where the link framework cannot be executed. Not just estimating the qualities of the diverse parameters of this framework can likewise show whether the qualities are preparing a predefined esteem. In the event that the information was transmitted or showed effectively, however the framework repeatability mistake rate was endured.

Utilizing such a strategy, it is conceivable to build up the estimation of various water parameters without depending on costly sensors and strategies. By having such strategies to gauge different parameters like BOD (biochemical oxygen request), COD (substance oxygen request), nitrogen, and so forth.

## References


1. Khurana, M.K., Singh, R., Prakash, A., Chhabra, R.: An IoT based water health monitoring system. Int. J. Comput. Technol. Appl. (IJCTA) **9**(21), 07–13 (2016)
2. Mo, D., Zhao, Y., Chen, S.: Automatic measurement and reporting system of water quality based on GSM. In: 2012 Second International Conference on Intelligent System Design and Engineering Application, pp. 1007–1010. IEEE (2012)
3. Wang, J., et al.: A remote wireless sensor networks for water quality monitoring. In: 2010 International Conference on Innovative Computing and Communication and 2010 Asia-Pacific Conference on Information Technology and Ocean Engineering, pp. 7–12. IEEE (2010)
4. Mabrouki, J., El Yadini, A., Bencheikh, I., Azoulay, K., Moufti, A., El Hajjaji, S.: Hydrogeological and hydrochemical study of underground waters of the tablecloth in the vicinity of the controlled city dump mohammedia (Morocco). In: International Conference on Advanced Intelligent Systems for Sustainable Development, pp. 22–33. Springer, Cham (2018)
5. Le Dinh, T., Hu, W., Sikka, P., et al.: Design and deployment of a remote robust sensor network: experiences from an outdoor water quality monitoring network. In: 32nd IEEE Conference on Local Computer Networks (LCN 2007), pp. 799–806. IEEE (2007)
6. Rao, A.S., Marshall, S., Gubbi, J., et al.: Design of low-cost autonomous water quality monitoring system. In: 2013 International Conference on Advances in Computing, Communications and Informatics (ICACCI), pp. 14–19. IEEE (2013)
7. Cloete, N.A., Malekian, R., Nair, L.: Design of smart sensors for real-time water quality monitoring. IEEE Access **4**, 3975–3990 (2016)
8. Aswale, P., Patil, S., Ahire, D., Shelke, S., Sonawane, M.: Water environment moritoring system based on WSN, vol. 4, no. 4, p. 4 (2015)
9. Hongpin, L., Guanglin, L., Weifeng, P., Jie, S., Qiuwei, B.: Real-time remote monitoring system for aquaculture water quality. Biol. Eng. **8**, 8 (2015)
10. Ranjbar, M.R., Abdalla, A.H.: Low-cost, real-time, autonomous water quality testing and notification system. Int. J. Comput. Sci. Netw. Secur. **17**(5), 277–282 (2017)
11. Samsudin, S.I., Salim, S.I.M., Osman, K., Sulaiman, S.F., Sabri, M.I.A.: A smart monitoring of a water quality detector system. Indones. J. Electr. Eng. Comput. Sci. **10**(3), 951–958 (2018)
12. PH meter (SKU: SEN0161) - Robot Wiki | arduino | Pinterest. https://www.pinterest.fr/pin/737112663982502209/. Accessed 20 Apr 2019
13. IR LED Datasheet. https://www.vishay.com/docs/81078/tshg6200.pdf. Accessed 12 Jan 2019
14. IR Photodetector Datasheet. https://www.vishay.com/docs/81530/bpw83.pdf. Accessed 12 Jan 2018

15. ESP8266WiFi library. https://arduino-esp8266.readthedocs.io/en/latest/esp8266wifi/readme.html. Accessed 20 Apr 2019
16. HM-10 Bluetooth 4 BLE Modules. http://www.martyncurrey.com/hm-10-bluetooth-4ble-modules/. Accessed 20 Apr 2019
17. pH of Water - Environmental Measurement Systems. https://www.fondriest.com/environmentalmeasurements/parameters/water-quality/ph/. Accessed 20 Apr 2019
18. Arrhenius, S.: Recherches sur la conductibilité galvanique des électrolytes. PA Norstedt & Söner (1884)
19. Sadar, M.: Turbidity measurement: a simple, effective indicator of water quality change. Hach Hydromet (2003)

# Using Ensemble Methods to Solve the Problem of Pulsar Search

Mourad Azhari[1(✉)], Altaf Alaoui[1], Abdallah Abarda[2], Badia Ettaki[3], and Jamal Zerouaoui[1]

[1] Laboratory of Engineering Sciences and Modeling, Faculty of Sciences, Ibn Tofail University, Campus Universitaire, BP 133, Kenitra, Morocco
azharimourad@yahoo.fr, altaf.alaoui@gmail.com, j_zerouaoui@yahoo.fr
[2] Laboratoire de Modélisation Mathématiques et de Calculs Economiques, FSJES, Université Hassan 1er, Settat, Morocco
abardabdallah@gmail.com
[3] Laboratory of Research in Computer Science, Data Sciences and Knowledge Engineering, Department of Data, Content and knowledge Engineering, School of Information Sciences, Rabat, Morocco
ettakibadia@yahoo.fr

**Abstract.** Ensemble methods are a machine learning technique that combines several base models in order to produce one optimal predictive model. In this paper, we compare accuracy metric of three ensemble methods: Bagging, Random Forest, and Boosting. Then, We use the "CARET package", implemented in R language, to experiment the Time Resolution Universe (HTRU2) dataset, obtained from UCI Machine Learning Repository.

**Keywords:** Ensemble methods · Bagging · Random Forest · Boosting · Imbalanced class · Accuracy · Pulsar

## 1 Introduction

Pulsars are very dense stars called neutron stars, detectable on Earth, rapidly spinning and emitting a beam of electromagnetic radiation. Scientists use pulsars to study condensed matter physics, measure cosmic distances [1], find gravitational waves [2], testing General Relativity of Albert Einstein theory [3] and testing black hole super-radiance [4]. In practice the pulsars candidate detection remains difficult for scientist as it requires modeling the pulsar signal. The purpose of this signal modeling is to separate the pulsars into active or passive by using classification methods. In this issue, many scholars highlight the significant results of using Machine Learning (ML)classification method to solve various problems in several areas of astronomy phenomena [5] such us study the galaxies morphology [6], understand the Milky Way structure, identify the supernovae type [7] and select pulsar candidates [8].


Y. Farhaoui (Ed.): BDNT 2019, LNNS 81, pp. 183–189, 2020.
https://doi.org/10.1007/978-3-030-23672-4_14

Breiman [9] used the Bagging predictors for generating multiple versions of a predictor, he carried out several tests on real and simulated data sets using classification and regression trees. Zheng [10] applied the Boosting and the Bagging techniques with neural networks on time series in the field of finance. Alfaro et al. [11] compared the accuracy performance of two Adaboost techniques: Decision Tree (DT) and Artificial Neural Network (ANN). Suganthan [12] employed various kernels (linear, polynomial, Radial basic function (RBF)) of the SVM method combined with the Bagging technique to test its performance on six data sets UCI (Australian, Breast, Heart, Liver, Mushroom, Pima data sets). Mordelet and Vert [13] proposed an implementation of the model Bagging (SVM) in the Pattern Recognition letters data to solving the problem of learning from positive and unlabeled examples. Bauer and Kohavi have conducted an empirical Study to compare three different voting methods (Bagging, Bosoting, and Adaboost) on classification error [14].

In this paper, we compare three ensemble methods: Bagging, Random Forest and Boosting in order to solve the problem of pulsar search using accuracy metric and we applied these methods to HTRU2 dataset, obtained from UCI[1] Machine Learning Repository. We use two techniques to evaluate our results: Split data and repeated cross validation with under-sampling and over-sampling.

## 2 Proposed Ensemble Methods to Detect Pulsar Candidates

The main of ensemble methods is to improve the performance of single classifiers. The approach involves building several classifiers from the original data and then aggregate their predictions [15].

In this section, we present three ensemble methods to detect pulsar candidates: Bagging, Random Forest, and Boosting.

### 2.1 Bagging Method

Bagging, refers to Bootstrap Aggregating, is a method for improving the stability and performance of machine learning algorithm [9]. It forms a series of classifiers that are connected by voting and by generating repeated bootstrap samples of the data [16]. Bagging reduces variance, aids to avoid over-fitting and it can be used with any type of method.

### 2.2 Random Forest Method

Random Forest is a supervised learning algorithm that combine the Bagging method and a particular decision tree classifier. This classifier creates a set of decision trees from a randomly selected subset of training set. Then, it aggregates

[1] https://archive.ics.uci.edu/ml/datasets.html.

the votes from different decision trees to decide the final class of the test object [17]. There are two stages in Random Forest algorithm: The Random Forest creation and the prediction from the Random Forest classifier created in the first stage.

### 2.3 Boosting Method

The AdaBoost algorithm was introduced by Freund and Schapire [18]. It is an ensemble method for improving the model predictions of a weak learning algoritm. The idea of Boosting is to genearte the classifier sequentially [19]. The predictions are then combined via a weighted majority vote to produce the final prediction.

## 3 Description of HTRU2 Data Set

In this paper, we use the High Time Resolution Universe (HTRU2) dataset, obtained from UCI Machine Learning Repository. The data collected for each instance consists of four continuous variables derived from the folded pulse profile, four continuous variables from the dispersion measure – signal to noise ratio (DM-SNR) curve and a qualitative variable (target attribution) with two classes (no signal and presence of signal).

This work presents the pulsar data with eight features: Statistics of the folded pulse profile: Mean of the integrated profile, Standard deviation of the integrated profile, Excess kurtosis of the integrated profile, and Skewness of the integrated profile.

Statistics based on the DM-SNR curve: Mean of the DM-SNR curve, Standard deviation of the DM-SNR curve, Excess kurtosis of the DM-SNR curve, and Skewness of the DM-SNR curve.

The pulsar class consists of 9% of presence of signal (pulsar candidates) and 91% of no signal (no pulsar candidates). Thus, we have face with an imbalanced data set. We proceed to experiment two resampling techniques to handling imbalanced data set: Split data and repeated cross validation k-fold with under-sampling and over-sampling.

### 3.1 Split Data Set

HTRU2 dataset is split into a training sample for 70% (12528 instances) and test -sample for 30% (5370 instances). The relative frequencies of the classes in the

**Table 1.** Relative frequencies of the classes in the initial sample

| No signal | Presence of signal |
|---|---|
| 0.90842552 | 0.09157448 |

**Table 2.** Relative frequencies of the classes in the test-sample

| No signal | Presence of signal |
|---|---|
| 0.90883669 | 0.09116331 |

test-sample are conformed with the initial relative frequencies. we can conclude that the sample test is representative (Tables 1 and 2).

# 4 Experimental Results and Analysis

## 4.1 Evaluation Function

The evaluation of a classification algorithm performance is measured by the Confusion Matrix which contains information about the actual and the predicted class (Table 3):

**Table 3.** Confusion Matrix of actual and the predicted class

| Actual | Predicted | |
|---|---|---|
| | Positive class | Negative class |
| Positive class | True positive (TP) | False negative (FN) |
| Negative class | False postive (FP) | True negative (TN) |

Where:

The true positives (TP) is the number of candidate patterns that are already pulsars and are also being classified as pulsars.

The true negative (TN) is the number of candidates which are non-pulsars and also being classified as non -pulsars.

The false negative (FN) is the number of actual pulsar candidates that are incorrectly being classified as non- pulsars.

The false positive (FP) is the number of non-pulsars candidates that are incorrectly being classified as pulsars.

The equations from (1) to (4) give important metrics that are computed for comparison [20,21]:

**Sensitivity or Recall** is the proportion of actual positive cases which are correctly identified (True positive rate).

$$Recall = \frac{TP}{TP + FN} = \frac{TP}{TP + FN} \tag{1}$$

**Specificity** is the proportion of actual negative cases which are correctly identified.

$$Specificity = \frac{TN}{TN + FP} = \frac{TN}{N} \tag{2}$$

**Precision** is a Positive Predictive Value, it is calculated as:

$$Precision = \frac{TP}{TP + FP} \tag{3}$$

**Accuracy** is the proportion of the total number of predictions that were correct

$$Accuracy = \frac{TP + TN}{TP + TN + FP + FN} \tag{4}$$

## 4.2 Experimental Results

We use the "CARET package" implemented in R language to experiment the Time Resolution Universe (HTRU2) data set. We compare three models using tuning parameters of train function: Bagged CART (method = 'treebag'), Random Forest (method = 'rf') and AdaBoost Classification Trees (method = 'Adaboost') and we use "repeatedcv" 10-folds method with under-sampling and over-sampling techniques.

Table 4 shows the summary results for repeatedcv and test set in the cases of under-sampling and Over-sampling:

**Table 4.** Performance of three ensemble methods with under-sampling and over-sampling techniques

| Resampling | | Bagged CART | Random Forest | AdaBoost |
|---|---|---|---|---|
| Under-sampling | Repeated cross validation | **0,9369957** | **0,945995** | **0,9445135** |
| | Test set | **0,9608501** | **0,9630872** | **0,9636465** |
| Over-sampling | Repeated cross validation | **0,9608501** | **0,9443797** | **0,9443828** |
| | Test set | **0,9535794** | **0,962528** | **0,9670022** |

## 4.3 Analysis and Conclusion

In this subsection, we analyze the accuracy metric of three ensemble methods with under-sampling and over-sampling techniques.

In the context of under sampling, The test performance of three methods on HTRU2 dataset is extremely well. It outperforms 96% and it is conformed with repeated cross validation 10-folds. The Fig. 1 shows that the Random Forest achieved higher accuracy in comparison to the bagged cart and adaboost classification trees. Thus, Random Forest representS a good way to resist to under-fitting particularly in the case of the imbalanced data set such as HTRU2.

In the context of over- sampling, The test performance of three methods is also good, the accuracy measure outperforms 96% and it close to repeated cross validation performance. The Bagged cart algorithm works better than Random Forest and Adaboost Algorithms (see Fig. 2).

In the context of the both (under-sampling and over-sampling), Adaboost classification tree and Radom forest algorithms run efficientlly and work better than the Bagged cart classifier (see Fig. 3). Thus, we conclude that the Adaboost and the Random Forest Algorithms run efficiently to adjusting the train set distribution specially on the case of imbalanced data set such as HTRU2.

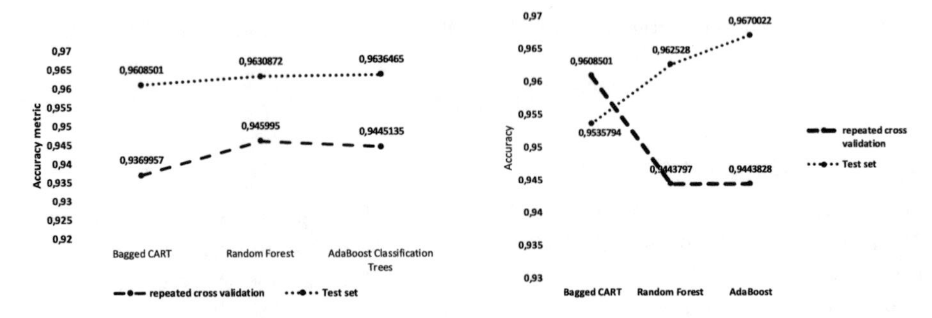


**Fig. 1.** Under-sampling performance of three Ensemble Methods.

**Fig. 2.** Over-sampling performance of three Ensemble Methods.

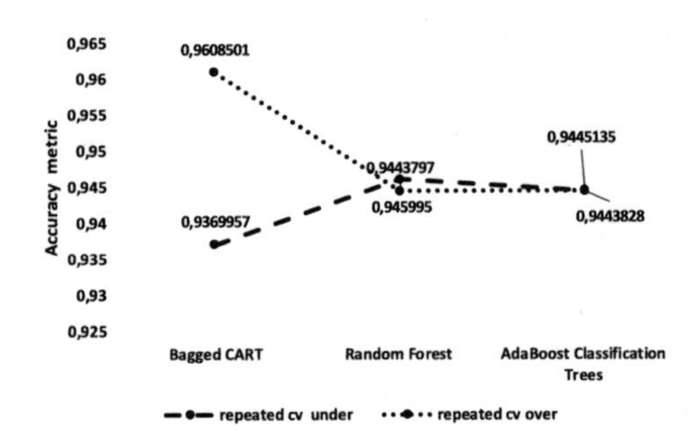


**Fig. 3.** Under and over-sampling performance of three Ensemble Methods.

## References

1. Thornton, D., Stappers, B., Bailes, M., Barsdell, B., Bates, S., Bhat, N.D.R., Burgay, M., Burke-Spolaor, S., Champion, D.J., Coster, P., D'Amico, N., Jameson, A., Johnston, S., Keith, M., Kramer, M., Levin, L., Milia, S., Ng, C., Possenti, A., van Straten, W.: A population of fast radio bursts at cosmological distances. Science **341**, 53–56 (2013). https://doi.org/10.1126/science.1236789
2. Hughes, S.A.: Gravitational wave astronomy and cosmology. Phys. Dark Univ. **4**, 86–91 (2014). https://doi.org/10.1016/j.dark.2014.10.003
3. Stairs, I.H.: Testing general relativity with pulsar timing. Living Rev. Relativ. **6**(2003). https://doi.org/10.12942/lrr-2003-5
4. Rosa, J.G.: Testing black hole superradiance with pulsar companions. Phys. Lett. B **749**, 226–230 (2015). https://doi.org/10.1016/j.physletb.2015.07.063
5. Ball, N.M., Brunner, R.J.: Data mining and machine learning in astronomy. Int. J. Modern Phys. D **19**, 1049–1106 (2010). https://doi.org/10.1142/S0218271810017160
6. Gauci, A., Adami, K.Z., Abela, J.: Machine Learning for Galaxy Morphology Classification (2010). arXiv:1005.0390 [astro-ph]
7. Möller, A.: Detection and classification of type Ia supernovae for cosmology in the complete data set of SNLS 187(n.d)

8. Eatough, R.P., Molkenthin, N., Kramer, M., Noutsos, A., Keith, M.J., Stappers, B.W., Lyne, A.G.: Selection of radio pulsar candidates using artificial neural networks: selection of radio pulsar candidates. Mon. Not. R. Astron. Soc. Lett. **407**, 2443–2450 (2010). https://doi.org/10.1111/j.1365-2966.2010.17082.x
9. Breiman, L.: Bagging predictors. Mach. Learn. **24**, 123–140 (1996). https://doi.org/10.1007/BF00058655
10. Zheng, Z.: Boosting and Bagging of Neural Networks with Applications to Financial Time Series 27 (2006)
11. Alfaro, E., García, N., Gámez, M., Elizondo, D.: Bankruptcy forecasting: an empirical comparison of AdaBoost and neural networks. Decis. Support Syst. **45**, 110–122 (2008). https://doi.org/10.1016/j.dss.2007.12.002
12. Ye, R., Suganthan, P.N.: A Kernel-Ensemble Bagging Support Vector Machine, pp. 847–852. IEEE (2012). https://doi.org/10.1109/ISDA.2012.6416648
13. Mordelet, F., Vert, J.P.: A bagging SVM to learn from positive and unlabeled examples. Pattern Recogn. Lett. **37**, 201–209 (2014). https://doi.org/10.1016/j.patrec.2013.06.010
14. Bauer, E.: An Empirical comparison of voting classification algorithms: bagging, boosting, and variants. Mach. Learn. **36**(36), 105–139 (1999)
15. Yang, P., Hwa Yang, Y., Zhou, B.B., Zomaya, A.Y.: A review of ensemble methods in bioinformatics. Curr. Bioinform. **5**, 296–308 (2010). https://doi.org/10.2174/157489310794072508
16. Dietterich, T.G.: Ensemble methods in machine learning. In: Multiple Classifier Systems, pp. 1–15. Springer, Heidelberg (2000). https://doi.org/10.1007/3-540-45014-9_1
17. Breiman, L.: Random forest. Mach. Learn. **45**(1), 5–32 (2001). https://doi.org/10.1023/a:1010933404324
18. Schapire, R.E.: Explaining AdaBoost. In: Schölkopf, B., Luo, Z., Vovk, V. (eds.) Empirical Inference, pp. 37–52. Springer, Heidelberg (2013). https://doi.org/10.1007/978-3-642-41136-6_5
19. Bühlmann, P.: Bagging, boosting and ensemble methods. In: Gentle, J.E., Härdle, W.K., Mori, Y. (eds.) Handbook of Computational Statistics, pp. 985–1022. Springer, Heidelberg (2012). https://doi.org/10.1007/978-3-642-21551-3_33
20. Kumar, R., Indrayan, A.: Receiver operating characteristic (ROC) curve for medical researchers. Indian Pediatr. **48**, 277–287 (2011). https://doi.org/10.1007/s13312-011-0055-4
21. Fawcett, T.: An introduction to ROC analysis. Pattern Recogn. Lett. **27**, 861–874 (2006). https://doi.org/10.1016/j.patrec.2005.10.010

# Pragmatic Text Mining Method to Find the Topics of Citation Network

Samaher Al_Janabi(✉), Mahdi Abed Salman, and Maha Mohammed

Department of Computer Science, Faculty of Science for Women (WSCI), University of Babylon, Babylon, Iraq
{samaher,mahdi.salman}@uobabylon.edu.iq, mememuhammed94@gmail.com

**Abstract.** Here is a large amount of data being generated every minute via websites, which are usually stored in corpus, and these data are either structured or unstructured or semi-structured. Each document in corpus often composed of a large number of words that are frequently repeated and which are insignificant. The traditional method to text preprocessing is following the sequence that is Tokenization, Stopwords Removal, Punctuation Removal and Stemming and finally computes TF-IDF to extract the score based on their frequency for each keyword. This paper focuses on building a pragmatic method to handle the documents of corpus and determined the topics for each document. This method involves (i) preprocessing the text documents using Rake that prove it extremely more efficient in term of time consuming (ii) Cleaning the document from some impurities (iii) building dictionary to keywords (iv) Apply LDA model to find the topics for each document. Two case study achieve in this paper, the first apply traditional method step on corpus while the second apply pragmatic method on the same corpus then compare the result of both methods. As result, we found pragmatic method given result based than traditional method in terms of accuracy and efficiency.

**Keywords:** Text classification · Preprocessing · Keywords extraction · Rake · LDA

## 1 Introduction

Since the data mining assume that the data is already stored in structural format so the overall preprocessing is focus on cleaning, normalize data and make number of tables join. In compare to text mining the preprocessing is focus on analysis and extraction the representative features based on natural language process (NLP). These preprocessing is responsible of convert the unstructured data of documents into more representative format so that improved subsequent processes [13].

Text mining is generally defined as the knowledge-intensive process by using set of analysis tool the user can be interacts over the time with document collection. The basic similarity between data mining and text mining are both seek to extract the useful knowledge by analysis, identification and representation of the data to obtain interesting


Y. Farhaoui (Ed.): BDNT 2019, LNNS 81, pp. 190–205, 2020.
https://doi.org/10.1007/978-3-030-23672-4_15

patterns. In case of text mining, the data sources are document collection so the useful information is not obtained by an analysis the database record but in unstructured textual data in documents collections [25]. There are many methods involves in text mining such as clustering, classification, information retrieval and so on.

Keywords can be considering as the smallest unit which used to identify and reviews the subject of a text [1]. The process of discovering the keywords within text is one of the basic functions of analysis large textual databases this process is known as keywords extraction. This task is considered core for several applications of text mining such as classification, clustering, indexing, topic detection, tracking and so on [6, 21].

The classification process of documents is the most important processes in text mining and because of the existence of millions of on-line documents, there must be a mechanism to organize these documents into classes to facilitate the process of retrieval of information and subsequent analysis. The classification process is used in automatic topic labeling, topic directory construction, identification of style that document is written, classify the document according to hyperlink associated with it. There are more classification methods such as Bayesian classification is one most popular method used for effective document classification. Support vector machine is also used for classification document if we assume the classes is number and create mapping function from the term space to class variable they preform effective classification since they appropriate for high-dimensional space [8]. The least-square linear regression is also used for this purpose [5, 26].

The traditional method to preprocessing text involves Several steps beginning of ***tokenization*** that convert text into list of tokens, ***remove punctuation*** such that symbol (, . > ? ! ...), ***remove stop words*** such that (the, an, a ..), ***stemming*** tokens where removing the suffix and prefix from it and ***lemmatize*** tokens where return it to roots [16]. Finally, apply the feature selection method to extract keywords (i.e., find weight to each token by apply TF_IDF). While the pragmatic method involves (i) preprocessing the text documents using Rake that prove it extremely more efficient in term of time consuming (ii) Cleaning the document from some impurities (iii) building dictionary to keywords (iv) Apply LDA model to find the topics for each document.

The reminder of this paper is organized as follows. Section 2 explain the related works on the text classification problem. Section 3 introduces the pragmatic method. Section 4 represent the two-case study and discussions the result. Finally, Sect. 5 show the conclusions.

## 2 Related Work

This section will explain the previous work handle the same problem. Dostal and Ježek [15]. They presented their experiment based on statistical methods and Wordnet-pattern evaluation on automatic key phrase extraction. They proposed new way to extract keypharse by the combination between TextRank (graph methods) and TF*IDF (statistical methods). The pragmatic method similar this work by find the same goal but differ in way to determine the topics.

Lau et al. [14]. They apply topic modeling to automatically induce word senses of a target word, and demonstrate that our word sense induction method can be used to automatically detect words with emergent novel senses, as well as token occurrences of those senses. We start by exploring the utility of standard topic models for word sense induction (WSI), with a pre-determined number of topics (=senses). We next demonstrate that a non-parametric formulation that learns an appropriate number of senses per word actually performs better at the WSI task. We go on to establish state-of-the-art results over two WSI datasets, and apply the proposed model to a novel sense detection task. The pragmatic method similar this work by using some of preprocessing stages such as lemmatize and remove stop words and the same goal.

Abu-Errub [2]. They proposed a modern strategy to classification the Arabic content to compared the pre-defined records categories with document based on its substance utilizing TF.IDF strategy (Term Frequency times Inverse Document Frequency) degree, The record is classified to the fitting sub-category utilizing Chi Square measure. The pragmatic method differ of this work by using rake as preprocessing method while this work used tokenization also, our work utilized English document while other utilized Arabic Tweet Extremity, finally we use the LDA model while other used Chi Square measure to classification.

Aaditya Jain and Mishra [9]. They presented modified Maximum Entropy based classifier. Most extreme Entropy classifiers give astonishing bargain of adaptability for parameter definitions and take after presumptions closer to genuine world situation. This classifier is at that point combined with a Naïve Bayes classifier. Naïve Bayes Classification is an exceptionally basic and quick method. The presumption show is inverse to that of the Greatest Entropy. The combination of classifiers is done through administrators that directly combine the comes about of two classifiers to foresee course of reports in query. The pragmatic method similar this work by find the same goal but differ in terms of initial preprocessing and classification technique used.

## 3 Main Tool

### 3.1 Tokenization

Tokenization is the process of dividing the text into sequence of pieces such as words, keywords, symbols and other element called tokens [24]. Tokens may be separate by whitespace or punctuation marks or line breaks and this may be remove from the list of the words in another stage of preprocessing. Finally, in last of this stage each input text is convert into list of tokens. The list of tokens is then turning into input for another text processing stage [11]. There are many type of tokenization methods such as Treebank this method is implement of Penn Treebank word tokenizer. The principle of her work depends on the regular expressions to fragment of the text into words. It considers that the text is already split into sentence. It split the English compressions such that don't tokenize to do and n't and they'll become they and 'll. They are considered punctuation marks as separator for splitting words. Split the text into words if the punctuation marks is followed by whitespace. The drawback of this method is the underscore and the dot that attached to the word [7]. The example below explains how it works:

| **Input:** | "Nowadays, telecom industry faces fierce competition in satisfying its customers". |
|---|---|

| **Output :** | " | Nowadays | , | telecom | industry | faces | fierce | competition | I n |
|---|---|---|---|---|---|---|---|---|---|
| satisfying | Its | customers | " | | | | | | |

### 3.2 Punctuation Removing

There are Thirty-two punctuation marks fourteen are generally used in English grammar. They are the question mark, exclamation point, semicolon, dash, comma, hyphen, parentheses, brackets, colon, braces, apostrophe, period, quotation marks, ellipsis etc. After the tokenization process this marks does not have any benefits meaning so its most remove from the bag of words. hyphenated word may be treated as a one word either separated words [5]. The table below (Table 1) explain the punctuation marks and their shapes.

**Table 1.** Puntuation marks and their shapes

| Mark | Shape | Mark | Shape |
|---|---|---|---|
| Number sign | # | Close parentheses | ) |
| At symbol | @ | Open parentheses | ( |
| Ampersand | & | Dollar sign | $ |
| Quotation mark | “” | Underscore | _ |
| Semicolon | ; | Period | . |
| Question mark | ? | Less than | < |
| Left brace | { | Greater than | > |
| Right brace | } | Dash | — |
| Left bracket | [ | Present | % |
| Right bracket | ] | Exclamation point | ! |
| Quotation marks | “ | Dash | – |
| Colon | : | Period | . |
| Apostrophe | ’ | Equal sign | = |
| Comma | , | Caret | ^ |
| Asterisk | * | Backslash | \ |

### 3.3 Stopwords Removing

Stop words are those that emerge in every text document and do not have any meaning in the document. in the same time, it adds noise to bag-of-words comparisons, so it most removes from document in order to simplest the Subsequent preprocessing step [25]. The most common stop word is the “a”, “is”, “an”, “the”, “across”, “after”, “afterwards”, "again", "against", "all", "almost", "alone", "along", "already", "also", "although", "always", "am", "among", and so on. The tokens should be removed when is match any features in the stop words list.

### 3.4 Stemming

Stemming is the preprocessing method used to removing the suffix and prefixes from the feature and return it to root form. This make many different forms of word in single forms and this reduce the number of features in feature space thus improve the performance of Subsequent operations. Each algorithm attempts to convert the morphological variants of a word like introduction, introducing, introduces etc. to get mapped to the word 'introduce'. The most common prefixes is un, im, pre, mis ..etc. and suffix such as ing, ed, es, ful, er ..etc. There are different type of stemming algorithms each of them is work in different way and some of them is works in the wrong way it give word that do not have any meaning such as Dawson, N-Gram, Corpus Based, Snowball and so [12, 23]. Porter stemming is the most popular stemming algorithms is consider the base and many algorithms is developed base on it. Many of the algorithms have been consider the porter stemming as a comparator for the efficiency of the algorithms. it has five step and inside each step the rules is apply is have 60 rules. When the conditions is satisfied the prefix or suffix is removed accordingly, and the next step is performed. This algorithm has many advantages (1) it gives best result comparable with the result of other stemming algorithms (2) The percentage of error is small (3) it framework to design Snowball stemmer. And in same time it has drawback (1) the stemmed words always not real words (2) more time consuming because it has five step and 60 rules [3].

### 3.5 Rapid Automatic Keyword Extraction (Rake)

The main goal is to produce a way to extract the keywords from text in a fast and useful way, so we use the rake method to this purpose that is extremely efficient. Operate on different type of document especially those that do not follow specific Grammar, can easily applied on new **fields**, applied on individual text to enable application to dynamic collections. The rake has three input parameters first the stop words list (or stoplist) [22] and the phrase delimiters set, and word delimiters set. Rake begin extract keyword from document by parsing its text into a set of candidate keywords. First it use the words delimiters to split the text into array of words then this array is split sequences of contiguous Words by using phrase delimiters and the stoplist position. Rake assigned the same position to words within a sequence into text considered a candidate keyword [10].

### 3.6 Lemmatization

Lemmatization is process to compilation together the inflected forms of a word so they can be considered as single item [17]. It depends on identify the part-of-speech and the meaning of the words in the sentence as well as within the larger context surrounding that sentence, such as neighboring sentences or even an entire document. In many languages the words appear in different forms for example in English language the verb walk may be appear as “walking”, “walked”, “walks”. The base form is “walk” that is look up in the dictionary it called the lemma of the word. another example lemma for word “better” is “good”, So that lemmatization attempts to select the correct lemma depending on the context [3].

## 3.7 TF-IDF

The tf-idf is abbreviated to term frequency-inverse document frequency is a numerical statistic used to compute the important of word in document in collection or corpus. It is used in most as information retrieval, text mining, and user modeling as weighting factor. The first part tf is used to compute the number of time that word occur in document and the second is the log of number overall document in corpus divided by the number of documents that contain the words. So, the words appear in many documents has lower weight than that occur in distinct document.

$$\mathrm{TF_{IDF}} = \mathrm{TFti} * \mathrm{IDFt} \tag{1}$$

$$IDFt = log2(N/DFt) \tag{2}$$

where DFt is the number of documents containing a word t, and N is the number of documents in the collection [1]

## 3.8 Latent Dirichlet Allocation (LDA)

LDA is a generative probabilistic model that is more generally used in topic models for the collection of documents [18–20].

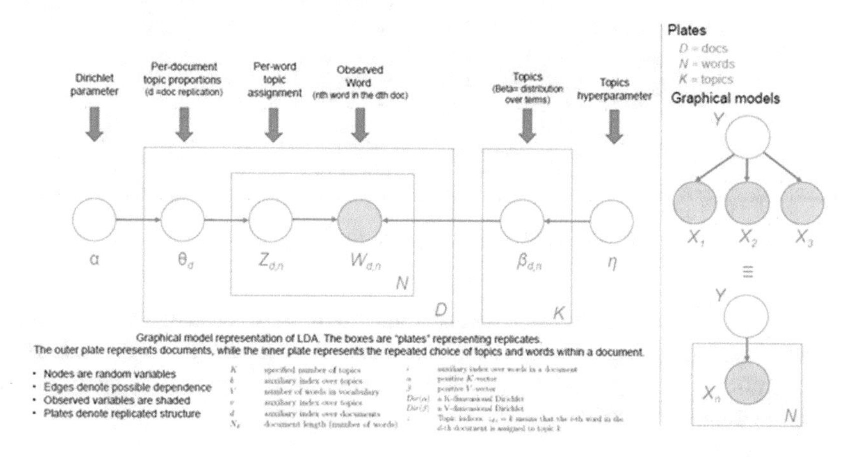

## 4 Pragmatic Method

At the first, we needed to explain the traditional method steps to compare with suggest method. Therefore Fig. 1 represents the steps of traditional method, while Fig. 2 shown algorithm of traditional methods for preprocessing corpus.

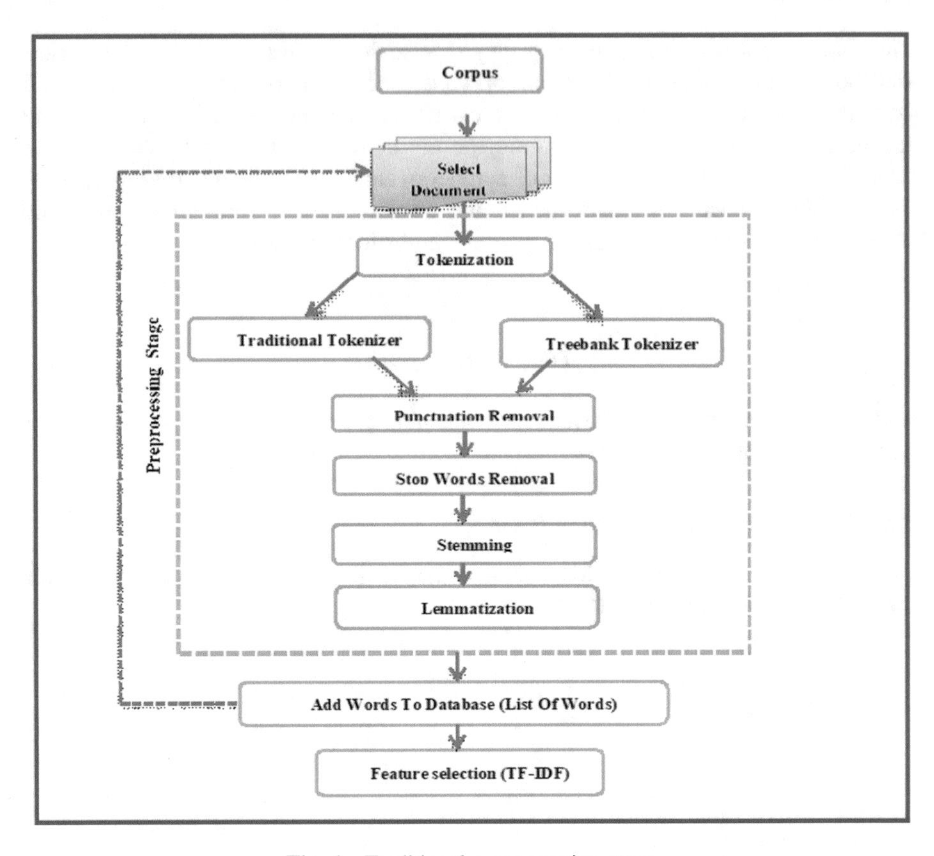


**Fig. 1.** Traditional preprocessing stage

While, in this section we will explain the main steps of pragmatic method to preprocessing corpus. Figure 3 represents the steps of pragmatic method, while Fig. 4 shown algorithm of pragmatic methods for preprocessing corpus. In general, we can have summarized as follow:

- First: Select a document from the corpus.
- Second: Apply Rake to split the document into set of sentences and determine the score for each sentence. remove sentence with lowest score.
- Third: Additional preprocessing apply to results obtained by the second step, because it may be containing many of the impurities such as symbols, stop words,

**Algorithm #1 : Traditional Method For Preprocessing Corpus**
**Input:** Corpus \\ Collection Of Documents
A:List Of Stopwords,
B: List Of Punctuation Symbols,
**Output**: Ks: List Of Keyword Of Corpus \\ Words With Scores
▪**Step1**: For Each Document ($D_i$)In Corpus
▪**Step2:** Call Tokenization Function \\ Split Document Into List Of Tokens(T)
T ← Tokenization ($D_i$)
▪**Step3:** Clean(T) Based On The Following:
- If Tj Belongs To A Remove (T, Tj) \\ J=1,2,,....,Length (T)
- If Tj Belongs To B Remove (T, Tj) \\ J=1,2,,....,Length (T)

▪**Step 4:** For Each Tokens In T (Stemming (Words))
▪**Step 5**: For Each Tokens In T (Lemmatize (Words))
▪**Step6**: scoring tokens using TF-IDF \\ see section 3.8
▪**Step7**: sort result based on score
▪**Step8**: chose top n as most important keywords

**Fig. 2.** Algorithm of traditional methods for preprocessing corpus

additions at the end or beginnings of tokens. Thus, we will get a bank of the legal words (i.e., keywords) of all the corpus.
- Forth: Build a dictionary for keywords that it obtained from the previous step. The purpose of this step getting the unique index for each keyword.
- Fifth: Aggregate all keywords and building digital corpus (i.e., index of sequence document, index of word in dictionary, frequency of that word).
- Sixth: Enter the digital corpus and dictionary to LDA model to determine the topics for each document.

# 5 Experimental Result

In this section, we will explain the results of two approaches to preprocessing text (traditional methods and Developed methods).

## 5.1 Collection Data

The first step in the beginning of the work we conducted extensive studies on the subject of citation research and then we collect the dataset of Citation. The data covers papers in the period from January 1993 to April 2003 (124 months). They include three folders the first is the citation graph second the date of each research was issued and finally the abstract of each research. Can get it from this link https://snap.stanford.edu/data/index.html#citnets.

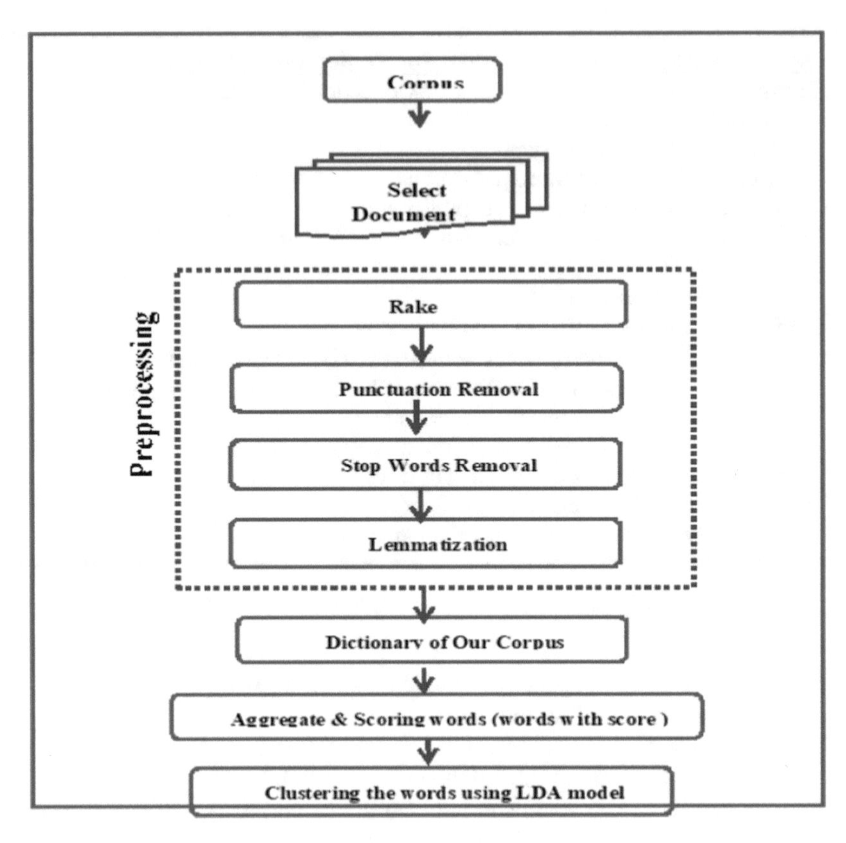


**Fig. 3.** Proposed method

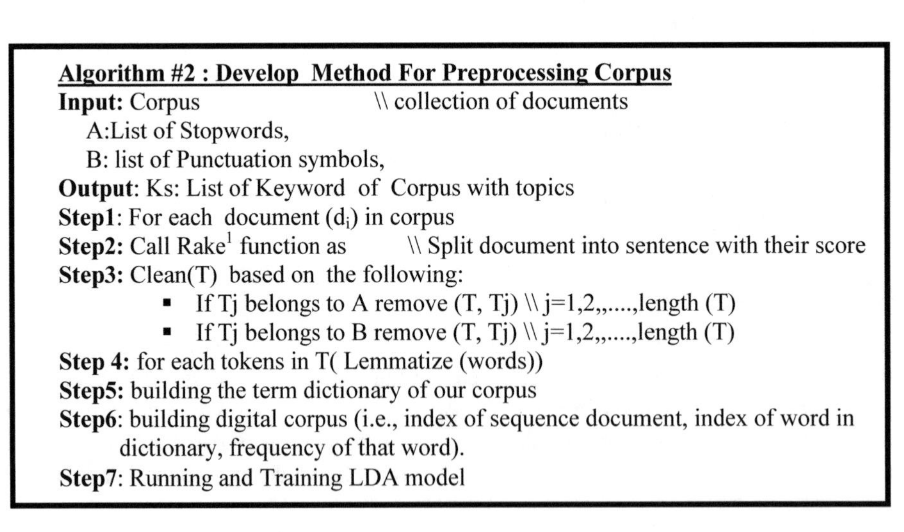

**Algorithm #2 : Develop Method For Preprocessing Corpus**

**Input:** Corpus \\ collection of documents

A:List of Stopwords,

B: list of Punctuation symbols,

**Output**: Ks: List of Keyword of Corpus with topics

**Step1**: For each document ($d_i$) in corpus

**Step2:** Call Rake[1] function as \\ Split document into sentence with their score

**Step3:** Clean(T) based on the following:

- If Tj belongs to A remove (T, Tj) \\ j=1,2,,....,length (T)
- If Tj belongs to B remove (T, Tj) \\ j=1,2,,....,length (T)

**Step 4:** for each tokens in T( Lemmatize (words))

**Step5:** building the term dictionary of our corpus

**Step6**: building digital corpus (i.e., index of sequence document, index of word in dictionary, frequency of that word).

**Step7**: Running and Training LDA model

**Fig. 4.** Algorithm of proposed method

## 5.2 Results of Traditional Method

When we apply the stages that explain in Fig. 1 on the sample abstract shown in Fig. 5, we get the following result:

"There are many words in documents that occur very frequently but in fact they meaningless and used to join the sentence to represent the meaning of text. So all text preprocessing focus on cleaning and normalize the text to extraction the representative features based on natural language process (NLP) to be appropriate for subsequent processes such as classification clustering information retrieval (IR)."

**Fig. 5.** Sample abstract for preprocessing stage

### 5.2.1 Preprocessing Phase

This phase considers very important to handle any corpus or document. it includes multi stages; first stage shown applying the two types of tokenization (i.e., Treebank Tokenizer and Traditional Tokenizer). Second stage applying Punctuation Removal, third stage Words Removal, Stemming the Words, and lemmatize The Words, the result for each stage in this phase explained below.

A. **Treebank Tokenizer**

**Table 2.** Tokenization text using Treebank Tokenizer

| there | are | many | words | in | documents | that | oc- cur |
|---|---|---|---|---|---|---|---|
| very | Frequently | but | in | fact | they | meaningless | and |
| used | to | join | the | sentence | to | represent | th e |
| meaning | of | text. | so | all | preprocessing | text | fo- cus |
| on | cleaning | and | normalize | the | text | to | represent |

Above table (Table 2) show the text is split based on all punctuation marks.

B. **Traditional Tokenizer**

**Table 3.** Tokenization text using Traditional Tokenizer

| there | are | many | words | in | documents | that | occur | | |
|---|---|---|---|---|---|---|---|---|---|
| very | Frequently | but | in | fact | they | meaningless | and | | |
| used | to | join | the | sentence | to | represent | the | | |
| meaning | of | text | . | so | all | preprocessing | text | focus | |
| on | cleaning | and | normalize | the | text | to | represent | | |
| features | Base d | on | natural | language | process | ( | nlp | ). | . |

Above table (Table 3) show the text is split based on white space.

- **Second Stage: Punctuation Removal**

**Table 4.** Remove punctuation marks

| there | are | many | words | in | documents | that | occur | | |
|---|---|---|---|---|---|---|---|---|---|
| very | Frequently | but | in | fact | they | meaningless | and | | |
| used | to | join | the | sentence | to | represent | the | | |
| meaning | of | text | | so | all | preprocessing | text | focus | |
| on | cleaning | and | normalize | the | text | to | represent | | |
| features | Based | on | natural | language | process | | nlp | | |

Above table (Table 4) show result after remove punctuation marks.

- **Third Stage: Stops Words Removal**

**Table 5.** Remove stop words

| | | many | words | | documents | | occur | | |
|---|---|---|---|---|---|---|---|---|---|
| | Frequently | | | fact | | meaningless | | | |
| used | | join | | sentence | | represent | | | |
| mean- ing | | text | | | | preprocessing | text | focus | |
| | cleaning | | normalize | | text | | represent | | |
| features | Based | | natural | language | process | | nlp | | |

Above table (Table 5) show result after remove stop words from text such that there, are, an.

- **Forth Stage: Stemming the Words**

**Table 6.** Stemming the tokens

| | | mani | word | | document | | occur | | |
|---|---|---|---|---|---|---|---|---|---|
| | Frequent | | | fact | | meaningless | | | |
| use | | join | | sentenc | | repres | | | |
| mean | | text | | | | preprocess | text | focu | |
| | Clean | | normal | | text | | repres | | |
| featur | Base | | natura | languag | process | | nlp | | |

Above table (Table 6) show result after remove the suffix and prefix from the token such as sentence → sentence.

- **Fifth Stage: Lemmatize the Words**

  Above table (Table 7) show result return the tokens to its root.

**Table 7.** Lemmatizing the tokens

| | | | | | | | | |
|---|---|---|---|---|---|---|---|---|
| | | many | word | | document | | occur | |
| | Frequently | | | fact | | meaning-less | | |
| us e | | join | | sentence | | represent | | |
| mean | | Text | | | preprocess | text | focus | |
| | clean | | normal | | text | | repre-sent | |
| feature | Base | | natural | language | process | | nlp | |

### 5.2.2 Feature Selection Phase

After we complete the text preprocessing, we apply TF-IDF on all corpus to extract the score based on their frequency for each keyword, Table 8 illustrate the top10 keywords in corpus:

**Table 8.** Keywords and scores

| Keyword | Score | Keyword | Score |
|---|---|---|---|
| Minor | 1.0 | Certain | 1.0 |
| Small | 1.0 | Various | 1.0 |
| Prove | 1.0 | Slightly | 1.0 |
| Hopefully | 1.0 | Contribution | 1.0 |
| Reference | 1.0 | Title | 1.0 |

## 5.3 Result of Pragmatic Method

### 5.3.1 Preprocessing Text Phase

We apply the stages that illustrate in Fig. 1 on the sample abstract in Fig. 5.

- **First Stage:** Apply the Rake function that depends on the fact that the keywords may be consists of multiple words and this words rarely may be contain punctuation marks or stop words such as and, the, to or other words without meaning. So for this reason it splitting the documents text according to this stop words and the phrase delimiters to candidate keywords. Then it calculates the score for each candidate keywords according to the word co-occurrences in document text.

   A. *After applying the Rake function on the sample abstract in Fig.* 5 *we have the following result* (Table 9):

   B. *Remove the sentence that score is less than or equal to 1* (Table 10).

**Table 9.** Candidate keywords with scores

| Keywords | Score | Keywords | Score |
|---|---|---|---|
| 'classification clustering information retrieval' | 16.0 | 'representative features based' | 9.0 |
| 'natural language process' | 9.0 | 'text preprocessing focus' | 0.8 |
| 'subsequent processes' | 0.4 | 'many words' | 0.4 |
| 'IR).' | 0.4 | 'text' | 2.0 |
| 'used' | 0.1 | 'meaning' | 0.1 |
| 'sentence' | 0.1 | 'join' | 0.1 |
| 'represent' | 0.1 | 'frequently' | 0.1 |
| 'occur' | 0.1 | 'fact' | 0.1 |
| 'normalize' | 0.1 | 'extraction' | 0.1 |
| 'nlp' | 0.1 | 'documents' | 0.1 |
| 'meaningless' | 0.1 | 'cleaning' | 0.1 |
| 'appropriate' | 0.1 | | |

**Table 10.** After remover the smallest sentence score

| Keywords | Score | Keywords | Score |
|---|---|---|---|
| 'classification clustering information retrieval' | 16.0 | 'representative features based' | 9.0 |
| 'natural language process' | 9.0 | 'text preprocessing focus' | 0.8 |
| 'subsequent processes' | 0.4 | 'many words' | 0.4 |
| 'IR).' | 0.4 | 'text' | 2.0 |

- ***Second Stage:*** Apply additional preprocessing to results obtained by the first stage because it may contain many of the impurities.

| **Keywords** |
|---|
| ['classification', 'clustering', 'information', 'retrieval'], |
| ['representative', 'feature', 'based'], |

| ['natural', 'language', 'process'], |
|---|
| ['text', 'preprocessing', 'focus'], |
| ['subsequent', 'process'], |
| ['many', 'word'] ,['IR'] ,['text'] |

- ***Third Stage:*** Building dictionary to our result from the previous stage, where each token get unique index, such that

[(0, 'classification'), (1, 'clustering'), (2, 'information'), (3, 'retrieval'), (4, 'based'), (5, 'feature'), (6, 'representative'), (7, 'language'), (8, 'natural'), (9, 'process'), (10, 'focus'), (11, 'preprocessing'), (12, 'text'), (13, 'subsequent'), (14, 'many'), (15, 'word'), (16, 'IR')]

- ***Forth stage:*** Building digital corpus (i.e., index of word in dictionary, frequency of that word).

[[(0, 1), (1, 1), (2, 1), (3, 1)], [(4, 1), (5, 1), (6, 1)], [(7, 1), (8, 1), (9, 1)], [(10, 1), (11, 1), (12, 1)], [(9, 1), (13, 1)], [(14, 1), (15, 1)], [(16, 1)], [(12, 1)]]

### 5.3.2 Classification Phase

*Applying* the LDA model to determine the topics to each document in all corpus (Table 11).

**Table 11.** Clusters for our result

| Cluster no. | Group Of Words |
|---|---|
| Cluster#1 | (1, '0.039*"quantum" + 0.030*"theory" + 0.030*"field" + 0.015*"action" + 0.014*"classical"'), |
| Cluster#2 | (2, '0.019*"method" + 0.018*"integral" + 0.015*"function" + 0.014*"expansion" + 0.014*"perturbative"'), |
| Cluster#3 | (3, '0.105*"string" + 0.050*"theory" + 0.026*"open" + 0.022*"field" + 0.016*"tachyon"'), |
| Cluster#4 | (4, '0.056*"boundary" + 0.035*"matrix" + 0.032*"model" + 0.030*"theory" + 0.019*"limit"'), |
| Cluster#5 | (5, '0.041*"effective" + 0.029*"field" + 0.026*"potential" + 0.022*"energy" + 0.019*"action"'), |
| Cluster#6 | (6, '0.026*"brane" + 0.025*"black" + 0.016*"cosmological" + 0.013*"can" + 0.012*"hole"'), |
| Cluster#7 | (7, '0.067*"gauge" + 0.023*"theory" + 0.017*"field" + 0.015*"model" + 0.013*"show"'), |
| Cluster#8 | (8, '0.031*"bps" + 0.020*"conformal" + 0.017*"charge" + 0.017*"topological" + 0.017*"field"'), |
| Cluster#9 | (9, '0.031*"algebra" + 0.018*"space" + 0.017*"quantum" + 0.016*"group" + 0.013*"lie"'), |
| Cluster#10 | (10, '0.043*"type" + 0.039*"noncommutative" + 0.029*"gauge" + 0.028*"supergravity" + 0.024*"theory"') |

## 6 Conclusion

As result, through experience by applying the traditional methods to preprocessing text that beginning by tokenization, punctuation removal, stop words removal, stemming, lemmatization and finally apply TF_IDF to feature selection and compare it to the proposed methods that following develop way to preprocessing text by apply rake algorithm to split text, done some cleaning, building dictionary then digital corpus and finally use LDA models to find topics to each document in corpus. we obtained that the proposed methods more powerful than the traditional way because it depends on rake that calculates the score for each candidate keywords according to the word co-occurrences in document text. Rather than compute the IF\IDF that based on frequency of words also take time less than traditional method. Finally, we approve the TF-IDF output not appropriate as input for LDA where the LDA required data in integer form. Rake reduced number stages where Merage two stages in single stage (i.e., preprocessing and feature selection) therefore, the results very suitable to LDA model to determine the topics of corpus.

## References


1. Abilhoa, W.D., De Castro, L.N.: A keyword extraction method from twitter messages represented as graphs. Appl. Math. Comput. **240**, 308–325 (2014)
2. Abu-errub, A.: Arabic text classification algorithm using TFIDF and chi square measurements. Int. J. Comput. Appl. **93**(6), 40–45 (2014)
3. Jivani, A.G., Anjali, M.: A comparative study of stemming algorithms. Int. J. Comp. Tech. Appl **2**(2004), 1930–1938 (2007)
4. Al_Janabi, S., Salman, M.A., Mohammad, M.: Multi-level network construction based on intelligent Big Data analysis. In: Farhaoui, Y., Moussaid, L. (eds.) Big Data and Smart Digital Environment. ICBDSDE 2018. SBD, vol. 53, pp. 102–118. Springer, Cham (2019). https://doi.org/10.1007/978-3-030-12048-1_13
5. Aggarwal, C.C.: Data Mining: The Textbook. Springer, New York (2015)
6. Meladianos, P., Tixier, A.J.P., Nikolentzos, G., Vazirgiannis, M.: Real-time keyword extraction from conversations. In: EACL, pp. 462–467 (2017)
7. Jurafsky, M.J.D.: Speech and Language Processing, 2nd edn. Pearson Education, Saddle River (2008)
8. Al-Janabi, S., Alkaim, A.F.: A nifty collaborative analysis to predicting a novel tool (DRFLLS) for missing values estimation. J. Soft Comput. (2019). https://doi.org/10.1007/s00500-019-03972-x. Springer
9. Jain, A., Mishra, R.D.: Text categorization: by combining Naïve Bayes and modified maximum entropy classifiers. Int. J. Adv. Electron. Comput. Sci., 122–126 (2016)
10. Rose, S., Engel, D., Cramer, N., Cowley, W.: Automatic keyword extraction from individual documents. Text Min. Appl. Theory, 1–20 (2010)
11. Vijayarani, S., Janani, R.: Text mining: open source tokenization tools – an analysis. Adv. Comput. Intell. Int. J. **3**(1), 37–47 (2016)
12. Shende, P.: Mining Text for Meaningful Words with Stemming Algorithm, pp. 13–16 (2016)

13. Feldman, R., Sanger, J.: Text Mining Handbook: Advanced Approaches in Analyzing Unstructured Data. United States of America by Cambridge University Press, New York (2006)
14. Lau, J.H., Cook, P., McCarthy, D., Newman, D., Baldwin, T.: Word sense induction for novel sense detection. In: Proceedings of the 13th Conference of the European Chapter of the Association for Computational Linguistics. Association for Computational Linguistics (EACL 2012), pp. 591–601 (2012)
15. Dostal, M., Ježek, K.: Automatic keyphrase extraction based on NLP and statistical methods. In: Proceedings of DATESO 2011 Annual International Workshop on Databases, Texts, Specification Object, pp. 140–145 (2011)
16. Sumathy, M., Chidambaram, K.L.: Text mining: concepts, applications, tools and issues - an overview. Int. J. Comput. Appl. **80**(4), 29–32 (2013)
17. Al_Janabi, S.: Smart system to create optimal higher education environment using IDA and IOTs. Int. J. Comput. Appl. (2018). https://doi.org/10.1080/1206212X.2018.1512460. Taylor & Francis
18. Wang, J., Bansal, M., Gimpel, K., Ziebart, B.D., Yu, C.T.: A sense-topic model for word sense induction with unsupervised data enrichment. Trans. ACL **3**, 59–71 (2015)
19. Al_Janabi, S., Mahdi, M.A.: Evaluation prediction techniques to achievement an optimal biomedical analysis. Int. J. Grid Util. Comput. (2019)
20. Brody, S., Lapata, M.: Bayesian word sense induction. In: Computational Linguistics, pp. 103–111 (2009)
21. Beliga, S.: Keyword extraction: a review of methods and approaches, pp. 1–9 (2014)
22. Dutta, A.: A novel extension for automatic keyword extraction. Int. J. Adv. Res. Comput. Sci. Softw. Eng. **6**(5), 160–163 (2016)
23. Salah, A., Babiker, A.: Improving Stemming Algorithm for Arabic Text Search, Sudan University Science and Technology (2014)
24. Kalajdzic, K., Ali, S.H., Patel, A.: Rapid lossless compression of short text messages. Comput. Stand. Interfaces **37**, 53–59 (2015). https://doi.org/10.1016/j.csi.2014.05.005
25. Meyer, D., Hornik, K., Feinerer, I.: Institutional Repository, October 2013
26. Baharudin, B., Lee, L.H., Khan, K.: A review of machine learning algorithms for text-documents classification. J. Adv. Inf. Technol. **1**(1), 4–20 (2010)

# Grey-Markov Model for the Prediction of the Electricity Production and Consumption

S. Elgharbi[1], M. Esghir[1(✉)], O. Ibrihich[2], A. Abarda[3], S. El Hajji[1], and S. Elbernoussi[1]

[1] Faculty of Sciences, Laboratory of Mathematics, Computing and Applications, Mohammed V University in Rabat, Rabat, Morocco
safaa.elgharbi@edu.uca.ac.ma, esghirmustapha@yahoo.fr, elhajji.said@gmail.com, souad.elbernoussi@gmail.com

[2] Département de génie réseaux et télécommunications, Sultan Moulay Slimane University in Beni Mellal, Ecole Nationale des Sciences Appliquées de Khouribga, Beni-Mellal, Morocco
wafaa.ibrihich@gmail.com

[3] Laboratoire de Modélisation, Mathématiques et de Calculs, Economiques, FSJES, Université Hassan 1er, Settat, Morocco
abardabdallah@gmail.com

**Abstract.** The electricity planning of a country requires accurate models for the prediction of the electricity production and consumption. It is an important process which guarantees a better future of the energy sector. In the field of predictive modeling and probabilistic forecasting, the Grey-Markov model is considered desirable since it gathers the GM(1,1) prediction model with the ability of the Markov chain to correct the errors related to the prediction. As an application, the historical data of the electricity production and consumption in Morocco from 1990 to 2017 was selected, and the predicted results from 2018 to 2030 show the efficiency of the Grey-Markov model comparing to the GM(1,1) model.

**Keywords:** Grey-Markov model · GM(1,1) model · Markov chain · Prediction · Electricity production and consumption · Morocco

## 1 Introduction

The electricity is the central technology in our epoch. In the last decade, the electricity production in Morocco knew a remarkable growth from 12.86 TWh in 2000 to 22.85 TWh in 2010, and this increase was due to the development of the heating and the air conditioning, the proliferation of electric machines and the policy of the rural electrification. As for the electricity consumption, it plays a primordial role in the economy of a developing country like Morocco. It grows much more rapidly than other energies consumption, more specifically from 14.11 TWh to 25.10 TWh in the period 2000–2010.


Y. Farhaoui (Ed.): BDNT 2019, LNNS 81, pp. 206–219, 2020.
https://doi.org/10.1007/978-3-030-23672-4_16

Lately, the liberalization tendency and changes in the Moroccan energy market required reducing environmental impacts of electricity production and consumption, using new techniques that are less harmful to the environment. The year 2017 was marked by a growth in the electricity production. In fact the electricity generated at the national level reached 31.89 TWh versus 30.83 TWh in 2016. Trend analysis shows a largest increase in electricity consumption in 2016, which is recorded especially by administrative (+9.8 %) and agricultural (+6.4 %) sectors. So to improve the efficiency of the Moroccan energy sector, developing new scenarios is required and the modeling of electricity planning is the simplest way.

Forecasting the electricity production and consumption can be seen as grey systems since they are affected by many factors such as climate, population, economic growth, which cause less accuracy and not enough data. To come up with solutions for these issues, many theories have been developed.

Fuzzy mathematics (Zadeh 1965), Grey systems theory (Deng 1982) and Rough set theory (Dawlak 1982) have been the most interesting works used for studying indeterministic systems with incomplete information. Energy [11, 17, 21], economy [2], agriculture [20], industry [1], tourism [3, 6, 15] are some of these theories applications.

Grey systems theory was first introduced by professor Julong Deng [4] in 1982, as a means of studying uncertain systems with a lack of information. This theory has gained a great popularity in many prediction applications, due to its reliability, efficacity and ability to deal with the poorness of the information and the uncertainty of the data sample.

The $GM(1,1)$ model is the best known and the most frequently used between all the Grey prediction models. It aims to transform a disordered discrete raw data into a regular series using the so-called Accumulated Generating Operation (AGO) which is important for the Grey modeling.

The Grey forecasting model $GM(1,1)$ is a first-order one variable model whose solution has an exponential nature which is not helpful if the data varies considerably and that makes the prediction accuracy poorer. Grey-Econometric model, Grey Cobb-Douglas model, Grey-Markov model and Grey time series model are some of the Grey-combined models [12] that were established to enhance the accuracy of the Grey model.

In this work, we shed light on the Grey-Markov model proposed by Liu et al. in 2004 in order to reduce the error and get better results in different fields. This model benefits from the prediction performance of the $GM(1,1)$ model and the capacity of the Markov Chain model to handle the high fluctuation of the data. Shengneng Hu in his paper [7] used the Grey-Markov chain model for the prediction of traffic accidents in Chinese Xinyang city. Suo Ruixia [16] applied the Grey-Markov model to forecast the China's energy consumption. Using the same model, Kazemi et al. [10] forecasted the energy demand of Industry sector in Iran. The Grey-Markov model was also used for the inflation prediction [9], the analysis of sport training injury measurement [18], the activity recognition [14] and for many other works. Other prediction methods were employed in the electricity sector. In a paper of Qingyou Yan et al. [19], a market share forecasting model was established based on Markov chain, and a system dynamics model was constructed to forecast the electricity demand. Madhugeeth [13] utilized artificial neural networks for electricity power load forecasting in Sri Lankan, India. Ding et al. [5] combined a new initial condition and rolling mechanism to modify the grey prediction model to accurately forecast China's overall and industrial electricity consumption.

The outline of this paper is given as follows: Sects. 2 and 3 introduce the GM(1,1) model and the Markov chain model, respectively. Section 4 presents the Grey-Markov model while Sect. 5 demonstrates some numerical experiments in which the Moroccan electricity production and consumption is predicted during 2018–2030 basing on some historical data. A concluding discussion is given in Sect. 6.

## 2 GM(1,1) Model

For a given sequence of raw data

$$X^{(0)} = \left(x^{(0)}(1), x^{(0)}(2), \ldots, x^{(0)}(n)\right), \tag{2.1}$$

with $x^{(0)}(k) \geq 0 \quad (1 \leq k \leq n)$, we generate a new sequence

$$X^{(1)} = \left(x^{(1)}(1), x^{(1)}(2), \ldots, x^{(1)}(n)\right) \tag{2.2}$$

basing on accumulating generation; that is

$$x^{(1)}(k) = \sum_{i=1}^{k} x^{(0)}(i), \qquad k = 1, \ldots, n. \tag{2.3}$$

Let $Z^{(1)}$ be the sequence of which the values are the mean of consecutive neighbors of $X^{(1)}$ given by

$$Z^{(1)} = \left(z^{(1)}(1), z^{(1)}(2), \ldots, z^{(1)}(n)\right) \tag{2.4}$$

with

$$z^{(1)}(k) = \frac{x^{(1)}(k) + x^{(1)}(k-1)}{2}, \qquad k = 2, \ldots, n \tag{2.5}$$

and $z^{(1)}(1) = x^{(1)}(1)$.

Consider now the $GM(1,1)$ model

$$x^{(0)}(k) + az^{(1)}(k) = b, \qquad k = 2, \ldots, n. \tag{2.6}$$

If we use the least-squares method, then the estimate sequence of parameters $\hat{a} = [a, b]^T$ verifies [12]

$$\hat{a} = (B^T B)^{-1} B^T Y, \tag{2.7}$$

where

$$Y = \begin{pmatrix} x^{(0)}(2) \\ x^{(0)}(3) \\ \vdots \\ x^{(0)}(n) \end{pmatrix} \quad \text{and} \quad B = \begin{pmatrix} -z^{(1)}(2) & 1 \\ -z^{(1)}(3) & 1 \\ \vdots & \vdots \\ -z^{(1)}(n) & 1 \end{pmatrix}. \tag{2.8}$$

Introduce

$$\frac{dx^{(1)}}{dt} + ax^{(1)}(t) = b \tag{2.9}$$

the whitenization equation of the Grey model $x^{(0)}(k) + az^{(1)}(k) = b$.

**Theorem 1** [12]. *If* $[a, b]^T = (B^T B)^{-1} B^T Y$, *then the following hold true:*

- *The solution (or time response function) of the whitenization function*

$$\frac{dx^{(1)}}{dt} + ax^{(1)}(t) = b \tag{2.10}$$

*is given by*

$$x^{(1)}(t) = \left( x^{(1)}(0) - \frac{b}{a} \right) e^{-at} + \frac{b}{a}. \tag{2.11}$$

- *The time response sequence of the GM(1,1) grey differential equation*

$$x^{(0)}(k) + az^{(1)}(k) = b \tag{2.12}$$

*is given by*

$$\hat{x}^{(1)}(k+1) = \left( x^{(1)}(0) - \frac{b}{a} \right) e^{-ak} + \frac{b}{a}, \qquad k = 1, 2, \ldots, n. \tag{2.13}$$

- *Let* $x^{(1)}(0) = x^{(0)}(1)$, *then*

$$\hat{x}^{(1)}(k+1) = \left( x^{(0)}(1) - \frac{b}{a} \right) e^{-ak} + \frac{b}{a}, \qquad k = 1, 2, \ldots, n. \tag{2.14}$$

- *The restored values of* $x^{(0)}(k)$*s can be given by*

$$\hat{x}^{(0)}(k+1) = \hat{x}^{(1)}(k+1) - \hat{x}^{(1)}(k), \qquad k = 1, 2, \ldots, n. \tag{2.15}$$

The parameters $(-a)$ and $b$ in the Grey model are called the development coefficient and grey action quantity, respectively.

## 3 Markov Chain Model

The Markov chain is a stochastic process that inhold the Markov property: The future state of the process can be predicted using only the present state without knowing the previous states:

$$P(x_{n+1} = i_{n+1}/x_0 = i_0, \ldots, x_n = i_n) = P(x_{n+1} = i_{n+1}/x_n = i_n) \tag{3.1}$$

where $\{x_n : n \in T \subset \mathbb{N}\}$ is a stochastic process, and $i_0$, $i_1, \ldots$, $i_{n+1}$ are the states of the system.

**Definition 1.** *Let $E$ be the set of states, $n \in T \subset \mathbb{N}$ and let $(x_0, x_1, \ldots, x_n)$ be a sequence of raw data.*

- *The number*
$$p_{ij} = P(x_1 = j/x_0 = i) \tag{3.2}$$
*is called transition probability from state $i$ to state $j$ after one step.*
- *The matrix*
$$P(1) = (p_{ij})_{(i,j)\in E^2} \tag{3.3}$$
*is called one-step state transition probability matrix.*

**Proposition 1.** *The following properties hold true:*

- $\forall (i,j) \in E^2, \qquad p_{ij} \geq 0$
- *The state transition probability matrix is stochastic, i.e.*
$$\forall i \in E : \sum_{j\in E} p_{ij} = 1. \tag{3.4}$$
- *The n-step state transition probability matrix is given by:*
$$P(n) = P(1)^n, \; \forall n \in \mathbb{N}. \tag{3.5}$$
- *Let $M_{ij}$ be the number of transitions from state $i$ to state $j$ in one step and $M_i$ be the number of times when the state $i$ occurs. Then the transition probability $p_{ij}$ is given by:*
$$p_{ij} = \frac{M_{ij}}{M_i}. \tag{3.6}$$

The Markov chain theory consists in:

- Defining the states space
- Creating the one-step state transition probability matrix $P(1)$
- Finding the future state of the system after one step, basing on the last state (present state) of the system.
- Creating the $k$-steps state transition probability matrix $P(k) = P(1)^k$ and finding the future state of the system after $k \geq 2$ steps.

# 4 Grey-Markov Model

The Grey-Markov model is used to forecast the information, basing on sequences of historical data, to get good and accurate results.

After establishing the $GM(1,1)$ model and getting the estimations $\hat{X}^{(0)}$ of the data, we divide the sequence of residual errors uniformly into $s$ states to get the state transition probability matrix.

Let $e = (e(i))_{i=1,\ldots,s}$ be the error sequence and let $E_i = [E_{1i}, E_{2i}]$ be a state interval where
$$E_{1i} = \hat{x}(k) + a_i \quad \text{and} \quad E_{2i} = \hat{x}(k) + b_i, \tag{4.1}$$

such that $a_i + b_i = 2e(i)$, for $i = 1, \ldots, s$.

Consider the state transition probability matrix $P(1)$, if the last element of the sequence is in the state $E_l$, then consider the row $l$ of the matrix. If $\max_{j \in \{1,\ldots,n\}} P_{lj} = P_{lm}$, then the state of the system will be $E_m$ in the next step. If the maximum is not unique, consider the state transition probability matrix $P(2) = P(1)^2$, else $P(k)$, $k \geq 3$. Proceed in the same way to get the future states in the next steps.

Finally, if the next state is $E_i$, then the predicted value by the Grey-Markov model is given by:

$$\hat{y}(k) = \frac{1}{2}(E_{1i} + E_{2i}) = \hat{x}(k) + \frac{a_i + b_i}{2}, \tag{4.2}$$

for all $k = 1, \ldots, n$ and $i = 1, \ldots, s$.

## 5 Numerical Experiments

In this section, some numerical experiments will be reported to examine the efficacity of the Grey-Markov model when applied to forecast the electricity production and consumption in Morocco. In these experiments, we will compare the performance of the Grey-Markov model with the GM(1,1) model. The historical data used in this study are collected from ONEE - Electricity Branch [22]. The algorithms were coded in Matlab R2016b. All the experiments were performed on a computer of Intel(R) at CPU 2.20 GHz and 4.00 GB of RAM.

### 5.1 Forecasting the Electricity Production in Morocco

In order to predict the total electricity production in Morocco, we use the Grey-Markov model which requires following the procedure below:

**Table 1.** The electricity production (GWh) in Morocco during the period 1990–2017.

| Year | Electricity production | Year | Electricity production | Year | Electricity production |
|---|---|---|---|---|---|
| 1990 | 9628 | 2000 | 12863 | 2010 | 22851.4 |
| 1991 | 9205 | 2001 | 15046 | 2011 | 24363.6 |
| 1992 | 9719 | 2002 | 15235 | 2012 | 26495.5 |
| 1993 | 9910 | 2003 | 16578 | 2013 | 26940.7 |
| 1994 | 11470 | 2004 | 16383.5 | 2014 | 28081.5 |
| 1995 | 12092 | 2005 | 16911 | 2015 | 29914.2 |
| 1996 | 12583 | 2006 | 19822.3 | 2016 | 30839.8 |
| 1997 | 13593 | 2007 | 20007 | 2017 | 31889.8 |
| 1998 | 12968 | 2008 | 20306.8 | | |
| 1999 | 12895 | 2009 | 20935.3 | | |

- **Establishing the GM(1,1) model:**
Table 1 shows the original data of the electricity production in Morocco during the period 1990–2017. Basing on these data and using the least-squares method, we get the GM(1,1) grey model:

$$x^{(0)}(k) - 0.048z^{(1)}(k) = 8521.6, \qquad k = 1, \ldots, n. \tag{5.1}$$

According to Theorem 1, the restored values of $x^{(0)}(k)$s are given as follows:

$$\hat{x}^{(0)}(k+1) = (1 - e^{a})(x^{(0)}(1) - \frac{b}{a})e^{-ak} = 8772.45e^{0.048k}, \qquad k = 1, \ldots, n. \tag{5.1}$$

- **Dividing the states:**
We divide the residual errors sequence uniformly into 5 states. Figure 1 shows the state division of the total electricity production.

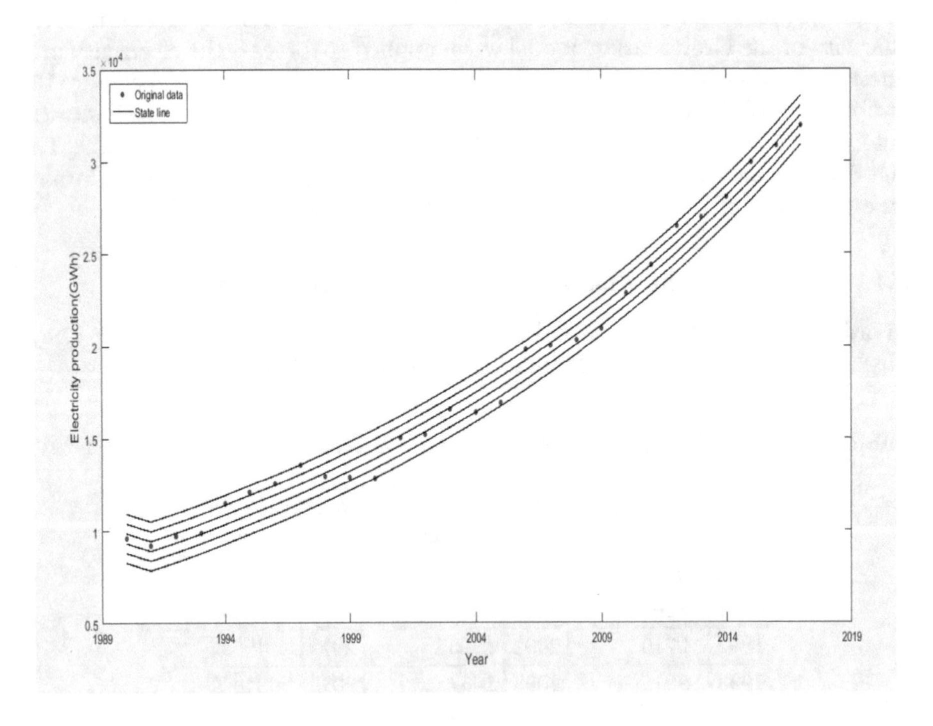


**Fig. 1.** State division by residual errors.

- **Generating the transition probability matrix:**

The state transition probability matrix after one step can be obtained according to Sect. 3 as follows:

$$\begin{pmatrix} 1/4 & 0 & 1/2 & 0 & 1/4 \\ 2/3 & 0 & 1/3 & 0 & 0 \\ 1/12 & 1/3 & 1/3 & 1/12 & 1/6 \\ 0 & 0 & 1 & 0 & 0 \\ 0 & 0 & 1/3 & 1/6 & 1/2 \end{pmatrix} \tag{5.2}$$

- **Obtaining the Grey-Markov predictive values:**
  Applying (4.2), we calculate the forecasting values of the Moroccan electricity production during 1990–2017. Figure 2 and Table 2 give the forecast values by the GM(1,1) model and the Grey-Markov model. For each model, the residual and relative errors are both reported.

**Table 2.** Prediction results of GM(1,1) model and Grey-Markov model for the electricity production (GWh)

| Year | GM(1,1) model | | | State | Grey-Markov model | | |
|---|---|---|---|---|---|---|---|
| | Forecast value | Residual error | Relative error (%) | | Forecast value | Residual error | Relative error (%) |
| 1990 | 9628 | 0 | 0 | $E_3$ | 9609 | 19 | 0.19 |
| 1991 | 9205.5 | −0.5 | −0.005 | $E_3$ | 9186.5 | 18.5 | 0.2 |
| 1992 | 9660 | 59 | 0.61 | $E_3$ | 9641 | 78 | 0.8 |
| 1993 | 10136.6 | −226.6 | −2.29 | $E_3$ | 10117.7 | −207.7 | −2.09 |
| 1994 | 10637 | 833 | 7.26 | $E_5$ | 11674.6 | −204.6 | −1.78 |
| 1995 | 11162 | 930 | 7.69 | $E_5$ | 12199.7 | −107.7 | −0.89 |
| 1996 | 11713 | 870 | 6.91 | $E_5$ | 12750.7 | −167.7 | −1.33 |
| 1997 | 12291 | 1302 | 9.58 | $E_5$ | 13328.8 | 264.2 | 1.94 |
| 1998 | 12897.8 | 70.2 | 0.54 | $E_3$ | 12879 | 89 | 0.68 |
| 1999 | 13534.5 | −639.5 | −4.96 | $E_2$ | 12987.3 | −92.3 | −0.71 |
| 2000 | 14202.5 | −1339.5 | −10.41 | $E_1$ | 13127 | −264 | −2.05 |
| 2001 | 14903.6 | 142.4 | 0.95 | $E_3$ | 14884.7 | 161.3 | 1.07 |
| 2002 | 15639.2 | −404.2 | −2.65 | $E_2$ | 15092 | 143 | 0.93 |
| 2003 | 16411.2 | 166.8 | 1.01 | $E_3$ | 16392.3 | 185.7 | 1.12 |
| 2004 | 17221.2 | −837.7 | −5.11 | $E_1$ | 16145.8 | 237.7 | 1.45 |
| 2005 | 18071.3 | −1160.3 | −6.86 | $E_1$ | 16996 | −85 | −0.50 |
| 2006 | 18963.3 | 859 | 4.33 | $E_5$ | 20001 | −178.7 | −0.90 |
| 2007 | 19899.3 | 107.7 | 0.54 | $E_3$ | 19880.5 | 126.5 | 0.63 |
| 2008 | 20881.6 | −574.8 | −2.83 | $E_2$ | 20334.4 | −27.6 | −0.13 |
| 2009 | 21912.3 | −977 | −4.67 | $E_1$ | 20837 | 98.3 | 0.47 |
| 2010 | 22994 | −142.6 | −0.62 | $E_3$ | 22975 | −123.6 | −0.54 |
| 2011 | 24129 | 234.6 | 0.96 | $E_3$ | 24110 | 253.6 | 1.04 |
| 2012 | 25320 | 1175.5 | 4.44 | $E_5$ | 26357.6 | 137.9 | 0.52 |
| 2013 | 26569.7 | 371 | 1.38 | $E_4$ | 27079 | −138.3 | −0.51 |
| 2014 | 27881.2 | 200.3 | 0.71 | $E_3$ | 27862.3 | 219.2 | 0.78 |
| 2015 | 29257.4 | 656.8 | 2.19 | $E_4$ | 29766.8 | 147.4 | 0.49 |
| 2016 | 30701.6 | 138.2 | 0.45 | $E_3$ | 30682.7 | 157.1 | 0.51 |
| 2017 | 32217 | −327.2 | −1.03 | $E_2$ | 31670 | 219.8 | 0.69 |

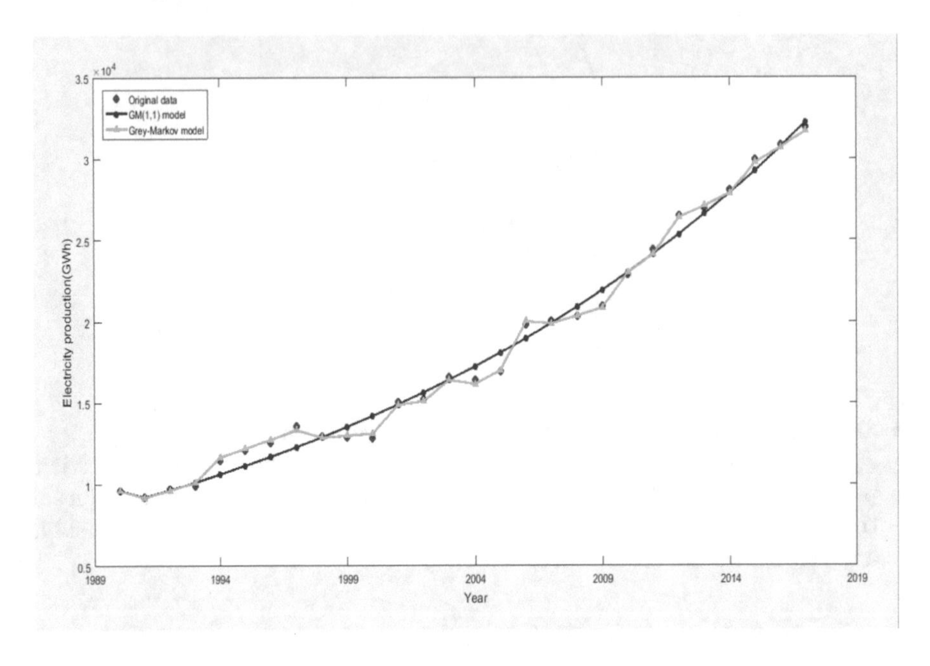


**Fig. 2.** Forecasting curve of the electricity production in Morocco.

Using the Grey-Markov model and basing on the original data, we can also predict the electricity production for the next 13 years. Table 3 presents the electricity production in Morocco during 2018–2030.

**Table 3.** Forecasting electricity production in Morocco from 2018 to 2030.

| Year | Electricity production (GWh) | Year | Electricity production (GWh) |
|---|---|---|---|
| 2018 | 33788.4 | 2025 | 47349 |
| 2019 | 35457 | 2026 | 49687 |
| 2020 | 37208.3 | 2027 | 52140.5 |
| 2021 | 39045.8 | 2028 | 54715 |
| 2022 | 40974 | 2029 | 57416.8 |
| 2023 | 42997.5 | 2030 | 60251.8 |
| 2024 | 45120.8 | | |

### 5.2 Forecasting the Electricity Consumption in Morocco

Table 4 presents the historical data of the electricity consumption in Morocco. Following the same steps, we get the GM(1,1) model

$$x^{(0)}(k) - 0.052z^{(1)}(k) = 8.61, \qquad k = 1, \ldots, n. \tag{5.3}$$

and the restored values are given by:

$$\hat{x}^{(0)}(k+1) = 8.84e^{0.052k}, \qquad k = 1, \ldots, n. \tag{5.4}$$

Using the Sturge formula [8], we can divide the residual errors sequence uniformly into $s$ states, where $s$ is the integer portion of $\frac{ln(n)}{ln(2)}$. In this case, the number of states is $s = 4$ and the state division of the total electricity consumption is the following:

$$E_1 = [\hat{x}^{(0)}(k) - 1.9506, \hat{x}^{(0)}(k) - 1.1483],$$
$$E_2 = [\hat{x}^{(0)}(k) - 1.1483, \hat{x}^{(0)}(k) - 0.3459],$$
$$E_3 = [\hat{x}^{(0)}(k) - 0.3459, \hat{x}^{(0)}(k) + 0.4565],$$
$$E_4 = [\hat{x}^{(0)}(k) + 0.4565, \hat{x}^{(0)}(k) + 1.2588].$$

Applying the Markov chain model, we get the one-step transition probability matrix as follows:

$$P(1) = \begin{pmatrix} 0 & 1 & 0 & 0 \\ 0 & 2/3 & 1/3 & 0 \\ 2/15 & 0 & 2/3 & 1/5 \\ 0 & 0 & 1/2 & 1/2 \end{pmatrix} \tag{5.5}$$

**Table 4.** Total electricity consumption in Morocco during 1990–2015.

| Year | Electricity consumption | Year | Electricity consumption | Year | Electricity consumption |
|---|---|---|---|---|---|
| 1990 | 8.91 | 1999 | 12.9 | 2008 | 23.13 |
| 1991 | 9.39 | 2000 | 14.11 | 2009 | 23.81 |
| 1992 | 10.34 | 2001 | 15.29 | 2010 | 25.10 |
| 1993 | 10.52 | 2002 | 15.79 | 2011 | 27.07 |
| 1994 | 11.2 | 2003 | 17.32 | 2012 | 29.15 |
| 1995 | 11.64 | 2004 | 18.44 | 2013 | 29.72 |
| 1996 | 11.88 | 2005 | 19.32 | 2014 | 30.93 |
| 1997 | 12.52 | 2006 | 21.20 | 2015 | 30.67 |
| 1998 | 13.49 | 2007 | 22.35 | | |

**Table 5.** Prediction results of GM(1,1) model and Grey-Markov model for the electricity consumption(TWh)

| Year | GM(1,1) model | | | State | Grey-Markov model | | |
|---|---|---|---|---|---|---|---|
| | Forecast value | Residual error | Relative error (%) | | Forecast value | Residual error | Relative error (%) |
| 1990 | 8.91 | 0 | 0 | $E_3$ | 8.96 | −0.05 | −0.56 |
| 1991 | 9.32 | 0.07 | 0.75 | $E_3$ | 9.37 | 0.02 | 0.21 |
| 1992 | 9.82 | 0.52 | 5.03 | $E_4$ | 10.67 | −0.33 | −3.19 |
| 1993 | 10.34 | 0.18 | 1.71 | $E_3$ | 10.40 | 0.12 | 1.14 |
| 1994 | 10.90 | 0.30 | 2.68 | $E_3$ | 10.95 | 0.25 | 2.23 |
| 1995 | 11.48 | 0.16 | 1.37 | $E_3$ | 11.54 | 0.10 | 0.86 |
| 1996 | 12.10 | −0.22 | −1.85 | $E_3$ | 12.15 | −0.27 | −2.27 |
| 1997 | 12.74 | −0.22 | −1.76 | $E_3$ | 12.80 | −0.28 | −2.24 |
| 1998 | 13.43 | 0.06 | 0.44 | $E_3$ | 13.48 | 0.01 | 0.07 |
| 1999 | 14.15 | −1.25 | −9.69 | $E_1$ | 12.60 | 0.30 | 2.32 |
| 2000 | 14.91 | −0.80 | −5.67 | $E_2$ | 14.16 | −0.05 | −0.35 |
| 2001 | 15.71 | −0.42 | −2.75 | $E_2$ | 14.96 | 0.33 | 2.16 |
| 2002 | 16.55 | −0.76 | −4.81 | $E_2$ | 15.80 | −0.01 | −0.06 |
| 2003 | 17.43 | −0.11 | −0.63 | $E_3$ | 17.49 | −0.17 | −0.98 |
| 2004 | 18.37 | 0.07 | 0.38 | $E_3$ | 18.42 | 0.02 | 0.11 |
| 2005 | 19.35 | −0.03 | −0.15 | $E_3$ | 19.41 | −0.09 | −0.46 |
| 2006 | 20.39 | 0.81 | 3.82 | $E_4$ | 21.25 | −0.05 | −0.23 |
| 2007 | 21.48 | 0.87 | 3.89 | $E_4$ | 22.34 | 0.01 | 0.04 |
| 2008 | 22.63 | 0.50 | 2.16 | $E_4$ | 23.49 | −0.36 | −1.56 |
| 2009 | 23.85 | −0.04 | −0.17 | $E_3$ | 23.90 | −0.09 | −0.38 |
| 2010 | 25.12 | −0.02 | −0.08 | $E_3$ | 25.18 | −0.08 | −0.32 |
| 2011 | 26.47 | 0.60 | 2.22 | $E_4$ | 27.33 | −0.26 | −0.96 |
| 2012 | 27.89 | 1.26 | 4.32 | $E_4$ | 28.75 | 0.40 | 1.37 |
| 2013 | 29.39 | 0.33 | 1.11 | $E_3$ | 29.44 | 0.28 | 0.94 |
| 2014 | 30.96 | −0.03 | −0.1 | $E_3$ | 31.02 | −0.09 | −0.29 |
| 2015 | 32.62 | −1.95 | −6.36 | $E_1$ | 31.07 | −0.40 | −1.3 |

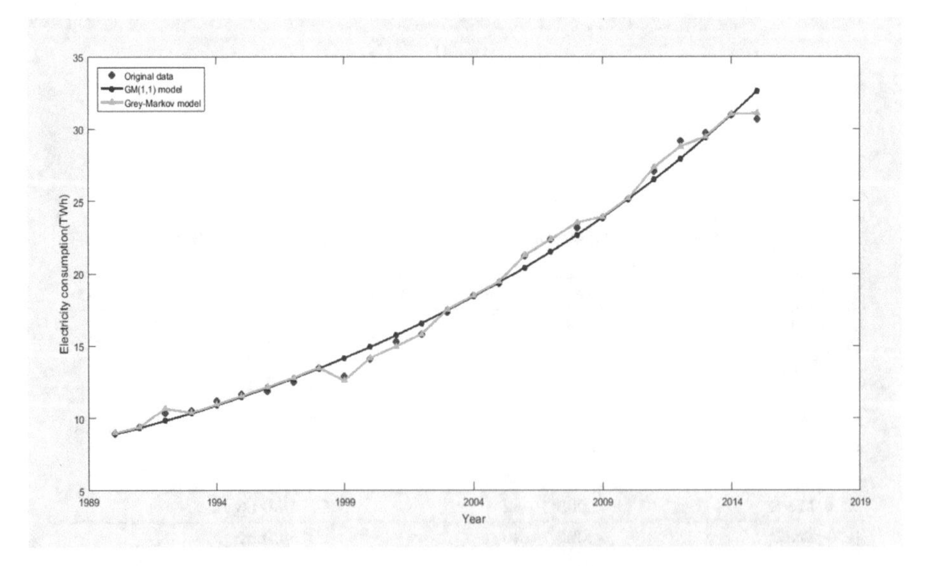


**Fig. 3.** Forecasting curve of the electricity production in Morocco.

**Table 6.** Forecasting electricity consumption in Morocco from 2016 to 2030.

| Year | Electricity consumption (TWh) | Year | Electricity consumption (TWh) |
|---|---|---|---|
| 2016 | 33.62 | 2024 | 51.44 |
| 2017 | 35.46 | 2025 | 54.24 |
| 2018 | 37.40 | 2026 | 57.18 |
| 2019 | 39.45 | 2027 | 60.29 |
| 2020 | 41.60 | 2028 | 63.56 |
| 2021 | 43.87 | 2029 | 67.01 |
| 2022 | 46.27 | 2030 | 70.64 |
| 2023 | 48.78 | | |

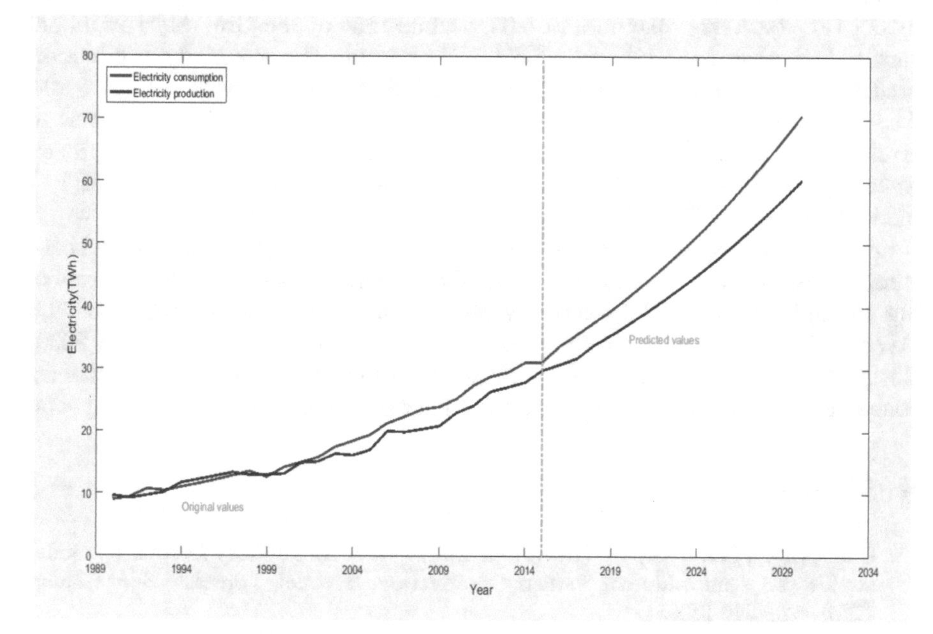


**Fig. 4.** Electricity production and consumption forecasts.

Therefore, using the matrix $P(1)$, we obtain the predicted values of the electricity consumption during 1990–2015, as listed in Table 5. The Grey-Markov model can also be applied to forecast the Moroccan electricity consumption from 2016 to 2030 (See Table 6).

## 6 Discussion and Conclusion

In the present work, we utilized the Grey-Markov model for forecasting the electricity production and consumption in Morocco. The Grey-Markov model is a combination of the GM(1,1) grey model and the Markov chain model, which gives more precision and better forecast results.

This research confirms the efficiency of the proposed model comparing to the GM(1,1) model, and that was obvious due to the good practical application results. This prediction was based on the analysis of the historical data from 1990 to 2017 in order to give forecast values for the next years, till 2030.

Examination of Tables 1 and 4 shows the high fluctuation of the data which affect negatively the forecasting performance and the accuracy of the model. According to Figs. 2 and 3, we can clearly deduce that the predicted values by the Grey-Markov model are very close to the original values. Tables 2 and 5 also highlight the difference between the two models which is indicated by the residual and relative errors of both models.

For the electricity consumption forecasting, the relative errors of the GM(1,1) model vary between −9.69% and 5.03% while that of the Grey-Markov model vary only between −3.19% and 2.32%. Concerning the electricity production prediction, the relative errors of the GM(1,1) model vary between −10.41% and 9.58 % and that of the Grey-Markov model don't exceed 1.94%. The average error is also reported to assure the great accuracy of the proposed approach since it is found to be 0.07 % for the Grey-Markov model while for the GM(1,1) model it equals 0.29% which makes the forecast precision lower.

Finally, it is easy to remark that the Moroccan electricity production and consumption are expected to know a noticeable growth form 2018 to 2030. According to Tables 2 and 4, the electricity production is predicted to reach 60259.8 GWh and the electricity consumption is expected to exceed 70 TWh in 2030. Basing on these results and the Fig. 4, we can obviously see that the electricity consumption will widely surpass the electricity production in the next few years.

## References


1. Chang, S.C., Lai, H.C., Yu, H.C.: A variable P value rolling Grey forecasting model for Taiwan semiconductor industry production. Technol. Forecast. Soc. Chang. **72**(5), 623–640 (2005)
2. Chen, C., Chen, H.L., Chen, S.: Forecasting of foreign exchange rates of Taiwan's major trading partners by novel nonlinear Grey Bernoulli model NGBM(1,1). Commun. Nonlinear Sci. Numer. Simul. **13**(6), 1194–204 (2008)
3. Claveria, O., Torra, S.: Forecasting tourism demand to catalonia: neural networks vs. time series models. Econ. Model. **36**, 220–228 (2014)
4. Deng, J.L.: Control problems of grey systems. Syst. Control. Lett. **1**(5), 288–94 (1982)
5. Ding, S., Hipel, K.W., Dang, Y.G.: Forecasting China's electricity consumption using a new grey prediction model. Energy **149**, 314–328 (2018)
6. Hu, Y.C., Jiang, P., Chiu, Y.J., Tsai, J.F.: A novel grey prediction model combining Markov chain with functional-link net and its application to foreign tourist forecasting. Information **8**, 126 (2017)
7. Hu, S.: Prediction of city traffic accidents based on grey Markov chain model. Revista de la Facultad de Ingenieria U.C.V. **32**(4), 144–151 (2017)

8. Imbusch, G.F., Yen, W.M.: The McCumber and sturge formula. J. Lumin. **85**(4), 177–179 (2000)
9. Chen, X., Jiang, K., Liu, Y.: Inflation prediction for China based on the Grey Markov model. IEEE (2015)
10. Kazemi, A., Modarres, M., Mehregan, M.R., Neshat, N., Foroughi, A.: A Markov chain grey forecasting model: a case study of energy demand of industry sector in Iran. In: 3rd International Conference on Information and Financial Engineering, vol. 12, Singapore (2011)
11. Kumar, U., Jain, V.K.: Time series models (Grey-Markov, Grey Model with rolling mechanism and singular spectrum analysis) to forecast energy consumption in India. Energy **35**, 1709–1716 (2010)
12. Liu, S., Lin, Y.: Grey Information: Theory and Practical Applications. Advanced Information and Knowledge Processing. Springer (2006)
13. Madhugeeth, K.P.M., Premaratna, H.L.: Forecasting power demand using artificial neural networks for sri lankan electricity power system. In: IEEE Region 10 Colloquium and the Third International Conference on Industrial and Information Systems, Kharagpur, India (2008)
14. Shouse, K.: Activity recognition using Grey-Markov model, thesis, Western Carolina University (2011)
15. Sun, X., Sun, W., Wang, J., Zhang, Y., Gao, Y.: Using a Grey-Markov model optimized by Cuckoo search algorithm to forecast the annual foreign tourist arrivals to China. Tour. Manag. **52**, 369–379 (2016)
16. Suo, R.: The energy consumption prediction model based on Grey-Markov model. Int. J. Acad. Res. Reflect. **5**(5) (2017). ISSN 2309-0405
17. Wei, S., Yanfeng, X.: Research on China's energy supply and demand using an improved Grey-Markov chain model based on wavelet transform. Energy, 1–16 (2016)
18. Wendong, Z.: Analysis of sports training injury measurement with Grey Markov model (GMM). In: 10th International Conference on Measuring Technology and Mechatronics Automation (2018)
19. Yan, Q., Qin, C., Nie, M., Yang, L.: Forecasting the electricity demand and market shares in retail electricity market based on system dynamics and Markov Chain. Math. Probl. Eng. (2018)
20. Yong, H.: A new forecasting model for agricultural commodities. J. Agric. Eng. Res. **60**, 227–35 (1995)
21. Zhou, D., Yu, Z., Zhang, H., Weng, S.: A novel grey prognostic model based on Markov process and grey incidence analysis for energy conversion equipment degradation. Energy **109**, 420–429 (2016)
22. http://www.one.ac.ma

# Smart Monitoring System for the Long-Term Control of Aerobic Leachate Treatment: Dumping Case Mohammedia (Morocco)

J. Mabrouki(✉), I. Bencheikh, K. Azoulay, M. Es-soufy, and S. El Hajjaji

Laboratory of Spectroscopy, Molecular Modeling, Materials, Nanomaterial, Water and Environment, CERNE2D, Faculty of Science, Mohammed V University in Rabat, Avenue Ibn Battouta, BP1014 Agdal, Rabat, Morocco
jamalmabrouki@gmail.com

**Abstract.** Intensive aeration can give an abnormal state of handling and conceivably present a lot of adaptability in their methods of activity, which on a fundamental level enables the administrator to set technique to work to convey the ideal treatment objective. By and by, be that as it may, Intensive Aeration is frequently abused utilizing basic stages with fixed time interims. Propelled control systems can be utilized to enable aeration to understand its full handling potential and advance its abilities, and this archive depicts a few online measures and control techniques that can be utilized for this reason. A contextual investigation is displayed. It indicates how a procedure demonstrate was first used to test three control alternatives for the city's leachate treatment plant in Mohammedia, Morocco. It was then trailed by two field preliminaries and the resulting execution of one of them to expand its preparing limit.

**Keywords:** Control · Dissolved oxygen · Respirometry · Intensive aeration

## 1 Introduction

Population growth, the improvement of the quality of life and the high density of urban areas, new forms of water pollution are generated. Indeed, the burial and storage of solid waste must not only allow efficient waste management, but also the treatment after drainage and recovery of the two effluents that are biogas and leachate. The storage center located along the P3313 road in the rural municipality of Chaâba El Hamra in Beni Yakhlef, Mohammedia, Morocco, receives household waste and similar waste from different activities. The Mohammedia landfill is based on the rational storage of solid waste in order to avoid any risk of harm to human health and the environment [1].

Intensive aeration is broadly utilized as an adaptable and economical procedure for the organic treatment of leachate. The escalated air circulation process is regularly worked by a fixed calendar involving a progression of stages: filling, response, decantation, sitting [2–4]. The type and duration of each phase (or combination of phases) may vary


Y. Farhaoui (Ed.): BDNT 2019, LNNS 81, pp. 220–230, 2020.
https://doi.org/10.1007/978-3-030-23672-4_17

depending on the treatment, which - in principle at least – provides almost unlimited flexibility in the way Intensive aeration treats incoming waste. This should allow the Intensive aeration to provide the optimal level of treatment for any given waste. Intensive aeration is the most controllable process option for wastewater treatment [5]. However, in practice, the phases and schedules are usually set, which may be partly attributed to the lack of measures and control algorithms that use this inherent "controllability".

Since Intensive aeration has the potential for great flexibility in its operation, what measures can be done to optimize their operation and how can they be used. This paper presents several ideas for advanced control options and provides a real case study for applying one of the ideas. The key procedure factors are the utilization of basic and reasonable online procedure estimations to derive the centralization of concoction factors, which are troublesome or costly to quantify straightforwardly (NH4 +, NOx). What's more, PO43-). It is commonly concurred that the exchanging grouping ought to be adjusted to the real burden utilizing pH, ORP, and broke up oxygen (DO) as backhanded procedure pointers [6, 7].

The leachate produced in the old landfills is an exceptionally safe wastewater portrayed by a low BOD/TKN proportion. Thusly, the evacuation of nitrogen must be accomplished if an outer wellspring of biodegradable COD is accommodated the denitrification procedure. The motivation behind this investigation is to build up a powerful and solid observing apparatus to change the stage term and expansion of COD in ROSs that manage landfill leachates.

## 2 Related Work

### 2.1 Parameters Control Approaches

The assessing the pressure driven limit is by and large comprehended by architects and in many reactors and Control frameworks are very much intended to adapt to the stream varieties. In any case, in not many cases are modifications made to the timings of the React stage for better utilization of organic limit or alterations made Task Settle and Decant to represent changes in sedimentation qualities of initiated muck. The creators propose the accompanying 7 control ideas for Thought for cutting edge control of Intensive air circulation, a significant number of which have substantiated themselves to the pilot or to the full scale:

In-situ optical solids test to gauge blended alcohol suspended solids (MLSS) focuses and as an input estimation to control the tap speed

- Oxygen Uptake Rates (OUR) to decide when vigorous treatment is finished;
- The utilization of Oxidation Reduction Potential (ORP) and pH to demonstrate finishing of different phases of treatment (anoxic, anaerobic, oxygen consuming);
- Online phosphate analyzer to progressively gauge phosphorus discharge and take-up;
- Disintegrated Oxygen (DO) estimation to by implication measure OUR;
- Rate of progress of DO to show total treatment;
- Online smelling salts and nitrate analyzers to legitimately measure nitrification and denitrification rates and to demonstrate when treatment stages are finished;

Fundamentally there are three restricting criteria in the working of an Intensive air circulation which are:

- Pressure driven limit. The Intensive air circulation must deal with the progression of leachate, including their fluctuation. The distinction between the least dimension of settling and the most elevated amount filling adequately directs the measure of leachate that can be leachate in each parcel. On the off chance that the pinnacle stream surpasses that can be taken care of in one clump then the sequencing of Concentrated air circulation must be moved quicker by utilizing shorter stage times or changing the quantity of stages enable all the more parcels to process.
- Organic limit. Contingent upon the sort and dimension of treatment required, adequate biomass must be kept up in the reactor (i.e., satisfactory slop age) with adequate response time in every one of the response stages to give expulsion of substrate and required supplements.
- Settlement rate. As recently showed, sedimentation attributes of initiated slop are hard to foresee and control. The sedimentation rate of initiated slop directly affects the time that must be took into consideration. Set up and draw stages. The sedimentation bend likewise confines how much the decanter can go down before the solids are drawn in the profluent.

## 2.2 Intensive Aeration Technology

The Intensive aeration process is effective for young leachates whose $BOD_5$/COD and COD/P ratios are respectively greater than 0.5 and 300 [8, 9]. The quantities of air to be injected have been studied by some authors. Thus, Labbe (1996) reported the use of 16.5 and 11.25 kg $O_2$/h in 500 $m^3$ basins installed in series for an initial COD of less than 10,000 mg/L. The reduction of this parameter was 40% in the first basin and 42.86% in the second. According to Rocher et al. [10], injection of 30 m3 of air per kilogram of BOD5 in wastewater is sufficient to remove 1.5 to 2 kg of BOD5 per cubic meter per day [11] recommended the use of 6.45 kg of oxygen per cubic meter per hour. The compilation of these data led to the use of a high flow of oxygen to achieve good purification yields. Thus for the test presented in this article, the choice was made to inject 7 L of air per minute into 20 L of leachates, corresponding to 6.3 kg of oxygen per cubic meter per hour.

Research facility scale treatment leachate (20 L work volume) created in a landfill in Mohammedia, Morocco, with a total cycle comprising of a progression of 4 sub-cycles (filling, anoxia). What's more, oxic respond), trailed by one hour of settling. The subcycle arrangement utilized in this examination is run of the mill for concentrated wastewater. In light of the leachate qualities utilized in this investigation (Table 1), sodium acetic acid derivation was added amid the anoxic stage to give a biodegradable COD to denitrification. The framework was outfitted with pH, ORP and DO sensors (WTW, Weilheim, Germany). More details of the treatment system are reported in Spagni et al. [12].

The fundamental tasks performed by the observation framework were as per the following:

- End-of-stage acknowledgment and exchanging: this spares time contrasted with the fixed time exchanging plan. End-of-stage tasks included on/off exchanging of air circulation and blending;
- The finish of the cycle incorporated the activity of the emanating extraction and muck treatment siphons;
- Addition of natural carbon (acetic acid derivation) for nitrification.

## 3 Results of Analysis and Discussion

### 3.1 Characterization Results of the Mohammedia Leachate

The results obtained at the Mohammedia city controlled landfill site were compared with those obtained at other landfills and showed that the metal composition of the landfill leachates is typical of a dominant household landfill. Moreover, the calculation of the BOD5/COD ratio provides information on the landfill fermentation step. Applied to the leachates observed at the Mohammedia landfill, the BOD5/COD ratio gives the value of 0.40 indicating a young discharge that is not yet stabilized, corresponding to the acid phase of anaerobic degradation (Table 1).

**Table 1.** Main leachate characteristics and variability, with n representing the number of samples.

| | Unit | Max | Min | SD | n |
|---|---|---|---|---|---|
| pH | – | 8.7 | 7.55 | 0.50 | 18 |
| CODt | mg/L | 9560 | 528 | 720 | 18 |
| CODf | mg/L | 3940 | 440 | 685 | 18 |
| BOD5 | mg/L | 1500 | 30 | 526 | 5 |
| TKN | mgN/L | 2250 | 252 | 452 | 17 |
| NH4+-N | mgN/L | 1519 | 167 | 525 | 18 |
| Ptot | mgP/L | 94 | 78 | 2.4 | 15 |

### 3.2 Result of Conventional Aeration Treatment

Aeration caused a significant and continuous decrease in total nitrogen. During the first ten days, total nitrogen marked an increased fall from 1.260 to 580 mg/L, a reduction of 55.2%. And since the concentration of ammonium ($NH^{4+}$), nitrites ($NO^{2-}$) and nitrates ($NO^{3-}$) remains insignificant, even negligible, compared to the total amount of nitrogen removed, organic nitrogen (N-Org) is probably transformed into volatile nitrogen elements (NH3, N2 …) [13, 14]. Given the acid pH value (4.6) during this period (first 10 days), the assumption of the transformation of organic nitrogen into ammonium $NH^{4+}$ then ammonia NH3 can be excluded [9, 15] because in acidic medium, there is absence of the OH- ions that react with ammonium to give ammonia (Fig. 1).

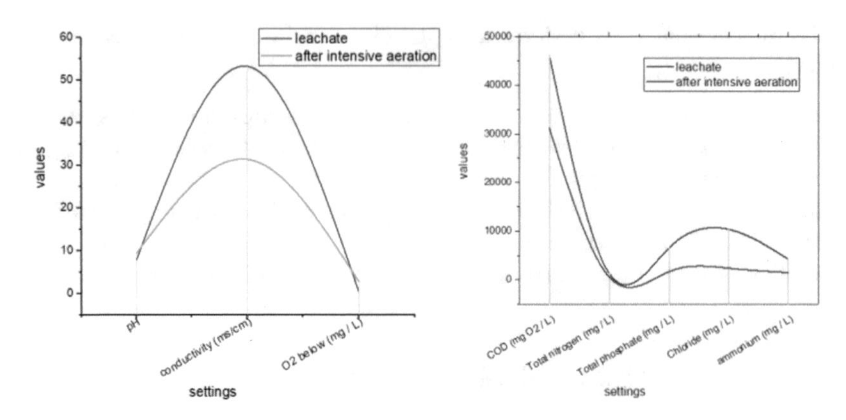


**Fig. 1.** Effect of intensive aeration of leachates

### 3.3 Developed and Structure of the Monitoring System

The Mohammedia landfill incorporates an organic treatment step intended to treat day by day stream rates, trailed by filtration and UV cleansing. The city of Mohammedia is a locale in full advancement, which causes the expansion of influent progression of the plant in a stupendous manner. Subsequently, the plant was drawing nearer to structure capacity. So as to meet these developing needs stream for the time being, studies have been directed to create alternatives to upgrade and augment treatment capacity of Intensive air circulation past their current plan ability utilizing the standards of "intelligent Intensive air circulation innovation".

The process signals (DO, pH and ORP) were acquired through an Arduino Uno board and monitored by a local PC with TCP/IP Internet connection. To close the feedback loop, a group of solid state switches, driven by the DAQ digital outputs, controlled the peristaltic pumps (feed, effluent extraction and sludge waste), agitator and aerator. The complete monitoring system is illustrated in Fig. 2. The monitoring and control system has been developed, which provides all the necessary data acquisition and processing functions and also enables remote operation.

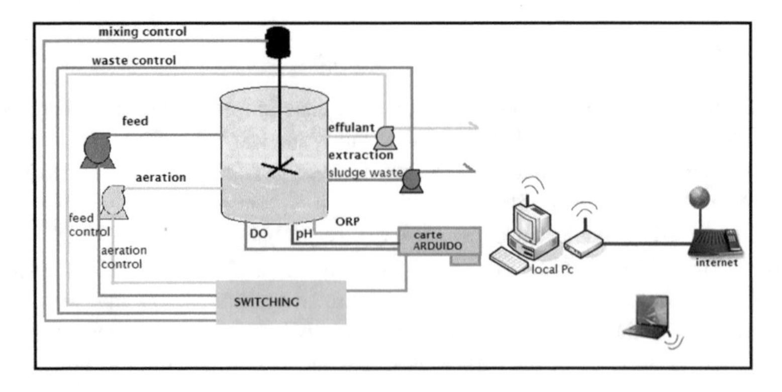


**Fig. 2.** Monitoring and control system designed around the pilot plan

### 3.4 Process Modeling

As a major aspect of the examination to misuse the Intensive Aeration of Mohammedia, the idea of integrant Intensive Aeration Process Control was tried out of the blue utilizing a dynamic procedure model of the plant on the Hydromantis GPS test system. - X. The recreations gave great outcomes and control approaches dependent on three distinct estimations were built up: ORP, disintegrated oxygen (DO) and rate of progress of DO (or “dO/dt”).

The presence of critical procedure designs in the Intensive Aeration cycle and its identification by computerized reasoning calculations have been broadly illustrated [7, 16–18]. Be that as it may, at no other time these highlights have been consolidated in an independent observing framework considered for long haul unattended activity. After an itemized examination the most pertinent practices showing the finish of the oxygen consuming stages on account of nitrogen expulsion were characterized [17]. On the off chance that phosphorus evacuation is likewise required an all-inclusive rundown of pointers can be characterized [19]. From Table 2 it creates the impression that all the applicable pointers are made out of flag subsidiaries, consequently the need to channel the procedure information and infer them in a numerically vigorous manner.

**Table 2.** Significant process indicators used for data programming.

| Phase | End of process | Indicator |
|---|---|---|
| Aerobic | Nitrification | • Sharp DO increase<br>• Ammonia valley<br>• ORP discontinuity<br>$\frac{dpH}{dt} \to 0 + \frac{dDO}{dt} \to 0$ |

The utilization of the second ORP subordinate is supported by the need of distinguishing the significant “nitrate knee” brokenness, denoting the consumption of nitrate and the progress from high-impact conditions. These conditions are featured in Fig. 4, appearing operational record from the pilot procedure considered in this investigation.

### 3.5 Monitoring System

The checking framework comprises of various progressive tasks on the information, as appeared in Fig. 4. Following the procurement, the information are approved and detached utilizing a wavelet channel, at that point an advanced induction is performed. The subsequent choice to end the stage enacts the comparing actuators, along these lines shutting the control circle. The arrangement of tasks in Fig. 4 is presently quickly surveyed, despite the fact that for reasons of room, not all strategies can be depicted in detail. The intrigued peruse will allude to Marsili-Libelli (2006) for a nitty gritty depiction of wavelet separating and standards of the fluffy induction framework, despite the fact that in this application a phonetic methodology was pursued rather than

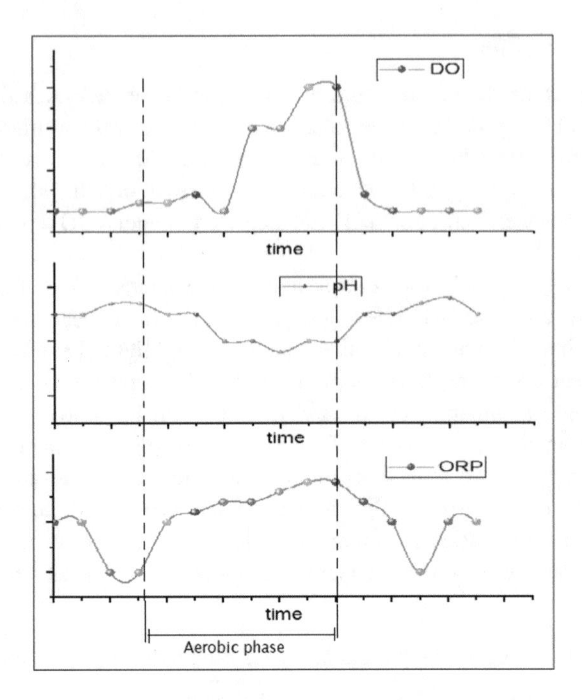


**Fig. 3.** Relevant diagrams indicating the end of the aerobic phase.

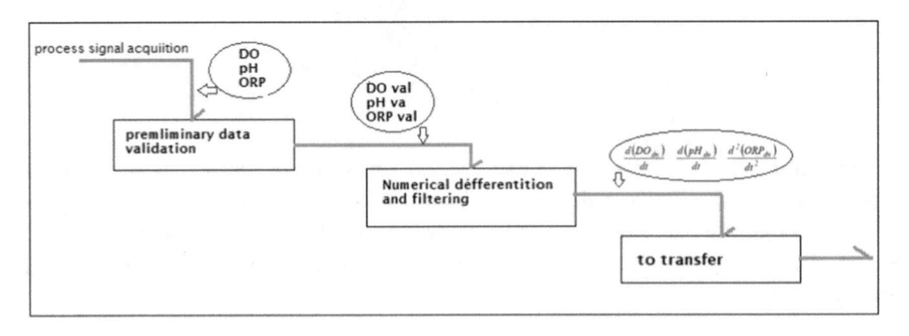


**Fig. 4.** Structure of the monitoring system.

the order system fluffy utilized in this article. Another fluffy application in control SBRs is portrayed in Bae et al. 2006.

### 3.6 Full Scale Testing

In the wake of demonstrating the Intensive Aeration control approach in a test system, it was then effectively tried for multi week on one of the Intensive Aeration of Mohammedia. Figure 3 demonstrates the aftereffects of one cycle of the trial. Amid the response period of this cycle, extra estimations were taken for smelling salts and the OD was physically approved utilizing the versatile test of the plant. Toward the finish of the response period of this cycle, it was obvious that smelling salts fixations were falling by about 20 mg/L and an extra 50 min of air circulation would most likely enable Intensive Aeration to achieve endogenous opposition; the response stage was along these lines reached out by 50 min to test the Intensive Aeration control which stops the fans when the OD surpasses 500 mg/L.

In Fig. 3, it tends to be seen that the quick increment in OD happens when the alkali fixation achieves zero, as anticipated by the model and depicted in the standards of Intensive Aeration. This figure likewise demonstrates that the ORP bend does not react as fast as they DO bend. At long last, it very well may be found in this assume the on-line estimation of the OD dependent on fluorescent innovation gave readings like those of the convenient galvanic test.

The full-scale test was exceptionally compelling utilizing either the DO approach or the dO/dt approach. The Intensive Aeration control was proceeded past the underlying one-week preliminary and both Intensive Aeration have since been designed to utilize this control framework. A tributary evening out tank and extra ventilation limit were introduced and, together with the Intensive Aeration control, diminished the plant to 30 ML/d at 40 ML/d.

### 3.7 Limitations and Constraints of Control

The thoughts introduced in this archive are straightforward and may appear glaringly evident to many. It is in this way important to inquire as to why not many expansive scale establishments utilize the thoughts exhibited here and the greater part of the examinations are constrained to research center tests.

A noteworthy issue in actualizing one of the thoughts is the limitations forced on a few Intensive Aeration by designers anxious to give savvy arrangements by sharing hardware between reactors, yet subsequently expelling a significant part of the adaptability inalienable in basic Intensive Aeration frameworks (e.g. in research facilities). Since bowls need to share and organize the utilization of fans or focal siphons, the operational stages are restricted to permit the movement of hardware between bowls. Lamentably, for frameworks with shared siphons and fans (most of substantial scale Intensive Aeration for residential applications), it is impossible that they will be intended to discharge the mutual hardware strings and the task of the gear. Propelled control will fundamentally be muddled and to some degree compelled.

Another entanglement is the trouble of dealing with a group ceaseless inundation stream. Whenever, somewhere around one of the Intensive Aeration bunches must get the approaching stream in the establishment, which restricts an opportunity to guarantee that the volume is as yet accessible for the approaching stream. Two methodologies can be embraced to conquer this confinement and permit more prominent operational adaptability. Either an influent leveling tank can be utilized to keep up info power, or Intensive Aeration can be planned with steady power. Leveling of tributaries is most likely the most ideal approach to boost ground-based limit and propelled control potential. In Mohammedia's landfill, for instance, it was conceivable to twofold the limit of Intensive Aeration by utilizing an influent leveling tank.

### 3.8 Preliminary Data Validation (PDV) Algorithm

The way to fruitful observing is the capacity to choose whether the information procured is noteworthy and conceivably to secure right obtaining or sensor blunders, to give a primer information approval calculation. This method frames the premise of the reconnaissance framework and depends on an educated correlation between the last example obtained and the recently approved information. PDV will in general regularize the information by expelling abrupt and unexplained varieties between nearby examples. A choice variable is characterized as the supreme distinction between the last example up and the effectively approved the past one.

Approval of the information as indicated by which one of the accompanying activities is embraced:

- The example up is acknowledged without adjustments.
- The approved example is acquired as a weighted normal.
- The example is dismissed and supplanted by the past approved.

This methodology is made with the PDV calculation as in it is desirable over maintain a strategic distance from PDV and straightforwardly forms the information with the wavelet channel in perspective on their deduction. Further, wavelet separating is over and again connected after every deduction, to smooth out the numerical commotion. The relationship among PDV wavelet sifting and numerical inference, executed with second-request focal contrasts.

## 4 Conclusions

A long haul checking arrangement of the Intensive air circulation process, in light of man-made reasoning ideas, was structured and created to control an Intensive Aeration pilot plant of 20 L both locally and remotely. The inferential controller depends on a calculation and starter condition of information approval and numerical separation and a fluffy inferential framework to decide the most suitable control activity, comprising of oxygen consuming stage exchanging.

A few estimation and control thoughts were displayed that utilize this adaptability and "Controllability", including a solid case of how the Intensive Aeration Control approach utilizing a straightforward DO, ORP and pH test was utilized on the Intensive air circulation of Mohammedia.

The framework has been actualized in the Hydromantis GPS-X programming stage and can be worked by a remote station utilizing the stage's local web distributing devices. One month of ceaseless unattended task with right stage identification more noteworthy than 85% exhibited the vigor of the observing framework, equipped for controlling the procedure through regular temperature changes and a few power changes.

## References


1. Mabrouki, J., El Yadini, A., Bencheikh, I., Azoulay, K., Moufti, A., El Hajjaji, S.: Hydrogeological and hydrochemical study of underground waters of the tablecloth in the vicinity of the controlled city Dump Mohammedia (Morocco). In: Ezziyyani, M. (ed.) AI2SD 2018. AISC, vol. 913, pp. 22–33. Springer, Cham (2019). https://doi.org/10.1007/978-3-030-11881-5_3
2. Wilderer, P. Irvine, R.L. Goronszy, M.C.: Sequencing batch reactors technology. IWA Scientific and Technical report n. 10, IWA Publ. London (2001)
3. Artan, N., Wilderer, P., Orhon, D., Morgenroth, E., Özgür, N.: The mechanism and design of sequencing batch reactor systems for nutrient removal – the state of the art. Wat. Sci. Tech. **43**, 53–60 (2001)
4. Artan, N., Orhon, D.: Mechanisms and design of sequencing batch reactors for nutrient removal. IWA Scientific and Technical report n. 19, IWA Publ. London (2005)
5. Shaw, A.R., Watts, J.B.: The use of respirometry for the control of sequencing batch reactors: principles and practical application. In: WEFTEC 2002, Proceedings of Water Environment Federation 75th Annual Conference, Chicago, Illinois, September 2002
6. Pavšeli, N., Hvala, N., Koijan, J., Roš, M., Šubelj, M., Mušič, G., Strmčnik, S.: Experimental design of an optimal phase duration control strategy used in batch biological wastewater treatment. ISA Trans. **40**, 41–56 (2001)
7. Spagni, A., Buday, J., Ratini, P., Bortone, G.: Experimental considerations on monitoring ORP, pH, conductivity and dissolved oxygen in nitrogen and phosphorus biological removal processes. Wat. Sci. Tech. **43**(11), 197–204 (2001)
8. Labbe, H.: Essai de caractérisation et de valorisation d'un lixiviats d'ordures ménagères. Thèse INP Toulouse-France, 263 (1996)
9. Ekama, G.A., Söte mann, S.W., Wentzel, M.C.: Biodegr adability of activated sludge organics under anaerobic conditions. Wat. Res. **41**, 244–252 (2007)
10. Rocher, J.L., Crimins, J.M., Amatrudo, M.S., Kinson, M.A., Todd-Brown, J., Lewis, J.I.: LuebkeStructural and functional changes in tau mutant mice neurons are not linked to the presence of NFTs. Exp. Neurol. **223**, 385–393 (2008)
11. Vasel, J.L.: Cours du transfert de masse et sonapplication dans le traitement de l'eau. Master spécialisé chimie et microbiologie de l'eau. Faculté des sciences d'Agadir (2012)
12. Spagni, A., Lavagnolo, M.C., Scarpa, C., Vendrame, P., Rizzo, A., Luccarini, L.: Nitrogen removal optimization in a Sequencing Batch Reactor treating sanitary landfill leachate. J. Environ. Sci. Health (2007, Submitted)
13. Anonymeb: Secrétariat d'Etat chargé de l'Environnement, Maroc. Rapport sur l'Etat de l'Environnement du Maroc, chapitre IV: Déchets (2001)
14. Jupsin, H., Praet, E., Vasel, J.-L.: Caractérisation des lixiviats de CET et modélisation de leur évolution. In: Proceedings of International Symposium on Environmental Pollution Control and Waste Management 7–10 January 2002, Tunis (EPCOWM 2002), pp. 884–896 (2002)
15. Hakkou, R.: La décharge publique de Marrakech: caractérisation des lixiviats, étude de leur impact sur les ressources en eau et essais de leur traitement. L'université Cadi Ayyad faculté des sciences et techniques Marrakech. Thèse de Doctorat d'État EsScience. N° d'ordre: 23/C271, 141 (2000)
16. Luccarini, L., et al.: Soft sensors for control of nitrogen and phoshorus removal from wastewaters by neural networks. Wat. Sci. Tech. **45**(5), 101–107 (2001)
17. Marsili-Libelli, S.: Control of SBR switching by fuzzy pattern recognition. Wat. Res. **40**, 1095–1107 (2006)

18. Sin, G., Insel, G., Lee, D.S., Vanrolleghem, P.A.: Optimal but robust N and P removal in SBR's: a model-based systematic study of operation scenarios. Wat. Sci. Tech. **50**(10), 97–105 (2004)
19. Bae, H., Seo, H.Y., Kim, S., Kim, Y.: Knowledge-based control and case-based diagnosis based upon empirical knowledge and fuzzy logic for the SBR plant. Wat. Sci. Tech. **53**(1), 217–224 (2006)

# The Reality and Future of the Secure Mobile Cloud Computing (SMCC): Survey

Samaher Al_Janabi(✉) and Nawras Yahya Hussein

Department of Computer Science, Faculty of Science for Women (SCIW), University of Babylon, Babylon, Iraq
samaher@uobabylon.edu.iq

**Abstract.** Mobile communication and mobile computing has revolutionized the way subscribers across the globe leverage and use a variety of services on the go. Mobile devices have evolved from mere simple handheld mobile phones that enabled voice calls only a few years back to smartphone devices that enable the user to access applications and value-added services anytime, anyplace. All of these, together with the recent reigniting of cloud computing technology, has spearheaded into the opportunistic and economically viable Mobile Cloud Computing (MCC) as the next innovative technology and ultimately a *utility* wave such as electricity or water. But this has created huge loopholes in facilitating reliable, scalable, performance oriented, safe, secure and auditable trusted services at affordable prices and profitability. Thus, this research proposal aims to create a fully-fledged Mobile Commerce-based Safe Mobile Cloud Computing (MCSMCC) framework to specifically accommodate Safe MCC (SMCC) with economic pricing models and e-trading schemes. As for the safety measures comprising of security, privacy, trust, identity management, audit and digital forensics among other related measures to suit SMCC, the objectives are to define and specify the necessary processes and protocols for secure e-trading, These objectives are further complimented by further defining and specifying SMCC pricing schemes that will be fair and equitable to users and vendors alike, and encourage critical mass market uptake and usage with confidence. Both of these central objectives will be experimented through simulation and prototyping to test and verify their purpose, effectiveness and efficiency.

**Keywords:** Cloud Computing (CC) · Mobile Cloud Computing (MCC) · Open Cloud Computing Federation (OCCF) · Mobile Commerce-based Safe Mobile Cloud Computing (MCSMCC)

## 1 Introduction

The merging, splitting and re-merging of Information Communication Technology (ICT) in a multi-faceted of ways, whether it is hardware, software and data communications or their combinations thereof, is so obvious in our lives. The growth of mobility with ICT has changed our lives fundamentally in an unprecedented way.


Y. Farhaoui (Ed.): BDNT 2019, LNNS 81, pp. 231–261, 2020.
https://doi.org/10.1007/978-3-030-23672-4_18

The ubiquitous mobile phone is an example of this phenomenon that it has become an external body part of every human being who has encountered it.

The market of mobile phones is on an all-time high. According to the International Data Corporation (IDC) [1], the premier global market survey and analyst firm, the worldwide smartphone shipments are forecast to grow 40.0% year over year to more than 1 billion units during 2014. According to Cisco Internet Business Solutions Group (IBSG) authors [2], an excess of 80% of the world's population has access to the mobile phone and new devices like the iPhone, Android smartphones, PDAs, palmtops and tablets which have brought a host of mobile computing applications (apps) at the palms of people's hands.

Simultaneously, Cloud Computing (CC) has emerged as a reenergized phenomenon that represents a rudimentary way by which IT services and functionality are charged for and delivered. It is seeing the beginning of "something" that is expected to grow in leaps and bounds. NIST *(National Institute of Standards and Technology, USA)* released in its September 2011 "Special Publication 800-145" [3], the definition of Cloud Computing as follows:

"Cloud Computing is a model for enabling convenient, on-demand network access to a shared pool of configurable resources (e.g. networks, servers, storage, applications and services) that can rapidly be provisioned and released with minimal management effort or service provider interaction".

A more formal definition that encapsulates the key benefits of cloud computing from a business-oriented perspective as well as its unique features from a technological perspective by Sean Martson et al. promulgated [4]:

"It is an information technology service model where computing services (both hardware and software) are delivered on-demand to customers over a network in a self-service fashion, independent of device and location. The resources required to provide the requisite quality-of service levels are shared, dynamically scalable, rapidly provisioned, virtualized and released with minimal service provider interaction. Users pay for the service as an operating expense without incurring any significant initial capital expenditure, with the cloud services employing a metering system that divides the computing resource in appropriate blocks".

Unlike conventional mobile computing paradigms and technologies, the resources in mobile cloud computing are virtualized and assigned in a group of numerous distributed computers rather than local computers or servers. To name a few, many applications based on Mobile Cloud Computing (MCC), such as Google's Gmail, Maps and Navigation systems for mobile, Google's Drive online storage, Voice Search, MobileMe from Apple, LiveMesh from Microsoft and Motoblur from Motorola, HomeSync from Samsung and various other applications on the popular Android platforms, have been developed and served to users at rates which even the user cannot cope with.

However, delivering cloud services in a mobile environment onsets numerous challenges and problems. Mobile devices cannot handle complicated applications due to their inherent characteristics and distinctive drawbacks, such as, it is impossible that a mobile device is always online because it is either powered off, the battery depleted, or out of the wireless communication range. In an MCC environment the offline solution of the mobile device needs to be considered if the service's offerings are to appear seamless. The absence of standards, business and pricing models, security and privacy, together with other safety measures such as trust, identity management, digital forensics and audit trails, system and operational management, and the ever-evasive elastic mobile application's requirements are to a large extend impeding the development of MCC. This paper is one tiny novel step in that direction to understand the challenges and to devise a program for producing working solutions which are essential on many fronts that potentially has many benefits and spin-offs.

This paper introduces the basic model of MCC, its background and related works comprising of key technologies, challenges and issues, current research status and identifies the research opportunistic perspectives. It is followed by the aims and objectives of the proposed research and key research questions to be answered, technical approach of how and what work will be carried out, expected results and other paper related details.

## 2 Background and Related Works

The idea of Cloud Computing (CC) was conceived by John McCarthy in 1961 [5]. It is a re-discovered technology topic that has gained popularity and it is expanding rapidly in service deployment environments that have business opportunities and economic benefits. Sharing resources, software and information via the Internet are the main functions of CC in order to reduce cost, underutilization and to satisfy demand. The cloud model as defined by NIST [3] promotes this principle and explicitly emphasizes availability as a core feature composed of five essential characteristics: (*On-demand self-service, Broad network access Resource pooling, Rapid elasticity and Measured Service*), three service models (*Software as a Service (SaaS), Platform as a Service (PaaS) and Infrastructure as a Service (IaaS)*) and four deployment models (*Private Cloud, Community Cloud, Public Cloud and Hybrid Cloud*).

The four deployment models are: Private (internal) cloud: enterprise owned or leased, behind a firewall, Public (external) cloud: sold to the public, mega-scale infrastructure (e.g. Amazon EC2), Hybrid cloud (virtual private cloud): composition of two or more clouds (e.g. Amazon VPC), Community cloud: shared infrastructure for a specific community (e.g. academic clouds).

Missing in the NIST declaration is what type of "*User Interface*" should be provided to all these service offerings? A suggestion is offered below.

The consumers of cloud computing can utilize applications without installation or configuration of special software and access their personal data from any computer over the Internet through centralized control but distributed management procedures. In cloud, everything is offered as a service (i.e.XaaS). The cloud computing services identify a layered system structure for cloud computing as shown in Fig. 1.

- ***Infrastructure as a Service (IaaS):*** at the infrastructure layer, enterprises lease equipment such as servers and network tools instead of purchasing from the service providers through the internet.
- ***Platform as a Service (PaaS):*** this layer provides services for the customers who can develop, test and run applications through the Internet, etc.

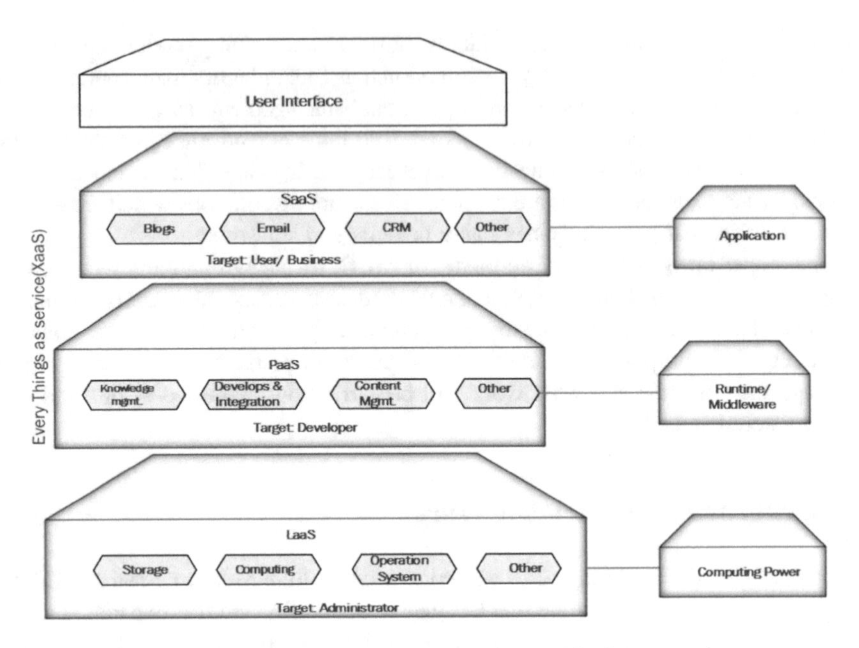


**Fig. 1.** Cloud computing emerging as "everything" as a service

- ***Software as a Service (SaaS):*** this layer is where the users or enterprises pay for software use and upgrades.
- ***User Interface: (yet to be universally defined):*** enables interaction with all the above layers in an open system style and functionality that would offer minimalistic man machine interfacing and interaction in a seamless non-intrusive manner.

Traditional mobile computing in juxtaposition with the revival and inheritance of cloud computing has resulted in the emergence of Mobile Cloud Computing (MCC) since 2009. It is predicted to be the "big player" in the years ahead as more and more users worldwide prioritise and take to sophisticated smartphones as their primary communication devices. While Hoang T. Dinh et al. in their comprehensive survey on mobile cloud computing emphasise that there is a significant paradigm shift by merging of mobile computing and cloud computing [6].

The Mobile Cloud Computing Forum [7] defines MCC as follows:

*"Mobile Cloud Computing at its simplest refers to an infrastructure where both the data storage and the data processing happen outside of the mobile device. Mobile cloud applications move the computing power and data storage away from mobile phones and into the cloud, bringing applications and mobile computing to not just Smartphone users but a much broader range of mobile subscribers".*

Alternatively, MCC can be defined as a combination of mobile Web and cloud computing [8], which is the most popular tool for mobile users to access applications and services on the Internet [9]. MCC, thus, provides mobile users with the data processing and storage services in clouds. The mobile devices do not need a powerful configuration (e.g., CPU speed and memory capacity) since all the complicated computing modules can be migrated and delegated to be processed in the MCC.

Market forces and economic opportunities show indications that XaaS will gambit and engulf the entire SaaS stack ranging from the lowest networking layer, storage, servers, virtualization, rich OS, middleware, dynamic runtime facilities, raw and meta data management to the highest application (apps) layer will be offered and managed as a set of services by vendors and service providers. This principle involves an entire paradigm shift, whereby user management of resources is completely moved away and catered for by the XaaS vendors and service providers. However, it is debatable whether this is positive or negative for the users who have to shed control of their intimate information and resources, and yet be prepared to pay for services that could be overpriced due to non-competitive cartelistic behaviour and agreements by the vendors and service providers, but it is a debate worthy of discourse certainly outside the scope of this proposal.

From a simple perspective, MCC can be thought of as an infrastructure where data and processing could happen outside of the mobile device, enabling a myriad of new types of applications, such as context-aware mobile social networking and semantic knowledge, coupled with ontolog driven Internet and Web of services based on Web 3.0/4.0 technologies [10]. As a result, many mobile cloud applications are not restricted to the powerful smartphones and their usage, but to a broad range of less advanced mobile phones due to their technologically inherent limitations. This is changing rapidly as the broader subscriber audience vie for smartphones which are increasingly becoming cheaper by the day and within their purchasing reach.

MCC can be simply divided into mobile computing and cloud computing, but it is their merged combination which is creating the buzz all around us. The mobile devices can be any on the go laptops, PDA, smartphones, tablets and so on, which connect with a base station or a hotspot by a radio link such as 3G/4G, WiFi, Edge or GPRS that typically connect wirelessly. Although the client is still mobile computing oriented today, the main thrust is rapidly evolving towards the uptake of the concept and availability of cloud computing which allows mobile users to send service requests to the cloud through a web browser or desktop application. The management component of the cloud then allocates resources to the request to establish connection, while the monitoring and calculating functions of mobile cloud computing are implemented to ensure Quality of Service (QoS) requirements are met in terms of Service Level Agreements (SLAs) until the connection and service is completed.

Typically, other than mobile devices which are connected to the mobile networks through base stations that establish and control the data communication connections between the networks and mobile devices, mobile users' requests and information are transmitted to the front-end computers to banks of servers providing a host of mobile network services. Here, based on Software Mobile Agent (SMA) computing, services like security (authentication, authorization and access control), privacy (identify management, settings for tracking user's usage behaviour, anonymity and non-disclosure of personal information) and account credentials (user validation, account management and billing) can be provided to the users via their Home Agent (HA) and subscribers' critical data stored in local databases. The subscribers' requests can then be delivered to a cloud vendor/service provider through the Internet through the SMA.

Cloud management controllers present in the cloud environment, process the requests to provide the mobile users with the corresponding cloud services. These services are developed based on the concepts of utility computing, virtualization and service-oriented architecture, which still have some way to go in their exposition [11].

What is important is that the details of cloud computing will be diverse in their offerings and usage contexts. Today, the major function of a cloud computing system/environment is storing data on the cloud and using technology by the client to access that data and other resources. This is likely to change in the future with new inventions, discoveries and end-user application usage patterns as progress is made towards eco-friendly MCC.

Some authors mentioned that CC is not entirely a new concept. For instance, Lamia Youseff et al. have stated in their position paper [12] that CC has manifested itself as a descendent of several other computing technology subject areas such as service-oriented architecture, grid and distributed computing, and virtualization, as well as inheriting their advancements and limitations, regarding the advantages and limitations of some of these technologies [11]. They introduced CC as a new paradigm in the sense that it presented a superior advantage over the existing under-utilized resources at the computing facilities and data centres. Several business models rapidly evolved to harness this technology by providing software applications, programming platforms, data-storage, computing infrastructure and hardware as services. R.Buyya et al. have introduced market oriented architecture [13] and economic modelling [14] by introducing CC as a type of parallel and distributed system consisting of a collection of interconnected and virtualized computers that offer computing resources from service providers to customers meeting their agreed SLA.

In terms of the *"XaaS"* layered stack architecture shown in Fig. 1 above, it illustrates the effectiveness of the CC model in terms of users' requirements in realising MCC as a *utility* service.

To exemplify this view in terms of *XaaS*, cloud computing is a large-scale distributed network system implemented on a number of servers in data centers, which is already bridging MCC to some extent. In the upper layers of this XaaS paradigm, Infrastructure as a Service (IaaS), Platform as a Service (PaaS), and Software as a Service (SaaS) are stacked. Data center's layer: this layer provides the hardware facility and infrastructure for the clouds. In the data center layer, a number of servers are linked together with high-speed networks to provide services for the customers. Typically, data centers are built in less populated places, with high power supply stability and a low risk of disaster. Infrastructure as a Service (IaaS): is built on top of the data center layer. IaaS enables the provision of storage, hardware, servers and networking components. The client typically pays on a per-use basis. Therefore, clients can save costs as the payment is only based on how much resource they really use. Infrastructure can be expanded or shrunk dynamically as needed. The examples of IaaS are Amazon EC2 (Elastic Cloud Computing) and S3 (Simple Storage Service). Platform as a Service (PaaS): offers an advanced integrated environment for building, testing and deploying custom applications. The examples of PaaS are Google App Engine, Microsoft Azure, and Amazon Map Reduce/Simple Storage Service. Software as a Service (SaaS): supports a software distribution with specific requirements. In this layer, the users can access application and information remotely via the Internet and pay only for what they use. Salesforce is one of the pioneers in providing this service model. Microsoft's Live Mesh also allows sharing files and folders across multiple devices simultaneously, so is Dropbox [15].

Advantages of Mobile Cloud Computing

Cloud computing is known to be a promising solution for mobile computing due to numerous reasons (e.g., mobility, communication, and portability) [16]. The following additions describe how the cloud can be used to overcome obstacles in mobile computing, thereby highlighting the advantages of MCC:

- ***Extending battery lifetime***: Battery is one of the main concerns for mobile devices. Several solutions have been proposed to enhance the CPU performance [17, 18] and to manage the disk and screen in an intelligent manner [19, 20] to reduce power consumption. However, these solutions require changes in the structure of mobile devices, or they require a new hardware that will result in an increase of cost and may not be feasible for all mobile devices. Computation offloading technique is proposed with the objective to migrate the large computations and complex processing from resource-limited devices (i.e., mobile devices) to resourceful machines (i.e., servers in clouds). This avoids taking a long application of execution time on mobile devices which results in large amounts of power consumption.
- ***Improving data storage capacity and processing power***: Storage capacity is also a constraint for mobile devices. MCC is developed to enable mobile users to store/access the large data on the cloud through wireless networks. For example, the Amazon Simple Storage Service (Amazon S3) [21] supports file storage service. Another example is Image Exchange which utilizes large storage space in clouds for mobile users [22].
- ***Improving reliability:*** Storing data or running applications on clouds is an effective way to improve the reliability since the data and application are stored and backed up on a number of computers. This reduces the probability of data and application lost on the mobile devices. In addition, MCC can be designed as a comprehensive data security model for both the service providers and users. For example, the cloud can be used to protect copyrighted digital contents (e.g., video, clip, and music) from being abused and unauthorized distribution [23].
- ***Dynamic provisioning:*** Dynamic on-demand provisioning of resources on a fine-grained, self-service basis is a flexible way for service providers and mobile users to run their applications without advanced reservation of resources.
- ***Scalability:*** The deployment of mobile applications can be performed and scaled to meet the unpredictable user demands due to flexible resource provisioning. Service providers can easily add and expand an application and service without or with little constraint on the resource usage.
- ***Multi-tenancy:*** Service providers (e.g., network operator and data center owner) can share the resources and costs to support a variety of applications and a large number of users.
- ***Ease of Integration:*** Multiple services from different service providers can be integrated easily through the cloud and the Internet to meet the users' demands.

***Mobile Commerce using Mobile Cloud Computing (MCC)***
Various mobile applications have incorporated some of the advantages of MCC, such as *mobile commerce*, which is a business model for commerce using mobile devices. Mobile commerce applications generally fulfill certain tasks that require mobility (e.g., mobile transactions and payments, mobile messaging, and mobile ticketing). These applications face various challenges (e.g., low network bandwidth, high complexity of mobile device configurations, and security) compelling them to be integrated into the MCC environment to address these issues. Yang et al. proposed a 3G E-commerce platform based on cloud computing [24]. This paradigm combines the advantages of both the 3G network and cloud computing to increase the data processing speed and security level based on PKI (public key infrastructure) [25]. The PKI mechanism uses an encryption-based access control and an over-encryption to ensure privacy of the user's access to the outsourced data. A 4PL-AVE trading platform utilizes cloud computing technology to enhance the security for users and improve the customer satisfaction, customer intimacy, and cost competitiveness [26].

Most of the current business and trading models that are available in the literature and in actual commercial use are biased in favour of the provider over the buyer in an unregulated trading environment. This in itself is a problem. One study with simulated results proposes a solution to this problem by proposing a new model called the Combinatorial Double Auction Resource Allocation (CDARA) model specifically for pricing of cloud computing service [27].

***Disadvantages of Mobile Cloud Computing***
Other than the business and commercial aspects of MCC across different applications, platforms, international trade regulations and trans-border information and money flows, in my humble believe and opinion, *per se,* one of the most important fundamental drawback to MCC, as understood today, is the whole area of safe and trusted computing encompassing the critical aspects of security, privacy, identity management, audit and digital forensics [28]. Although many contemporary researchers are tackling many of these topics and issues, it is a formidable challenge which has to be overcome through more advanced designs and developments of new frameworks, architectures and secure open system protocols and processes which are harmonized at a global level. Success in these areas can result in huge dividends and payoffs at different technical vertical levels across different horizontal enterprise and business sectors in mobile cloud computing.

**Looking Ahead - Challenges and Solutions**
The last decade brought with it several advancements in the way we perceive computing and mobility. Computing with the advent of CC and MCC is increasingly regarded as the *5$^{th}$ utility*, albeit, critical infrastructure utilities like water, electricity, gas and telephony, and to some extent, it is already providing the basic level of computing service that is considered essential to meet every day needs of the general community at a global level and context. CC is the latest paradigm proposed to deliver this vision with the effective merging of MC that is proving to be a promising solution for mobile computing for many reasons (e.g. mobility, communication, portability and availability). There are many challenges ahead of us as exemplified below:

***Resource Insufficiency:*** As processors are getting faster, screens are becoming sharper and devices are equipped with more sensors, a smartphone's ability to consume energy far outstrips the battery's ability to provide it. Thus, battery life of mobile devices remains a key limiting factor in the design of mobile applications. The two main contributors are (i) limited battery capacity and (ii) an increasing demand from users for energy-hungry applications. User demand is increasing by the day for resource intensive applications, like video games, streaming video and sensors equipped on mobile devices that produce continuous streams of data about the user's environment. Several solutions have been proposed to enhance the CPU performance and to manage the resources available optimally in order to reduce power consumption [17–20]. These solutions, however, require changes in the structure of mobile devices or require new hardware capabilities and configurations that would necessitate (re-)engineering resulting in higher cost.

Although outside the scope of this proposal, *computation offloading techniques* migrating large and intensive computations and complex processing from resource limited devices to resourceful facilities and applications *to the cloud is a research opportunity and challenged not to be missed.* Through these migrations, mobile devices avoid consuming large execution times. Several experiments have been conducted that evaluate the effectiveness of offloading techniques. Alenxey Rudenko *et al.* have demonstrated that remote execution of large tasks can reduce their power consumption by up to 50% [29]. Eduardo Cuervo *et al.* have shown that using Memory Arithmetic Unit and Interface (MAUI) structures which transform and migrate mobile components to servers in the cloud can save on the average 27% of energy consumption for computer games and 45% for the chess game [30].

***Data Storage Capacity and Processing Power:*** Storage is also a major concern for mobile devices. MCC is developed to enable mobile users to store and access large amounts of data on the cloud. Amazon Simple Storage Service (S3) is one such example [21]. It provides a simple web services interface that can be used to store and retrieve any amount of data, at anytime from anywhere on the web. Dropbox permits access to all kinds of document exchanges irrespective of which device created it that can be accessed from any smart device hovering on the cloud [15]. Flickr [31] is almost certainly the best photo sharing application based on MCC. It allows users to upload and share photos through mobile devices and web. Facebook [32] is the most successful social network application today and is also a typical example of using cloud in sharing images. MCC also helps reduce the running cost for compute-intensive applications. Cloud computing efficiently supports various tasks for data-warehousing, managing and synchronizing multiple documents online. Thus, mobile devices are no more constrained by storage capacity because their data is now stored on the cloud. Microsoft has embarked upon developing new office software [33] that will embrace cloud computing to fully integrate with all types of mobile devices. It will enable users to save, publish and share their work with other users, as well as their desktop computers and mobile devices.

***Division of Application Services:*** Mobile devices have inherently limited resources. Thus, the applications have to be *"divided"* in order to achieve a particular performance target such as low latency, minimum data transfer rates, fast response time etc. Considering the demands of MCC, the essential factors for delivering near optimum cloud services for mobile devices and users require:

- Optimum use of SMA technology.
- Optimal partition of application services across cloud and mobile devices.
- Low network latency in order to meet application and code offload interactivity.
- High network bandwidth for faster data transfer between cloud and mobile devices.
- Adaptive management and monitoring of real-time network conditions to optimize network usage and device costs against user-perceived performance of the cloud application.

And the following strategies can be adopted by vendors/service providers to address the above issues:

- ***Network bandwidth strategy:*** Using regional data centres or other means to bring content closer to mobile broadband as can be observed the how and where Google deploys them around the world.
- ***Network latency strategy:*** Application processor nodes to be moved to the edge of mobile broadband infrastructures to reduce delay like currently noticed with VoIP and video/audio streaming services.
- ***Battery saving strategy:*** Cloning the device in the network for performing energy consuming intensive compute and ancillary background management tasks such as automatic virus/malware scanning of mobile devices through the use of autonomic computing capabilities such as self-protecting and self-healing [34].
- ***Mobile cloud application elasticity:*** Dynamic optimization of application delivery and execution between the mobile device and the network through autonomic computing facets such as self-learning, self-optimization and self-configuration.

As a concrete example adopted from the Alcatel-Lucent report [35], Table 1 shows the mapping and values between applications and MCC critical infrastructure attributes with their levels of computing intensity, network bandwidth and network latency assignments.

**Table 1.** Application and mobile cloud computing infrastructure mapping (*Adapted from Source: Alcatel-Lucent* [36])

| Applications | MCC critical infrastructure attributes | | |
|---|---|---|---|
| | Compute intensity | Network bandwidth | Network latency |
| Web-mail (Yahoo!, Gmail) | Low | Low | High |
| Social networking (Facebook, Twitter) | Low | Medium | Medium |
| Web browsing | Low | Low | High |
| Online gaming | High | Medium | Low |
| Augmented reality | High | Medium | Low |
| Face recognition | High | Medium | Low |
| HD video streaming | High | High | Low |
| Language translation | High | Medium | Low |

- Compute intensity: High, required for compute-intensive apps
- Network bandwidth: High, required for content, heavy, large data transfer apps
- Network latency: Low, required for high interactivity

There are many other issues related to the implementation of MCC which have an impact on the pace of development and deployment. Some of the important ones are listed below:

**A. Absence of Standards**

Inspite of the various advantages of perceived or otherwise CC over the conventional computing techniques, there is no accepted open standard available to go by at this point of time. Interworking, interoperability and portability are also impossible between different CC Vendors & Service Providers (CCV&SP), which are impinging on the development of MCC. This prevents them to widely deploy and quickly develop cloud computing, *per se* MCC. Customers are reluctant to transform their current datacentres and ICT resources to cloud platforms owing to a number of unsolved technical, operational and business problems that exist in these platforms due to a lack of open standards such as:

- ***Limited scalability:*** Owing to the rapid growth, none of the CCV&SPs can meet all the requirements of all the users.
- Unreliable availability of a service: Dependence on a single CCV&SP's service can result in a bottleneck in the event of a breakdown of a service.
- ***Vendor/Service provider lock-in:*** Absence of portability makes it impossible for data and application transfer among CCV&SPs, consequently the customer is locked to a CCV&SP.
- ***Unable to deploy service over multiple CCV&SPs:*** Absence of interoperability makes it impossible for applications to be serviced and scaled over multiple CCV&SPs across diverse mobile devices and their supporting OSs.

In view of these disadvantages, B. Rochwerger *et al.* introduced a solution-based pathway called Open Cloud Computing Federation (OCCF) [43] that solves the problems of interworking, interoperability and portability among various CCV&SPs. Currently, however, the move to a common cloud standard, including that of MCC is impossible because most of the CCs and MCCs firms have their own APIs that have cost huge amounts of funds to develop and implement. The OCCF, in the absence of a full-hearted accommodation of Mobile Cloud Computing Forum (MCCF), lacks the necessary *"clout"* for harnessing the major players in delivering a practical realization mechanism. A possible suggestive approach is to have a Mobile Agent Based Open Cloud Computing Federation (MABOCCF) mechanism as advocated by Chetan S. *et al.* [37]. It should be advocated that these fora should unite into a single entity to overcome various administrative and technical issues, if the SMCC is to move rapidly ahead into practical solutions and business propositions.

**B. Access Schemes**

MCC will be deployed in a heterogeneous access scenario in terms of Wireless Network Interfaces. Mobile nodes access the cloud through different radio access technologies viz. GPRS, WLAN, 4G/LTE, WiMAX, CDMA2000, WCDMA etc. MCC requires the following features:

- An ***"always-on"*** connectivity for a low data rate cloud control signalling & management channel
- An ***"on-demand"*** available wireless connectivity with a scalable access link bandwidth upgrade option
- A ***"QoS conscious"*** network selection and use that takes energy-efficiency and costs into account to enhance lowest pricing offering to the subscribers while optimising vendor/service provider profitability

Access management is a very critical aspect of MCC. A possible solution is to use context and location information to optimize mobile access as proposed by A. Klein *et al.* [38]. Deploying MCC by utilizing the context information, such as device locations and capabilities and user profiles, this can be used by the mobile cloud server to locally optimize access management of their users.

**C. Elastic Application Models**

CC services are scalable via dynamic provisioning of resources on a fine grained modelling, self-service access basis in real-time, and without users having to imagine and/or engineer for peak loads and threshold violations. This requirement particularly manifests in MCC due to the intrinsic limitations of mobile devices, including the more sophisticated smart ones. For example, the iPhone 5 is equipped with a 1.3 GHz dual core swift CPU, minimum 16 GB RAM, iOS 7 platform allowing about 8 h of talk time and 14.4 Mbps speed on HSDPA 3G/4G networks [39]. Compared to today's PC and server platforms, these devices still cannot run compute-intensive applications. Thus, an elastic application model is required to solve the fundamental processing problems. Beyond the scope of this proposal, many contemporary researchers [40–47] have proposed a host of possible solutions to overcome such constraints.

**D. Security - Safety Measures**

One cannot underestimate the challenge facing MCC safety measures which goes beyond the mundane "security and privacy" title, when in effect it could be, to name several, any or all of these entities:

- Access control
- Applied cryptography
- Availability, resilience, and usability
- Certification of safety measures
- Cloud security
- Cryptanalysis
- Cybercrime
- Cyber security, cyber safety measures
- Database security
- Digital forensics and law enforcement
- Digital rights management
- Embedded system security
- Identity management

- Information security in vertical applications
- Multimedia security
- Network security
- Mobile client device and server cloud platform security, IDPS and malware,
- Mobile security for cyber-physical and virtual systems: collaborative, cooperative and contextual
- Privacy and anonymity
- Regulatory and legal issues pertaining to safety measures
- Risks, threats, countermeasures, evaluation and security certification
- Safety measures: frameworks, policies, management and operational strategies, processes and protocols
- Security intelligence, policies, protocols and systems and wares
- Security of smart cards, NFC and RFID systems
- Smart grid security
- Smartphone security
- Trust model and management
- Trusted computing
- Verification and validation of safety measure components, processes and protocols

It would be unrealistic to attempt to satisfy all the above entities in a single work, no matter how huge the availability of the amount of manpower, financial resources and timeline. The important point to realise is that some aspects from these entities are critical for establishing a safe and viable MCC that can be deployed and used with confidence.

Mobile devices today have all the functionalities of a standard computer plus total communication access. This poses a security threat to the mobile devices as well as the CC facilities. The threat detection and prevention services run on the mobile devices to combat these security threats, warrant intensive usage of resources, both in terms of computation and power [48]. A possible solution is to move a large proportion of these detection and prevention services to the cloud that seamlessly engage the mobile device to provide overall protection all round for the participating entities. It potentially saves the mobile device CPU and memory requirements with increased data transfer exchanges and bandwidth usage as the price to be paid. This approach has the following potential benefits:

- Better detection of intrusions, anomalies, masquerades and multifarious malicious software
- Reduced on-mobile-device resource consumption to a minimum
- Reduced on-mobile-device software hosting and complexities

E-marketplace as a concept and business model is a flexible and efficient approach to assist companies or corporations to extend their businesses to reach larger markets without regional boundaries via ad hoc networks. MCC easily fits and sits in such an e-marketplace environment with the proviso that this must be performed in a safe, secure, trusted and auditable manner to ensure customer (buyer) and supplier (seller) confidence against all kinds of intrusive attacks and masquerades. It requires a new paradigm, improved integration architectures and services to bring this to fruition.

Such e-marketplace is typically a cooperative distributed system composed of economically motivated software agents that interact cooperatively and/or competitively, find and process information, and disseminate it to humans and to other agents. In addition, it must provide support for common economic services and transactions, such as dynamic pricing, negotiation, automated supply chains, as well as other e-marketplace service infrastructure to ensure secure, trusted and reliable transactions [49]. Therefore, the Software Mobile Agent (SMA)-based e-marketplaces for MCC with strong mobility will cause the high risk security threats; in contrary, low mobility causes low security threats. Although much have been dedicated to the design of SMA-based marketplaces, the lack of standards for SMA-based MCC e-marketplace framework incorporating safe, secure, trusted and auditable services reflects that many critical technical, operational and business issues need to be resolved.

Security of mobile agents is a moving target problem because of the nature of hacking and malware that is introduced on a regular daily basis. The most important techniques to overcome this problem in the e-marketplace and mobile computing applications are still in their infancy, and challenging research area with many topics. There are different security approaches for mobile agents that have been proposed to protect the platform, host, agents and route. The security protocols, such as the Secure Socket Layer (SSL) [50] and Secure Electronic Transaction, (SET) [51] are used for confidentiality and integrity to secure the communication between the agents on different hosts. The SSL channel may be insecure for the mobile agents since a mobile agent may move to a corrupted platform or host to communicate with other agents. Also, it is dangerous for the mobile agents to exchange their sensitive information without the use of cryptography techniques as their information can be stolen or contaminated/corrupted by malicious attacking agents. SET on the other hand offers far superior r security than SSL since it uses the Public Key Infrastructure (PKI) for privacy and X.509 digital certificates to authenticate participants in an e-marketplace [52]. More importantly, the merchant cannot visualize or access the sensitive information on his server. This ensures the buyer's confidentiality, privacy and safety.

One of the major weaknesses in today's mobile agent-based e-marketplace system's usage is the failure to accommodate the end-user or system processes to use mobile trading and its security, audit and forensic features in a seamless and transparent way. For instance, Chan [53] developed a shopping system based on a mobile agent which is called the Shopping Information Agent System (SIAS). His research concentrated on the security and reliability issues of mobile agents and he discussed about the possible security attacks by malicious hosts against the mobile agents in the system. Specific solutions to prevent these attacks were devised. Another mobile agent-based e-commerce framework proposed by Zhao et al. [54] focused on the security, integrity and confidentiality of mobile agents to secure the trading transactions. They used the secure message digest and encryption to illustrate the Mobile Agent-based E-Commerce Model (MA-ECM).

High risk security threats appear with strong mobility. To reduce the security threats of mobile agents, Zhang et al. [55] proposed a framework for e-commerce that uses both single hop and multi-hop to reduce the risks. The first phase uses multi-hop for information collection where the mobile agent does not take any sensitive information. The second phase uses single hop to take the sensitive information to reduce the security risks. Wang Rui-Yun et al. [56] proposed a different solution to protect the mobile agent in e-commerce through a secure authentication infrastructure using PKI and combining it with integrity protection and confidentiality protection with regards to PKI limitations.

It is important to ensure that the personal or sensitive information carried by a mobile agent or stored on a platform is accessible only to authorized parties [57]. To address such privacy and trust issues from both the user and the system perspectives, Yu Bin and Yi Xiang [58] proposed a secure transmission mechanism for the mobile agent-based system to overcome the weakness of traditional communications for the distribution of data encryption key without going through a third-party. While Wang et al. [59] proposed a migration mechanism of security and fault tolerance for mobile agents that include trust and travel services. The trust service can be duplicated as a backup at several different trusted nodes in the e-marketplace environment. When the trust service expires, the system can select one backup from the trusted nodes as the trust service via the *election algorithm* [60]. However, this mechanism has some drawbacks such as the expiry of trust services and trusted nodes which may affect the backup of the trust service, therefore, causes the travel transmission problem of the mobile agent which still needs an effective solution to the problem.

The right of an Internet user acting anonymously conflicts with the rights of a server victim identifying the malicious user as exposed by Giannakis Antoniou et al. [61]. They proposed a security protocol for privacy and forensics investigation process which dramatically increases the level of the protocol's reliability that can address the reliability issue efficiently by defining what is considered as malicious and what is not at the beginning of a communication. But the application of this protocol has not been tested in real-life practical e-marketplace trading environments.

The frameworks and systems described above SMA e-marketplace trading include some aspects of security, trust, privacy and forensic issues. Also, one research proposed an extended secure and auditable agent-based framework for mobile users. They investigated the requirements and analysed the architecture needed to define a secure and auditable agent-based framework via the auditable and forensic investigations. They exemplified the framework and protocols with an auction e-marketplace as the commerce application and presented scenarios to prove their point. They defined the essential requirements for the secure and auditable agent-based e-marketplace as well as the requirements for a mobile commerce environment which an agent-based e-marketplace framework should include features such as security, privacy, identity management, trust, audit and digital forensics as a comprehensive set of safety measures.

The need to observe the benefits, constraints and limitations from the SMA e-marketplace experiences and knowledge for applicability in MCC requires a fresh visit to all these issues but without reinventing the wheel. It would require proposing a MCC framework, protocols, processes and de facto standards for contributing to international standards for a set of safe measures comprising security, privacy, identity management, trust, auditable service and digital forensic mechanisms for reply and proofing against illegal activities from outside or within the system.

**E. Economic Modelling and Pricing**

In a cloud environment, providers compete to offer their services, whilst the users seek to receive resources to run their tasks on the basis of their Quality of Service (QoS) and pricing requirements [62]. Usually providers and users are independent in their policies and resource allocation strategies [63]. Market-based techniques and economic models which regulate supply and demand for resources, provide an incentive for providers to optimize their returns and motivate the users to select the lowest cost service provider based on their budget and level of QoS [64]. The crucial aim of a resource allocation procedure is to find an allocation that is feasible and optimal for a given service. Due to the heterogeneous nature of cloud computing, resource allocation is often based on two approaches; that is, system centric and user centric. The system centric, a traditional approach, optimizes system-wide measures of performance such as an overall throughput of the system. The user centric approach mainly concentrates on providing maximum utilization for users on the basis of their QoS requirements and budgets. In cloud computing, where the availability of resources and workloads are changing dynamically, meeting the QoS constraints together with maintaining an acceptable level of system performance and utilization are the primary problems to tackle. This dynamic behaviour in turn creates challenges between the users and their service providers to maximize their required usage of resources or tasks as efficiently as possible, but coming from the opposite sides of the spectrum. Therefore, the resource allocation mechanism in this market environment needs to be optimal for both sides to meet in the middle of the spectrum.

Providing, however, a market-based resource allocation model that is economically efficient and state-of-the art and which considers equitable benefits for both the users and the providers, needs more thorough research to bridge the gap to meet these specific requirements for MCC, particularly, SMCC.

- **Open Research Issues as a Summary to Related Works**
  From the foregoing background and related works, it could be succinctly summarised that there are many open research issues, questions and unknowns. But this proposed proposal SMCC hinges on:
- ***Energy efficiency in the MCC eco-system:*** Owing to the limited resources such as battery life, available network bandwidth, storage capacity and processor performance, on the mobile devices, researchers are always on the lookout for solutions that result in optimal utilization of available resources through a variety of engineering and computational means.
- ***Task division:*** Researchers are always on the lookout for strategies and algorithms to offload computation tasks from mobile devices to cloud. However, due to differences in computational requirements of numerous applications available to the users and the variety of mobile devices available in the market, an optimal strategy is an area to be explored.
- ***Better service:*** The original motivation behind MCC was to provide PC-like services to mobile devices. However, owing to the varied differences in features between fixed and mobile devices and features between wired and wireless data communication, transformation of services from one to the other may not be as

direct as one may perceive. This requires QoS and SLAs to be re-explored with the vision of new eco-friendly business models.

- *Security:* The absence of standards poses a serious issue specifically with respect to security and privacy of data being delivered to and from the mobile devices to the cloud.
- ***Economic modelling & pricing schemes for better SMCC service:*** The original motivation behind MCC was to provide PC-like services to mobile devices. However, owing to the varied differences in features between fixed and mobile devices and features between wired and wireless data communication, transformation of services from one to the other may not be as direct as one may perceive. This requires smart economic models and pricing schemes that are synchronised with QoS offerings and SLAs to be re-explored with the vision of new eco-friendly business and services models. This also requires the research to answer a multi-disciplinary problem at the nexus of ICT services, Computer Science, Economics and Law: that is:

*"How to design a market-based resource allocation model to address the requirements of a SMCC environment which satisfies both the users and the providers as optimally as possible coming from opposite sides of the supply and demand costing/pricing spectrum?"*

***SMCC Summary Conclusion for a Better Service***

Mobile Cloud Computing, as a development and extension of Cloud Computing and Mobile Computing, is the most rapidly emerging and accepted technology with fast growth from every side of the equation of consumers, vendors and service providers. There is a huge economic potential in this arena. CC services are scalable via dynamic management which somehow nicely fits with the MCC phenomenon. The combination of cloud computing, mobile computing, wireless communication infrastructures, portable computing devices, transient location-based services, mobile Web context and meaningful application dispositions has laid the foundation to harness at novel SMCC and business opportunities provided outstanding challenges and issues are resolved.

## 3 Precise Problem Statement Recapped

From the foregoing it should be obvious that mobile technology also has become one of the most vibrant and stimulating areas driving forward the ICT society of the 21$^{st}$ century. Mobile communication networks have evolved into the new delivery platform and services consumption models for many institutions, industries and government organizations worldwide. Mobile devices, such as smart phones or tablets or their combined integrated devices, open the door to a great assortment of new applications and services. This may also include new types of specific application settings such as medical environments, support for disabled people, industrial workspaces, process monitoring, automatics and diagnostics or measurement activities in remote and distributed environments. Location, other context information and personal data are used to tailor services directly to the needs of the user. This leads to pervasive environments in which users can always access the Internet through cloud computing. While this seems to be an exciting development it also raises substantial safety measure issues such as security and privacy concerns which need to be properly addressed.

Mobile commerce (a.k.a. m-commerce) comprises applications and services that are accessible from Internet and Web enabled mobile devices. It involves new technologies, services and business models. Whilst it is different from traditional e-commerce it can also be considered as an extension of it since, among other reasons, it makes e-commerce available, in a modern way, to new application areas and to a new set of customers with appropriate economic models and pricing schemes. This includes the incorporation of web services and cloud-based services, autonomic mobile computing, and the integration and interplay with urban systems, industrial settings, medical environments and other contexts. Economic models and pricing schemes together with safety measures are deemed to be the most important factors for the success of mobile communications, commerce and mobile applications that fit into the affordable and secure mobile cloud computing (SMCC) mould.

This proposal addresses the research pertaining specifically to economic models and pricing schemes together with safety measures for SMCC.

## 4 Gaps in the Literature to Be Addressed by This Proposal

The main goal is to reassess the gaps in the literature and determine the path to provide mobile pricing and secure payment solutions in the proposed SMCC framework that mainly caters for:

- Equitable and fair economic models and pricing schemes for SMCC that satisfies both the users and vendors/service providers alike. Although such models and schemes as solutions for grid computing and to some limited extent for CC exist, there is none to the proposers' knowledge for SMCC.
- A payment model that suits clients ranging from using non-sophisticated mobile devices to very smart mobile devices, as well as supported by MCC servers with varying degree of software and hardware sophistication. But it is not clear how safety measures in the form security and privacy is assured between the client and the MCC service provider, if it is non-SMCC, or a SMCC of a lesser degree.
- Defining a 3G/4G mobile e-commerce platform based on SMCC requirements with the characteristics of high Web bandwidth, high data processing speed, and high security. Although claims have been made that PKI systems and infrastructures would be able to cope to deliver against these characteristics, it's not clear how the PKI limitations, e-commerce trading negative aspects and lack of appropriate safety measures should be overcome.
- Defining and specifying the sub-frameworks and related e-commerce trading and payment protocols of the principal SMCC framework that would ensure that all the safety measures are accounted for (considered). Although claims are made that it would be easy to "transfer" existing mobile computing technology for secure e-trading payments to SMCC, their implications and limitations are not known.

- Mobile payments. Although mobile payments solutions for CC exist to some limited extent, it is not clear where in the SMCC infrastructure and associated protocol hierarchy client's credentials should be generated and stored in the face of open and unreliable communication. The most obvious choice would be at the end points of the application layer, but the implications and limitations of this choice needs serious study and assessment for SMCC.
- Battling intrusions, anomalies and masquerades. It is still not clear how to combat intrusions and prevent them in *hostile* and *open* Wireless Area Networks (W-LANs) and Wireless Sensor Networks (W-LSNs) in the context of MCC, even if SMCC is operative. It requires a new paradigm other than the collaborative Intrusion and Prevention System (IDPS) strategies.
- Some verification and validation techniques and tools are available for protocols, but their application to the suite of SMCC *protocols* is virtually zilch.
- Given the above (g) *protocols*, similar verification and validation techniques and tools have to be applied to the suite of SMCC *software* which is produced to support such protocols and other management functions.

## 5 Research Questions to Be Answered

Some of the key critical questions to be answered by the proposed technical work for SMCC are detailed below.

***Security for SMCC Client Side of Mobile Device***

Security for SMCC should be assured internally within the client's devices and during the transmission of data from the client to the MCC servers.

*The first question is: Where on the client side (i.e.in the memory of the mobile device or SIM card) should the SMCC be implemented?*

Our proposed mobile commerce framework based on MCC will be implemented in UICC (Universal Integrated Circuit Card) which is a Secure Element (a generic platform for smart card applications). It has been standardized by ETSI EP SCP (ETSI Project Smart Card Platform). The UICC can host a number of different applications, each defining and controlling its own application(s). The architecture of our proposed mobile e-commerce framework based on MCC has three layers of security: they are the Physical Infrastructure, Communication and Application layers. Physical Infrastructure layer security is about GSM and GPRS security which is vulnerable to many attacks. Secure and reliable end to end communication between UICC and Cloud Service Provider (CSP) is ensured using SSL/TLS and TCP at the Communication layer. Security at the application layer is ensured using HTTPS and our proposed mobile payment protocol. Provisioning is the process of installing a payment application on a UICC. Personalization is the process of logging data specific to a client into the mobile payment application. This includes providing the necessary cryptographic material required by the UICC or application in order to allow installation or personalization. It is also responsible for providing a chain of trust between the CSP and UICC, including appropriate logging to assist in audit, repudiation and forensic.

**Internal Security of SMCC**

- How the SMCC provider will provide authentication and integrity over the user's data?
- How the SMCC provider will protect the stored user's data in cloud storage servers from attackers. And how to protect private data from hackers whose aim is to hack the servers for access over private data?

**During the Transit of Data From the Client to the SMCC**
SMCC relies on service concept to deliver computing. In other words, when a user requests computing resources he does not need to buy these products but rather can rent a service from a service provider to meet his objectives. Security shows up as a main concern in mobile cloud computing as data should be transferred from the mobile client to the cloud service provider. The GSM network is the most popular environment for mobile commerce.

How an application layer security is guaranteed and made fail-safe in mobile payment and mobile commerce?

**Flaws in the Design of Security Protocols**
Secure protocols need to be designed for transferring data between the client and MCC. A protocol is a set of rules that follows the defined conventions to establish semantically correct communications between the participating entities. A security protocol is an ordinary communication protocol in which the message exchanged is often encrypted using the defined cryptographic mechanisms. The network is assumed to be hostile as it contains intruders with the capabilities to encrypt, decrypt, copy, forward, delete, and so forth. Considering an active intruder with such powerful capabilities, it becomes extremely difficult to guarantee proper working of a security protocol. Several examples show how carefully designed protocols were later found out to have security breaches.

So, how to detect flaws in the protocols?

How can we trust that the protocols designed for mobile e-commerce in SMCC are free from flaws?

**Privacy/Anonymity**
Data, information, computer and network security and trusted computing are the cornerstones of the Internet, the Web and the Information Society in which we live. Integrity of financial transactions, accountability for electronic signatures, and confidentiality within a virtual enterprise, privacy of personal information and dependability of critical infrastructure, all depend on the availability of strong trustworthy security mechanisms which can obviously provide anonymity but at the same time it allows for cyber criminals to use to hide their tracks. Ensuring the availability of these mechanisms requires solving a multitude of substantial research problems cutting across many critical domains ranging from economics, engineering, security, privacy, trust, audit and digital forensics.

Security and privacy are intertwined and have serious economic and societal consequences. The argument about security versus privacy is not clear-cut. When a government engages into a policy making exercise to balance the security requirements and privacy needs, it is pertinent to escape from the confinement posed by thinking on a micro level, and to consider the aggregate macro trends and consequences. In most attempts to study and analyze security and privacy, they seem to be pulling in opposite directions; privacy on the one hand, considered to be a right or a need by people is an integral aspect of the modern society, and its place within that society may be unchallenged and unquestionable. Yet, security in many cases has shown to attempt to waive the right to privacy, which is accompanied with the rationale of protecting the citizens' safety and welfare. More particular, privacy that relates to anonymity of actions is targeted by identification, non-repudiation and indirectly by accountability. Furthermore, the security services are fundamentally different from the well-known privacy criteria of unobservability, pseudonymity, unlinkability and anonymity, so they come with different agendas which can result in conflicting goals.

A common mistake when attending to privacy is to maintain a narrow focus on the so called Privacy Enhancing Technologies (PETs) based on security technologies but failing to observe the wider, global picture of overall security processes. It is of little benefit to establish what is regarded as "good" security by establishing the security services of confidentiality, integrity, and non-repudiation, but failing to invoke risk assessment exercises to uncover the darker side of security consisting of hacking, vulnerability scanning, malware and possible exploitation of backdoors in the hardware and software computer systems and network infrastructure services. Indeed, when designing the security of a system, one needs to consider the threats which may range from the very practical hacking technologies to social engineering. Therefore, it is essential during risk analysis to consider the security processes as a whole instead of the one-dimensional "good" security technologies that will guard us against attacks, simply because the adversarial technologies are created by "good guys" recruited in the camp by "bad guys". Often the good use of malicious technologies is termed as "ethical hacking". This creates a dilemma as one cannot always distinguish whether hacking – which is referred to as "adversarial technologies" – is used for good or malicious purposes.

- Will mobile cloud computing ensure privacy/anonymity in addition to security?
- How to stop the bad guys using Privacy-Invasive Technologies (PITs) from conducting sinister things in all kinds of guises in all types of e-business?
- Can PETs overcome PITs to reverse their alarmingly growing trend by directly assisting in the protection of the privacy interest of entities/agents on behalf of their legitimate clients through countermeasures which are designed to defeat or neutralize PITs?

**Cybercrimes in SMCC**

Several threats may compromise the service or the contract between the clients and SMCC/MCC providers. Despite the use of traditional security defense mechanisms, cybercrimes on SMCC/MCC infrastructure may always occur. It is therefore crucial to implement forensic techniques to help investigate cybercrimes when they do happen.

How to collect data, where and how to store metadata for each transaction, how to analyze log files, and how to identify attacks on cloud infrastructure?

**Autonomic Computing Systems in SMCC**

Autonomic Computing Systems are systems which are capable of adapting themselves to changes in their working environment in order to maintain required service level agreements, protect the execution of the systems from external attacks or prevent and recover from failures. Characteristics of Autonomic computing are self-healing, self-configuring, self-optimization and self-protection.

How can SMCC be self-healable, self-configurable, self-optimized and self-protectable?

**Verification and Validation of SMCC Protocols and Software**

Without applying proper verification and validation techniques and tools to ensure that the suite of SMCC protocols and software execute as intended, will be the biggest source of challenges and problems in deploying them.

- How to ensure that the suite of SMCC protocols and software are correctly certified in terms of verification and validation?

## 6 Technical Approach of How and What Work Will Be Performed

**Framework**

Before commencing with the technical aspects of the paper, the economics and business models for SMCC ranging from basic to sophisticated ones would be selected and studied. This initiative will consider different technical vertical levels of integration of SMCC services across different horizontal enterprise and business sectors availing of mobile cloud computing. This aspect of the work has the pronounced possibilities and opportunities to enhance the work to be multidisciplinary in nature that might require input from the economics, business, legal and cyber security and digital forensics disciplines offered by the participating institutions. Some of these models will be specified by looking at the MCC businesses themselves, and followed by specifically considering what is available in the state of the art (as already presented in the Background and Related Works Section), and either possibly selecting appropriate fair-trading pricing schemes like the combinatorial double auction for mobile users and cloud computing service providers or devising more sophisticated and equitable ones.

Subsequently, given the framework and understanding the economics and regulatory business issues, the principal driver of the core of the technical research of this proposal will be to define and specify the safety measures (security, privacy, identity management, trust, audit & digital forensic, etc.). It will be the major research effort since the safety measures themselves would require a comprehensive framework in specifying the protocols and the verification methods for SMCC. As an example of the type of research that is being considered by this proposed work is outlined below.

**SMCC Pricing Scheme**

After carrying out a systematic review and analysis of existing state of the art of market-based resource allocation in grid and cloud computing environments, a new market-based resource allocation model based possibly on the combinatorial double auction will be specified and designed for SMCC. Other competitive schemes will also be considered as well as simulated to test their effect. These schemes will be initially implemented as *rapid prototypes* using CloudSim, a Java based simulator for simulating cloud computing environments, but adjusted and tailored to meet SMCC requirements in order to test their respective pricing mechanisms under varying QoS e-trading and operating conditions. The system design of these pricing mechanisms would be based on object oriented technology and illustrated using UML use cases and sequence diagrams to comfort the research team. The data generated from the simulator will be analyzed and the results evaluated to form the final conclusion of which one is the most optimum.

For instance, the two popular categories for economic models in resource management are the commodity market model and the auction model. Providers in a commodity market model specify their resource price according to the amount of resources that a user would consume. In an auction model, both the providers and the users act individually and independently. They may agree surreptitiously on a selling price [65]. Auctions are mainly about products that have no standard price values, and their values might be affected by the supply and demand in a particular time. Auctions require little global price information, but the price can be implemented easily in a grid setting [66]. Single-sided and double-sided auctions are the two main categories of auctions. A single-sided auction, such as an English auction, is a mechanism for one-to-many price negotiations, whereas the double-sided auctions allow many-to-many price negotiations, like a combinatorial double auction. In the combinatorial double auction, both sides submit their bids for multiple items. This double auction is considerably more effective and is suitable for cloud computing than several one-sided combined auctions because it prevents monopoly situations by the service providers colluding together [67].

Fujiwara et al. [68] stated that in a cloud environment, the providers compete to offer services and users compete to achieve such resources to run their tasks on the basis of their Quality of Service (QoS) requirements. He also expressed that an efficient allocation mechanism among service providers and users has yet to be devised. Izakian et al. [69] expressed that distributed systems include a set of providers and users. Since these entities are independent and making decisions autonomously based on their policies and resource allocation, this is a challenging problem. In such systems, using market-based techniques for this problem regulates the supply and demand for resources, provides an incentive for providers, and motivates the users to trade-off between budget, and the required level of quality of service. Thus, providing a market-based resource allocation model that is economically efficient and state-of-the art and which considers equitable benefits for both the users and the vendors and service providers, needs more thorough research to bridge the gap to meet these specific requirements for SMCC. This also requires the research to answer a multi-disciplinary problem at the nexus of ICT services, Computer Science, Economics and Law: that is, "how to design a market-based resource allocation model to address the requirements of a SMCC environment which satisfies both the users and the providers as optimally as possible coming from opposite sides of the supply and demand costing/pricing spectrum?".

**Safety Measures**

Wireless technology has achieved great development during the last several years and it is extremely complex, compared with wired networks. There are many security issues with wireless networks, which make hackers consider them as the weakest point in the communication chain of mobile financial transactions. GSM network is the most popular environment for mobile transactions. The migration from e-commerce to m-commerce is not as easy as it first appears because all the existing e-commerce technologies were developed for wired networks, which were more reliable, more secure and faster than wireless and mobile networks. Therefore, without major revisions the current e-commerce technologies cannot be applied directly to m-commerce. In order to fully deploy m-commerce for business, there are two levels of security requirements that must be satisfied. The lower level requirement is the need for a secure wireless infrastructure to protect each individual wireless communication and the higher level requirement is for a secure protocol with which to conduct mobile payment and business transactions, thus protecting the legitimate security concerns of all the parties involved, namely, the customer, the merchant, issuer, acquirer and payment gateway. This is a precursor to the SMCC.

Wireless communication security is a serious problem for all wireless applications that must transmit data securely through an open airwave communication medium. IEEE 802.1x defines the standard for wireless authentication, key distribution, network monitoring, and similar issues. This standard uses Extensible Authentication Protocol (EAP) and it is supported by EAP are MD5 (Message Digest 5), TLS (Transport Layer Security), TTLS (Tunnelled TLS), LEAP (Lightweight EAP), and PEAP (Protected EAP) algorithms to authenticate exchanged messages.

The security community has agreed that cryptography is the only solution to the problem of ensuring authenticity, privacy and integrity for communications through insecure media and many encryption algorithms have been developed over the past few decades. However, in a wireless environment with limited physical resources, most existing encryption algorithms are computationally too intensive, sluggish and irritating to use. A lightweight encryption algorithm with an acceptable degree of security strength is a possible solution to this dilemma. Although the lower level security requirement of wireless communication security is the topic of considerable ongoing research and is a vital preliminary to the deployment of all SMCC applications, this proposal will focus on the higher level security requirements of mobile payment and transaction security through the mechanisms and protocols of safety measures.

Based on the above, we can conclude that current security solutions at the communication layer in mobile environments are not adequate. Therefore, security at the application layer must be added in order to achieve an end-to-end protection especially for mobile financial e-marketplace trading environments, of which SMCC will be. In order to ensure end-to-end security, PKI should be adopted for SMCC. Adopting PKI for mobile payments is a nontrivial task because the mobile phone has the fundamental limitation of performance such as less memory and a less powerful CPU, but it also has other disadvantages of being offline, which an SMCC platform can overcome. Also, wireless data network presents a more constrained communication environment such as less bandwidth and has different protocol compared to wired internet protocol. So it is very difficult to apply wired PKI technology for a wireless environment. At first, a mobile phone must generate a public key pair and compute digital signature using the key. A public key certificate could be issued to a mobile user through the wireless internet. The public key certificate provides a method to bind the public key and its owner. Using the certificate, the mobile user must authenticate itself by securing a trusted channel for internet service such as M-commerce. Generating and storing a client's credentials are a challenging task because it needs a tamper resistant device. In existing mobile payment schemes client's credentials are stored in the memory of the mobile phone; the mobile phone with PKI functionality is personalized (generation of the private key) by the issuer (usually by MNSP). This solution faces many disadvantages such as the generated signature cannot be considered as equivalent to the handwritten signature because the credentials are not stored in the tamper resistant device. The client's credentials are stored in the SIM; SIM with PKI functionality is personalized (generation of the private key) by the issuer (usually by MNSP). SMCC integrates the cloud computing into the mobile environment to overcome the obstacles related to the performance (e.g., battery life, storage, and bandwidth), environment (e.g., heterogeneity, scalability, and availability), and foster safety measures (e.g., security, reliability and privacy) which are very important for mobile commerce. But, for instance, security is the main concern in SMCC which includes confidentiality, authentication, integrity, non- repudiation and fraud detection.

Fraud is an intentional deception accomplished to secure an unfair gain, and an intrusion which are any set of actions that attempt to compromise the integrity, confidentiality or availability of a resource. Several threats may compromise the service or the contract between the users and the providers. Despite the use of traditional security defence mechanisms, cybercrimes on cloud computing infrastructure may always occur. It is therefore crucial to implement forensics techniques to help investigate cybercrime when they do happen. Several challenges pose questions such as how to collect data, where and how to store metadata for each transaction, how to analyse log files, and how to identify attacks on cloud infrastructure. Digital forensic computing should be enforced and implemented to cater for both pre and post events concerning cybercrime and other offending activities used to overcome the greatest challenges faced in dealing with frauds in mobile commerce as we know it today. As such, there is a need for designing secure mobile commerce protocols to accommodate SMCC.

A protocol is a set of rules that follow the defined conventions to establish semantically correct communications between the participating entities. A security protocol is an ordinary communication protocol in which the message exchanged is often encrypted using the defined cryptographic mechanisms. The mechanisms, symmetric key cryptography or asymmetric key cryptography, are used to obtain various cryptographic attributes such as confidentiality, entity authentication, message integrity, non-repudiation of a message, and message freshness, to name a few [70]. However, merely using cryptographic mechanisms, does not guarantee from the view point of security that semantic operation of the protocol is secure, even if it is verified and validated as correct. There indeed have been reported breaches in a few security protocols after they have being published and accepted as safe protocols. Therefore, the design of a security protocol is an intuitive process which is severely error-prone so a more rigid framework is required within which we can safely design secure protocols, specifically tailored to SMCC operations and e-trading requirements.

The network is assumed to be hostile as it contains intruders with the capabilities to encrypt, decrypt, copy, forward, delete, and so forth. Considering an active intruder with such powerful capabilities, it becomes extremely difficult to guarantee proper working of a security protocol. Several examples show how carefully designed protocols were later found out to have security breaches [71]. In order to gain confidence in the cryptographic protocol employed, it has been found desirable that the protocol be subjected to an exhaustive analysis that verifies and validates its security properties through reverse engineering by parties other than the inventors. Some of the tools developed for this purpose are Scyther [72] and AVISPA (Automated Validation of Internet Security Protocols and Applications) [73] and [74] to name a few. These tools differ in their input language and also in the way they verify the protocols and provide the output. Though Scyther and AVISPA tools eliminate the possibility of human error, but still the selection of these automated tools are very important in verifying the correctness of security protocols. Only very few automated tools explore all the possible behaviours, whereas others explore strict subsets. Ignoring these kinds of differences leads to completely wrong interpretations of the output from such tools. Cas Cremers and Pascal Lafourcade [75] conducted a study of state space relations in performance comparison of several well-known automatic tools for security protocol verification, After the analyses of the performances of the tools over comparable state spaces, they concluded that the Scyther [76] and ProVerif [77] tools are the most efficient and fastest, and their approximation techniques are effective, and both of them can handle unbounded verification. The Scyther tool has the added advantage of not using approximations.

One of the approaches would be to design secure mobile commerce protocols (bringing together the economic and safety measures aspects together) based on MCC by adopting Wireless Public Key Infrastructure (WPKI) using UICC as a secure element from the client's side. The designed mobile commerce protocols must ensure end-to-end security, that is, from the mobile client's device to the SMCC. SMCC based payment solutions will assist in securing merchants, protecting end-users and developers of applications requiring payments from the risk of data leakages and breaches through better designed mechanisms and platform operation by never allowing sight or storing payment card data in the clear. Our proposed mobile commerce protocols based on SMCC will ensure authentication, integrity, confidentiality and non-repudiation, identity protection from the merchant and eavesdroppers, transaction privacy protection from eavesdroppers and payment gateways. We will go further to ensure payment secrecy, goods order secrecy, goods forwarding secrecy, and prevent double spending, overspending and money laundering with the possibility of designing a new schema to enforce this automatically. In addition to these, our proposed suite of protocols will be designed to withstand replay, Man-in-the-Middle and impersonation attacks.

Our proposed safety measure framework will be designed to overcome security threats, privacy violations and cybercrime activities in an SMCC environment.

SMCC in our proposed framework and protocol suite will resort to autonomic computing principles to be able to self-monitor, self-heal, self-configure, self-optimize and self-protect [78, 79].

In our experimental development, it is envisaged that virtual servers run on Sun Enterprise T5240 based on Apache Hadoop and Google AppEngine is a suitable option in order to provide services for the consumption on various mobile clients, includes the Android devices such as Motorola Droid/Milestone and HTC Hero or the Apple iPhone. A variation of technologies will be used, such as Android OS, SDK, GAE SDK, JavaScript, HTML, virtualization technologies, and Amazon Web Services.

Verification and validation techniques will be applied to ensure that the suite of SMCC prototype software conforms to the functional specification, and to develop the necessary local skill sets to produce quality and certifiable software.

Application-specific SMCC (ASSMCC) involves developing specific applications for mobile devices which use cloud computing. While both can potentially allow a mobile device to perform more intensive operations than it could by using only local execution, ASSMCC has the added benefit that it allows for users of cloud computing who require more than simply increased computational power. For example, chat or e-mail clients require ASSMCC because the internet is used as a communication resource and not simply for storage or additional computational power (although such applications may leverage these resources as well). Several methods and systems have been proposed which aim to specifically facilitate mobile cloud computing for applications.

Writing SMCC applications using RESTful web services are helpful because RESTful web services (unlike standard web services) are easy to create, are not processor or time intensive, do not have continuous TCP connections and produce simple XML responses that can be easily parsed. So, by leveraging mobile applications that use these services it will be more easily possible to create solid cloud applications. In order to achieve elasticity in MCC applications our proposed framework divides a full application into pieces called Weblets. These Weblets have the important feature of portability. Any given Weblet can be switched between both the mobile and stationary devices.

## References

1. http://www.idc.com/getdoc.jsp?containerId=prUS24302813
2. IBSG Cisco: (i) "Mobile Consumers reach for the Cloud", 2012 and (ii) "The Road to 'Cloud Nine" - "How Service Providers Can Monetize Consumer Mobile Cloud", Henky Agusleo and Neeraj Arora (2013). http://www.cisco.com/web/about/ac79/docs/sp/Mobile-Cloud.pdf
3. Mell, P., Grance, T.: The NIST definition of Cloud Computing. Special Publication 800-14, vol. 15, September 2011
4. Marston, S., Li, Z., Bandyopadhyay, S., Zhang, J., Ghalsasi, A.: Cloud computing – the business perspective. Decis. Support Syst. **51**(1), 176–189 (2011)
5. McCarthy, J.: Life in the Cloud, Living with Cloud Computing. http://computinginthecloud.wordpress.com/2008/09/25/utility-cloud-computingflashback-to-1961-prof-john-mccarthy/
6. Dinh, H.T., Lee, C., Niyato, D., Wang, P.: A survey of mobile cloud computing: architecture, applications, and approaches. Wirel. Commun. Mob. Comput. **13**(18), 1587–1611 (2013)
7. http://www.mobilecloudcomputingforum.com/
8. Christensen, J.H.: Using RESTful web-services and cloud computing to create next generation mobile applications. In: The proceedings of the 24th ACM SIGPLAN Conference Companion on Object Oriented Programming Systems Languages And Applications (OOPSLA), pp. 627–634, October 2009

9. Liu, L., Moulic, R., Shea, D.: Cloud service portal for mobile device management. In: The proceedings of the IEEE 7th International Conference on e-Business Engineering (ICEBE), p. 474, January 2011
10. Ahamad, S., Al-Shourbaji, I., Al-Janabi, S.: A secure NFC mobile payment protocol based on biometrics with formal verification. Int. J. Internet Technol. Secur. Trans. **6**(2), 103–132 (2016). https://doi.org/10.1504/ijitst.2016.078579. https://www.inderscienceonline.com/doi/pdf/10.1504/IJITST.2016.078579
11. Al-Janabi, S., Al-Shourbaji, I., Shojafar, M., Abdelhag, M.: Mobile cloud computing: challenges and future research directions. In: IEEE, 2017 10th International Conference on Developments in eSystems Engineering (DeSE), Paris, pp. 62–67 (2017). https://doi.org/10.1109/dese.2017.21. http://ieeexplore.ieee.org/stamp/stamp.jsp?tp=&arnumber=8285798&isnumber=8285458
12. Youseff, L., Butrico, M., Da Silva, D.: Toward a unified ontology of cloud computing. In: Grid Computing Environments Workshop, pp. 1–10. IEEE (2008)
13. Buyya, R., Yeo, C., Venugopal, S.: Market-oriented cloud computing: vision, hype, and reality for delivering IT services as computing utilities. In: The proceedings of the 10th IEEE International Conference on High Performance Computing and Communications, (HPCC 2008), pp. 5–13 (2008)
14. Buyya, R., Yeo, C.S., Venugopal, S., Broberg, J., Brandic, I.: Cloud computing and emerging IT platforms: vision, hype, and reality for delivering computing as the 5th utility. J. Futur. Gener. Comput. Syst. **25**(6), 599–616 (2009)
15. https://www.dropbox.com/
16. Forman, G.H., Zahorjan, J.: The challenges of mobile computing. IEEE Comput. Soc. Mag. **27**(4), 38–47 (1994)
17. Kakerow, R.: Low power design methodologies for mobile communication. In: Proceedings of IEEE International Conference on Computer Design: VLSI in Computers and Processors, p. 8, January 2003
18. Paulson, L.D.: Low-power chips for high-powered handhelds. IEEE Comput. Soc. Mag. **36** (1), 21 (2003)
19. Davis, J.W.: Power benchmark strategy for systems employing power management. In: The proceedings of the IEEE International Symposium on Electronics and the Environment, p. 117, August 2002
20. Mayo, R.N., Ranganathan, P.: Energy consumption in mobile devices: why future systems need requirements aware energy scale-down. In: The proceedings of the Workshop on Power-Aware Computing Systems, October 2003
21. http://aws.amazon.com/s3/
22. Vartiainen, E., Mattila, K.V.-V.: User experience of mobile photo sharing in the cloud. In: The Proceedings of the 9th International Conference on Mobile and Ubiquitous Multimedia (MUM), December 2010
23. Zou, P., Wang, C., Liu, Z., Bao, D.: Phosphor: a cloud based DRM scheme with sim card. In: The proceedings of the 12th International Asia-Pacific on Web Conference (APWEB), p. 459, June 2010
24. Yang, X., Pan, T., Shen, J.: On 3G mobile e-commerce platform based on cloud computing. In: The proceedings of the 3rd IEEE International Conference on Ubi-Media Computing (U-Media), pp. 198–201, August 2010
25. Dai, J., Zhou, Q.: A PKI-based mechanism for secure and efficient access to outsourced data. In: The Proceedings of the 2nd International Conference on Networking and Digital Society (ICNDS), vol. 1, p. 640, June 2010

26. Leina, Z., Tiejun, P., Guoqing, Y.: Research of mobile security solution for fourth party logistics. In: The Proceedings of the 6th International Conference on Semantics Knowledge and Grid (SKG), pp. 383–386, January 2011
27. Al-Janabi, S., Al-Shourbaji, I.: A study of cyber security awareness in educational environment in the middle east. J. Inf. Knowl. Manag. **15**(01), 1650007 (2016). https://doi.org/10.1142/s0219649216500076
28. Al_Janabi, S.: Smart system to create optimal higher education environment using IDA and IOTs. Int. J. Comput. Appl. (2018). https://doi.org/10.1080/1206212x.2018.1512460
29. Rudenko, A., Reiher, P., Popek, G.J., Kuenning, G.H.: Saving portable computer battery power through remote process execution. In: Journal of ACM SIGMOBILE on Mobile Computing and Communications Review, vol. 2, no. 1 (1998)
30. Cuervo, E., Balasubramanian, A., Cho, D., Wolman, A., Saroiu, S., Chandra, R., Bahl, P.: MAUI: making smartphones last longer with code offload. In: The Proceedings of the 8th International Conference on Mobile systems, applications, and services, pp. 49–62 (2010)
31. http://www.flickr.com/
32. http://www.facebook.com/
33. http://seekingalpha.com/article/729681-microsoft-s-new-office-software-to-use-cloud-computing-adapt-for-mobile-appsand-devices
34. Ali, S.H.: Novel approach for generating the key of stream cipher system using random forest data mining algorithm. In: IEEE, 2013 Sixth International Conference on Developments in eSystems Engineering, Abu Dhabi, pp. 259–269 (2013). https://doi.org/10.1109/dese.2013.54
35. Mun, K.: Mobile Cloud Computing Challenges, September 2010. http://www2.alcatel-lucent.com/techzine/mobile-cloud-computing-challenges/#sthash.0BAMDI9B.dpuf
36. http://www2.alcatel-lucent.com/techzine/mobile-cloud-computing-challenges/
37. Rochwerger, B., Breitgand, D., Levy, E., Galis, A., Nagin, K., Llorente, L., Montero, R., Wolfsthal, Y., Elmroth, E., Caceres, J., Ben-Yehuda, M., Emmerich, W., Galan, F.: The RESERVOIR model and architecture, for the open federated cloud computing. IBM J. Res. Dev. **53**(4), 4 (2009)
38. Chetan, S., Kumar, G., Dinesh, K., Mathew, K., Abhimanyu, M.: Cloud computing for mobile world. chetan.ueuo.com
39. Klein, A., Mannweiler, C., Schneider, J., Schotten, H.D.: Access schemes for mobile cloud computing. In: The 11th International Conference on Mobile Data Management (MDM), pp. 387–392 (2010)
40. http://reviews.cnet.com/8301-6452_7-57602215/iphone-5s-vs-iphone-5c-vs-iphone-5-specs-compared/
41. Qureshi, S.S., Ahmad, T., Rafique, K., Shuja-ul-Aslam: Mobile cloud computing as future for mobile applications – implementation methods and challenging issues. In: The Proceedings of the IEEE CCISS 2011 (2011)
42. Chun, B.G., Maniatis, P.: Augmented smartphone applications through clone cloud execution. In: The Proceedings of the 12th Conference on Hot Topics in Operating Systems, Berkeley, CA, USA (2009)
43. Chun, B., Ihm, S., Maniatis, P., Naik, M., Patti, A.: CloneCloud: elastic execution between mobile device and cloud. In: The Proceedings of the 6th Conference on Computer Systems, pp. 301–314. ACM (2011)
44. Wen, Y., Zhang, W., Luo, H.: Energy-optimal mobile application execution: taming resource-poor mobile devices with cloud clones. In: INFOCOM IEEE Proceedings, pp. 2716–2720 (2012)

45. Samimi, F.A., McKinley, P.K., Sadjadi, S.M.: Mobile service clouds: a self-managing infrastructure for autonomic mobile computing services. In: The Proceedings of the 2nd IEEE International Workshop on Self-Managed Networks, Systems and Services (SelfMan), pp. 130–141 (2006)
46. Zhang, X., Jeong, S., Kunjithapatham, A., Gibbs, S.: Towards an elastic application model for augmenting computing capabilities of mobile platforms. In: The 3rd International ICST Conference proceedings on Mobile Wireless Middleware, Operating Systems, pp. 161–174 (2010)
47. Kosta, S., Aucinas, A., Hui, P., Mortier, R., Zhang, X.: ThinkAir: dynamic resource allocation and parallel execution in cloud for mobile code offloading. In: The INFOCOM IEEE Proceedings, pp. 945–953 (2012)
48. Yang, L., Cao, J., Tang, S., Li, T., Chan, A.T.S.: A Framework for partitioning and execution of data stream applications in mobile cloud computing. In: The Proceeding of the 5th International Conference on Cloud Computing (CLOUD), pp. 794–802. IEEE (2012)
49. Al-Janabi, S., Al-Shourbaji, I.: A hybrid Image steganography method based on genetic algorithm. In: 2016 7th International Conference on Sciences of Electronics, Technologies of Information and Telecommunications (SETIT), Hammamet, pp. 398–404 (2016). https://doi.org/10.1109/setit.2016.7939903
50. Al-Janabi, S., Al-Shourbaji, I., Shojafar, M., Shamshirband, S.: Survey of main challenges (security and privacy) in wireless body area networks for healthcare applications. Egypt. Inform. J. **18**(2), 113–122 (2017). https://doi.org/10.1016/j.eij.2016.11.001. http://www.sciencedirect.com/science/article/pii/S1110866516300482, ISSN 1110-8665
51. Al-Janabi, S.: Pragmatic miner to risk analysis for intrusion detection (PMRA-ID). In: Mohamed, A., Berry, M., Yap, B. (eds.) Soft Computing in Data Science, SCDS 2017. Communications in Computer and Information Science, vol. 788. Springer, Singapore (2017). https://doi.org/10.1007/978-981-10-7242-0_23
52. Freier, A.O., Karlton, P., Kocher: The SSL Protocol. PC (1996). http://home.netscape.com/eng/ssl3/ssl-toc.hml
53. Drew, G.N.: Using SET for Secure Electronic Commerce. Prentice-Hall Inc., Upper Saddle River (1999)
54. Liu, D.: Research of the two electronic commerce payment protocols: SSL and SET. Secur. Saf. Magzine **4**, 61–63 (2003)
55. Chan, H.-W.: Mobile A gent Security and Reliability Issues in Electronic Commerce. Master thesis. The Chinese University of Hong Kong (2002)
56. Zhao, S., Xin, F.-Q., Ma, J.-Z.: Research on secure mobile agent-based electronic commerce. China Acad. J. Electron. Publ. House, 10–11 (2007)
57. Zhang, D.-L., Lin, C.: Security model of mobile agent in e-commerce. China Acad. J. Electron. Publ. House **25**(6), 1271–1273 (2005)
58. Wang, R.-Y., Liang, L., Liu, W.-F.: Research of mobile agent-based security in the e-commerce. J. Appl. Res. Comput. **11**, 137–141 (2004)
59. Kannammal, A., Iyengar, N.: A model for mobile agent security in e-business applications. Int. J. Bus. Inf. **2**(2), 185–198 (2007)
60. Yu, B., Yi, X.: Mobile agent-based encryption transmission system. Masters thesis. The GuiZhou University, China (2006)
61. Wang, Y., Wang, Z., Wei, L.-F.: A migration mechanism of mobile agent system supporting security and fault-tolerance. J. Comput. Technol. Dev. **17**(3), 169–175 (2007)
62. Krzyzanowski, P.: Process synchronization and election algorithms. Lecture Notes on Distributed Systems (2000). http://www.cs.rutgers.edu/~pxk/rutgers/notes/content/06-mutex.pdf

63. Antoniou, G., Sterling, L., Gritzalis, S., Udaya, P.: Privacy and forensics investigation process: the ERPINA protocol. J. Comput. Sci. Interface. **30**(4), 229–236 (2008). http://www.w3.org/P3P/Platform-for-Privacy-Preferences-(P3P)-Project
64. Al_Janabi, S., Mahdi, M.A.: Evaluation prediction techniques to achievement an optimal biomedical analysis. Int. J. Grid Util. Comput. (2019)
65. Son, S., Jung, G., Jun, S.: An SLA-based cloud computing that facilitates resource allocation in the distributed data centers of a cloud provider. J. Supercomput. **64**, 606–637 (2013)
66. Izakian, H., Abraham, A., Ladani, B.T.: An auction method for resource allocation in computational grids. Futur. Gener. Comput. Syst. **26**, 228–235 (2010)
67. Buyya, R., Stockinger, H., Giddy, J., Abrams, D.: Economic models for management of resources in grid computing. J. Concurr. Comput. **14**, 1507–1542 (2002)
68. Hausheer, D., Stiller, B. (eds.): Implementation of economic grid traffic management and security mechanisms. In: Deliverable D4.1 of the EC-GIN Project, May 2009. http://www.csg.uzh.ch/research/ec-gin/publications/D4.1.pdf
69. Wang, X., Sun, J., Huang, M., Wu, C.: A resource auction based allocation mechanism in the cloud computing environment. In: The 26th IEEE International Parallel and Distributed Processing Symposium Workshops & PhD Forum (IPDPSW), pp. 2111–2115 (2012)
70. Fujiwara, I., Aida, K., Ono, I.: Applying double-sided combinational auctions to resource allocation in cloud computing. In: 2010, The 10th IEEE/IPSJ International Symposium Applications and the Internet (SAINT), 14 July 2010
71. Muhammad, S., Furqan, Z., Guha, R.K: Understanding the intruder through attacks on cryptographic protocols. In: The Proceedings of the 44th ACM Southeast Conference (ACMSE2006), pp. 667–672, March 2006
72. Stallings, W.: Cryptography and Network Security: Principles and Practices, 5th edn. Pearson Education (2011). ISBN-10: 0131873164, ISBN-13: 9780131873162
73. Cremers, C.J.F.: Scyther - Semantics and Verification of Security Protocols, Ph.D. thesis, Eindhoven University of Technology (2006)
74. Armando, A., et al.: The AVISPA tool for the automated validation of internet security protocols and applications. In: The Proceedings of Computer Aided Verification (CAV 2005). Lecture Notes in Computer Science, vol. 3576, pp. 281–285. Springer (2005). http://dl.acm.org/citation.cfm?id=2153265
75. Smyth, B.: Formal verification of cryptographic protocols with automated reasoning, Ph.D. thesis, School of Computer Science, College of Engineering and Physical Sciences, University of Birmingham, March 2011
76. Cremers, C., Lafourcade, P.: Comparing state spaces in automatic security protocol verification. Electronically published in Electronic Notes in Theoretical Computer Science from AVoCS 2007. http://www.lsv.ens-cachan.fr/Publis/PAPERS/PDF/CL-avocs07.pdf
77. Blanchet, B., Smyth, B., Cheval, V.: ProVerif 1.88: Automatic Cryptographic Protocol Verifier, User Manual and Tutorial 1, INRIA Paris-Rocquencourt, Paris, France (2013). http://prosecco.gforge.inria.fr/personal/bblanche/proverif/manual.pdf
78. Nogueira, J.H.M.: Mobile intelligent agents to fight cyber intrusions. Int. J. Forensic Comput. Sci. (2006)
79. Strassner, J., Raymer, D.: Implementing next generation services using policy based management and autonomic computing principles. In: The proceedings of the 10th IEEE/IFIP Network Operations and Management Symposium – NOMS (2006)

# A Multilayer Perceptron Classifier for Monitoring Network Traffic

Azidine Guezzaz[1,2](✉), Ahmed Asimi[2], Azrour Mourade[3], Zakariae Tbatou[2], and Younes Asimi[2,4]

[1] Technology High School, M2SC Team Essaouira, Cadi Ayyad University, Marrakesh, Morocco
a.guzzaz@gmail.com
[2] Faculty of Sciences, Ibn Zohr University, 8106, Dakhla, Agadir, Morocco
[3] Faculty of Science and Technology, IDMS Team Errachidia, Moulay Ismail University, Meknes, Morocco
[4] High School of Technology Guelmim, Ibn Zohr University, Agadir, Morocco

**Abstract.** The network security is the process of preventing and detecting unauthorized use of the networks and stopping unauthorized users from accessing any part of systems. So, the network monitoring is demanding task. It is an essential part in the use of network administrators who are trying to maintain the good operating of their networks and need to monitor the traffic movements and the network performances. The needs of data analysis and classification are significantly increased to categorize data and make decisions by implementation of standardization and information extraction techniques. The design of an efficient classifier is one of the fundamental issues of automatic training. The actual network monitoring systems suffer from many constraints at the level of analysis and classification of data. To overcome this problem, it was necessary to define reliable methods of analysis to implement a relevant system to monitor the circulated traffic. The main goal of this paper is to model and validate a new traffic classifier able to categorize the collected data within the networks. This new proposed model is based on multilayer perceptron that composed of three layers. An efficient training algorithm is proposed to optimize the weights but also a recognition algorithm to validate the model.

**Keywords:** Security · Monitoring · Perceptron · Training · Recognition

## 1 Introduction and Notations

The classification of data is one of the research axes which aims an automatic resolution of problems of the real world in this case the recognition of forms, classification and generalization using techniques of data processing [1, 2]. The large dimension and heterogeneity represent a specific challenge to data analysis methods.

The second section presents a background of some training algorithms and recognition methods of data which are more used in practice. A new classifier model of traffic based on a multilayer perceptron is presented in the third part.

The article is accomplished with a conclusion and perspectives. In this document, we use the following notations (Table 1):


Y. Farhaoui (Ed.): BDNT 2019, LNNS 81, pp. 262–270, 2020.
https://doi.org/10.1007/978-3-030-23672-4_19

**Table 1.** Used Notations.

| $f$ | : Sigmoid function |
|---|---|
| $(X_i)_{i=1\ldots n}$ | : Presented inputs |
| $X_i = \left(x_{i,j}\right)_{j=1\ldots m}$ | : The presented occurrences to input $X_i$ |
| $W^{(0)} = \left(w_{i,0}\right)_{i=1\ldots n}$ | : The initialized weights |
| $W_i = \left(w_{i,j}\right)_{j=1\ldots m}$ | : The model weights initialized randomly and associated to input $X_i$ |
| $w_{0,i}$ | : The initialized Bias to 1 and associated with the input $X_i$ |
| $a_i$ | : The weighted sum associated to input $X_i$ |
| $y(a_i) = f(a_i)$ | : The calculated output associated to input $X_i$ |
| $\varepsilon_i$ | : The calculated error associated to an input $X_i$ |

# 2 Background

The training is a developmental phase in which the behavior changes until a desired one is achieved by optimization of weights, the examples are presented to establish new connections or to modify the existing ones. The calculated result is compared with the expected response in the output [1, 3–7].

The unsupervised training extracts information from the learning observations. The obtained result is analyzed in order to be retained or rejected. In this type of training, the data are not complete. The choice of a training algorithm depends on the specification of the problem to be solved [8]. To simplify analysis, a preprocessing is often applied to the data presented to the training. Any element of the training basis is represented by an input / output pair. In fact, the classification of data becomes a very useful discipline that helps to solve complex problems [9, 10]. The data is not always in a ready format for analysis. So, the standardization of unstructured data requires an additional work.

## 2.1 Data Classification Methods

A classification method is a set of precise rules to classify objects based on the quantitative and qualitative variables characterizing these objects.

An artificial neural network is a collection of interconnected processing elements that transform inputs into outputs. The result of the transformation is determined by the characteristics of the elements and the weights associated with their interconnection. The weights are modified so that the network adapts to the desired output. The neural networks are proposed as alternatives to statistical analysis methods. A network of neurons is a set of mathematical methods used to model processes and recognize patterns [4, 7, 9, 11, 12]. It can classify activities from non-linear or incomplete data. A formal neuron is a processing unit of several inputs performing complex logic, arithmetic or symbolic functions. The output corresponds to weighted sum of inputs:

$$a_i = \sum_{i=1}^{n} w_i^{(j)} x_i^{(j)} + w_{0,i} \tag{1}$$

For recognition problems, it is useful to determine the number of classes to which the inputs belong. Each class is determined to receive an input. Two databases are used, one for training and other for validation [12, 13].

The more the training base is significant and contains a large amount of data, the more accurate the prediction rate, but the calculations will be long and complex.

### 2.2 Support Vector Machines

SVM are the machine learning methods derived from early work of Vapnik [14–17]. They are supervised training techniques designed to solve problems of discrimination and regression. The principle is to transform the input variables into a large characteristic space and then find a better hyper plane that models the data in space and separates two groups so that the linear boundary produces a maximum margin.

```
Algorithm: SVM Algorithm
   1. Find the hyper plane as a solution to a constrained
      optimization problem.
   2. The search for nonlinear separating surfaces is
      introduced using a kernel that encodes the
      nonlinear transformation of the data.
   3. The equations are obtained based on some scalar
      products using the kernel and some database
      weights.
```

### 2.3 Multilayer Perceptron

Multilayer perceptron (Rosenblatt 1957) is a neural network that composed of successive feed forward layers connecting neurons by weighted links [5, 18–20]. The input layer is used to collect the input signals and the output layer provides responses. One or more hidden layers are added for transfer. Multilayer perceptron learning is performed by the error gradient propagation algorithm [4, 5, 8] (Figs. 1, 2 and 3).

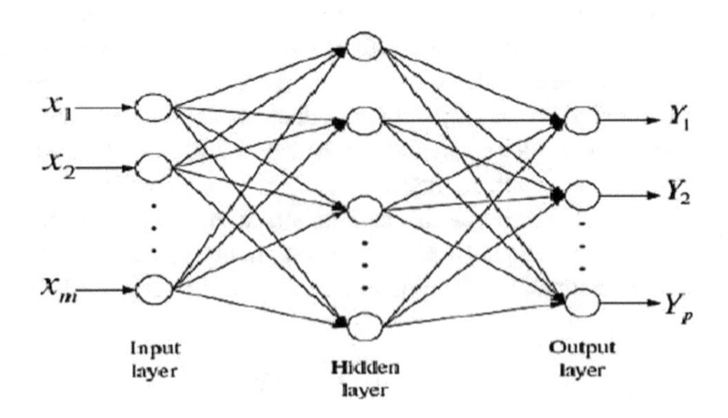


**Fig. 1.** Multilayer Perceptron Structure

```
Algorithm: Back Propagation Training
DBA : Training Base.
X_i = (x_{i,j})_{j=1...m} : Inputs.
C_i = (c_{i,j})_{j=1...m} : Desired Results for X_i.
W_i = (w_{i,j})_{j=1...m} : Weights for X_i.
θ_i : Calculated Results.
λ_i : Training rate.
BEGIN: Calculate (W_i)_{i=1...n} for Inputs (x_i)_{i=...n}.
        for i from 1 to n do
        Initialize randomly the wights
          Optimization of wights :
             For j from 1 to m do
                 w_{i,j} = w_{i,j} + λ_i(c_{i,j} - θ_i)x_{i,j};
             End For
        End For
END
```

# 3 Our Work

This section describes the proposed solutions to validate our new model of a heterogeneous traffic classifier based on the multilayer perceptron.

## 3.1 The Proposed Model

The contribution aims to propose a rigorous algorithm for the training and classification of the events that constitute the collected traffic. Incorporation of the multilayer perceptron is suggested. The proposed model is easy to implement, works with unstructured data and presents a high accuracy.

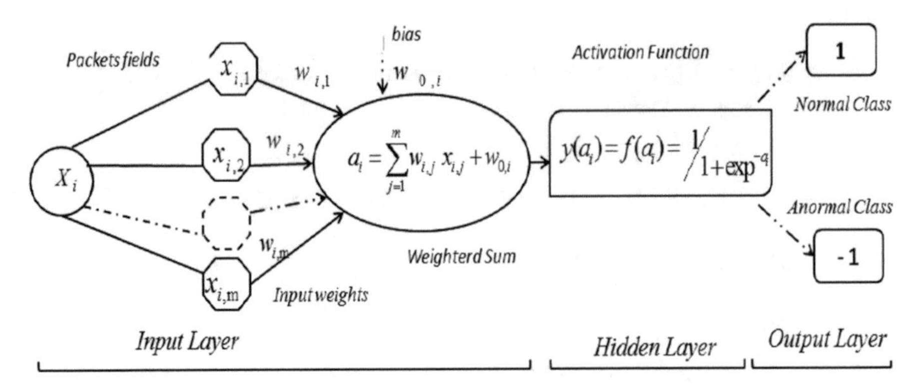


**Fig. 2.** Structure of the proposed model

The validation of such a model requires static data with supervised methods or structured data presented in sequence. Our architecture consists of three layers; each layer has neurons directly linked to the neurons of the next layer. The direct connections between the input layer and the output layer don't exist. We add a single layer hidden to achieve our goal.

### 3.2 Description of Solutions

The new design defines a supervised method using a three layer perceptron. The training basis is intercepted in a period when the network is offline. Over this period is longer, the base will be the most significant. The collection is shared horizontally between the training and testing base:

- 80% To implement the model (Training basis).
- 20% To evaluate the model performances (Test basis).

We find ourselves faced with an optimization model containing modifiable variables that describe the problem, together with constraints representing limits on these variables. We define a function that assigns a value to each iteration of adjusting and thus optimizes the model's weights. The cost function to minimize is:

$$1 - y(a_i) \text{ With } a_i = \sum_{j=1}^{m} w_{i,j} x_{i,j} + w_{0,i} \text{ for } i = 1 \ldots n \tag{2}$$

We are looking in the training phase to adjust the weights of the proposed model for reliable recognition.

```
Algorithm: Training Algorithm
Initialize the weights W^(0) = (w_{i,j})_{j=1...m} such as 0 ≤ w_{i,0} ≤ 10^-3 and w_{0,i} = 1 for
i = 1... n.
For i from 1 to n do
        Present the inputs: X_i = (x_{i,j})_{j=1...m}  and  a_i = Σ_{j=1}^{m} w_{i,j} x_{i,j} + w_{0,i} ;
        Calculate W_i^(op), ε_i ≠ 0 and a_i^(op) :
          For k from 1 to m do
             ε_i = 1 - y(a_i);
                 For j from 1 to m do
                 w_{i,j} = w_{i,j} + [1 - y(a_i)] x_{i,j};
               End For
                  a_i = Σ_{j=1}^{m} w_{i,j} x_{i,j} + w_{0,i};
                   w_{0,i} = w_{0,i} + [1 - y(a_i)];
        End For
  End For
```

The recognition phase consists in validating our model using the weights $\left(w_j^{(\max)}\right)_{j=1...m}$ obtained in the training phase:

```
Algorithm: Recognition Algorithm
New   Inputs X = (x^(j))_{j=1...m} ,   final   output   d, activation   state
a = Σ_{j=1}^{m} w_j^(max) x^(j) + w_0^(max);

Classification of Activities:
        if (1 - y(a) ≤ ε) then
            d = 1 // Normal activity.
        else
            d = -1 //Intrusion.
        End if
```

In this part, an experimental study is realized to validate the different tasks of the model, but also the constraints confronted during implementation.

To collect traffic, we use Wireshark [21, 22] in a Windows 7 operating system. It is a Libpcap based network traffic sniffer that controls wired or wireless interfaces in a specific time. It allows filtering and storing of the collection for later analysis. We use XML technology to standardize traffic.

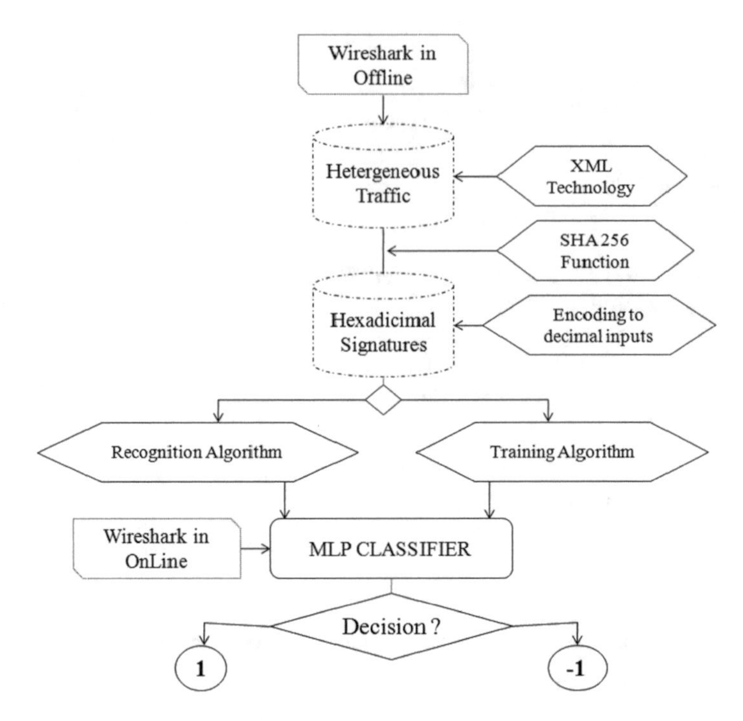


**Fig. 3.** Implementation Solutions

After the normalization of traffic occurrences, a coding process is applied on the basis. Each occurrence is signed by SHA256 [23], one of the most powerful hash functions and is not complex to code. Thus, a 256-bit hexadecimal signature is produced. The importance of this function is represented at the level of the constant size of the output which allows us to fix the number of neurons of the input layer supporting all the occurrences.

## 4 Conclusion and Perspectives

In fact, the classification of data becomes a very useful discipline that helps to solve complex problems. The data is not always in a ready format for analysis. The recognition is an indispensable task that associates an activity to one or more classes by applying a suitable algorithm on an audit base and generating a classifier able to distinguish the normal behavior from the new data. It becomes a necessary task that intervenes in applications that require making of decision. In this paper, we present some used techniques of training and classification to solve problems of real world and we deduce that the majority of the classifiers suffer from the limits at level of the large dimension of the data which have an impact on the classification task. The validation of the new traffic classifier is achieved by proposed solutions that guarantee an efficient and fast analysis.

The classifier represents a perceptual analysis layer which can be integrated into monitoring systems to make control of the data flow more reliable. The next task is to validate a relevant monitoring approach that implements this layer to analyze data.

## References


1. Malik, N.: Artificial neural networks and their applications. In: National Conference on 'Unearthing Technological Developments & their Transfer for Serving Masses' GLA ITM, Mathura, India, 17–18 April 2005
2. Dietterich, T.G.: Approximate statistical tests for comparing supervised classification learning algorithms. Neural Comput. **10**, 1895–1923 (1998). Massachusetts Institute of Technology
3. Kotsiantis, S.B., Zaharakis, I.D., Pintelas, P.E.: Machine Learning: A Review of Classification and Combining Techniques, Published online: 10 November 2007, Springer Science+Business Media B.V. (2007)
4. Dyboski, R., Gant, V.: Clinical applications of artificial neural networks, king's college London, published by the press Syndicate of the university of cambridge the Pitt Building, Trumpington Street, Cambridge, United Kingdom.
5. Schiffmann, W., Joost, M., Werner, R.: Optimization of the Backpropagation Algorithm for Training Multilayer Perceptrons, University of Koblenz Institute of Physics Rheinau 1 56075 Koblenz First edition published in 1992.
6. Malouche, D.: Méthodes de classications, essai-u2s-enit http://essai.academia.edu/Dhafer Malouche/dhafer.malouche@me.com
7. Rochaa, M., Cortezb, P., Nevesa, J.: Evolution of neural networks for classification and regression. Neurocomputing **70**, 2809–2816 (2007). Elsevier
8. Parizeau, M.: réseaux de neurones, Livre Automne (2004)
9. Lapôtre, R. : Faire parler les données des bibliothèques:du Big Data à la visualisation de données Diplôme de Conservateur des Bibliothèques, Sous la direction de Julien Velcin Maître de Conférence en informatique – Université Lumière Lyon 2
10. Bouveyron, C., Girard, S., Schmid, C.: Classification des données de grande dimension: application à la vision par ordinateur.
11. Bivens, A., Palagiri, C., Smith, R., Szymanski, B., Embrechts, M.: Network-based intrusion detection using neural networks
12. Suthaharan, S.: Machine Learning Models and Algorithms for Big Data Classification. ISIS, vol. 36. Springer, Boston, MA (2016). https://doi.org/10.1007/978-1-4899-7641-3
13. Solla, S.A.: Capacity control in classifiers for pattern recognition. AT&T Bell Laboratories, Holmdel, New Jersey 07733, USA and connect, The Niels Bohr Institute Blegdamsvej 17, DK-2100 Copenhagen, Denmark
14. Vapnik, V.N.: An Overview of Statistical Learning Theory. IEEE Trans. Neural Netw. **10** (5), 988–999 (1999)
15. Dom, T.N.*, Fekete, J.D.**, Poulet, F.***: Algorithmes rapides de boosting de SVM, *Equipe Aviz, INRIA Futurs, LRI Bât.490, Université Paris Sud 91405 Orsay Cedex. **IRISA TexMex, Université de Rennes I Campus de Beaulieu, 35042 Rennes Cedex.
16. Gadat, S., Algorithmes de support vector machines, laboratoire de statistique et probabilités umr 5583 cnrs-ups, www.lsp.ups-tlse.fr/gadat.
17. Lauer, F., Bloch, G. : Méthodes SVM pour l'identication, Centre de Recherche en Automatique de Nancy (CRAN UMR 7039), Nancy Université, CNRS CRAN-ESSTIN, rue Jean Lamour, 54519 Vand÷uvre Cedex.

18. Comon, P., Classification supervisée par réseaux multicouches, Supervised Classification by Multilayer Networks. THOMSON-SINTRA, Parc de Sophia Antipolis, BP 138, F-06561 albonne Cedex. Soumis à la revue Traitement du Signal le 7/9/91; révisé le 17 March 1992
19. Beheshti, M., Berrached, A., de Korvin, A., Hu, C., Sirisaengtaksin, O.: On Interval Weighted Three-layer Neural Networks. Department of Computer and Mathematical Sciences University of Houston-Downtown Houston, TX 77002
20. Mendes, R., Cortez, P., Rocha, M., Neves, J.: Particle Swarms for Feedforward Neural Network Training. 0-7803-7278-6/02/$10.00 02002 IEEE.
21. Guezzaz, A., Asimi, A., Sadqi, Y., Asimi, Y., Tbatou, Z.: A new hybrid network sniffer model based on Pcap language and sockets (Pcapsocks). Int. J. Adv. Comput. Sci. Appl. (IJACSA), **7**(2) (2016)
22. Asrodia, P., Patel, H.: Analysis of various packet sniffing tools for network monitoring and analysis department of computer science and engineering. Int. J. Electric., Electron. Comput. Eng. **1**(1), 55–58 (2012)
23. Drissi, A., Asimi, A.: One-way hash function based on goppa codes « OHFGC ». Appl. Math. Sci. **7**(143), 7097–7104 (2013)

# Design (More-G) Model Based on Renewable Energy & Knowledge Constraint

Samaher Al_Janabi(✉), Samah Alhashmi, and Zuhal Adel

Department of Computer Science, Faculty of Science for Women (SCIW), University of Babylon, Babylon, Iraq
samaher@itnet.uobabylon.edu.iq

**Abstract.** In recent years, there has been a widespread trend towards generating electricity based on environmentally friendly sources of energy that do not cause carbon dioxide emissions during generation, such as solar, wind and water energy. The main point in the process of generating electricity from these sources is the cost and effectiveness of the technology used to generate them. This proposal will focus on building a new multi-objectives optimization approach that considers all the real determinants of the Iraqi environment to generate renewable energy from the previously mentioned environment-friendly sources. Where we will seek to design an integrated system in terms of cost and efficiency simulates the reality, the drama of all its parameters, positive points and how to benefit from it in practice. In addition to identifying the most important obstacles that prevent the implementation of that model on the ground and try to overcome them.

**Keywords:** Multi objectives optimization · Knowledge constraints · Wind energy · Cuckoo optimization · Horizontal combination

## 1 Introduction

Renewable energy is a kind of inexhaustible energy that is not depleted, and is called renewable energy because it comes from natural resources (i.e., wind, water, sun), the most important characteristic of a clean and environmentally friendly energy that does not lose harmful gases such as carbon dioxide, which does not adversely affect the surrounding environment. It does not play a role that affects the temperature level. Renewable sources of energy are completely incompatible with their non-renewable sources, such as natural gas and nuclear fuel. These sources lead to global warming and the release of carbon dioxide when used, there are several types of renewable energy: Solar energy, bioenergy, wind power, hydroelectric power, sustainable biofuel energy, geothermal or geothermal energy, tidal energy. For example, Iceland is one of the leading countries in renewable energy, it provides 100% of its electricity needs by generating renewable sources, particularly geothermal energy for heating homes, Lighting and electricity generation for industrial use and the like, the importance of renewable energy comes from the multiplicity of its use in the various areas of human life and its role in meeting the human needs and daily requirements as they enter the military, domestic, industrial and agricultural fields [10], the main advantage of


Y. Farhaoui (Ed.): BDNT 2019, LNNS 81, pp. 271–295, 2020.
https://doi.org/10.1007/978-3-030-23672-4_20

renewable energy is (available in most countries of the world, do not pollute the environment and maintain the general health of living organisms, economic in many uses, ensure their continued availability and use uncomplicated techniques).

Renewable energy technologies are effective and efficient solutions for clean and sustainable energy development in most countries, taking into account the geographical location of those countries where intensive use of most renewable energy resources is most important.

Data is one of the most valuable treasures on the world and is the basis of computer science in different branches [7]. Data refer to any object have set of features recognition it, or specific feature characteristic that object or collection of objects with their features. It takes different types and can get by observation, search or recording. In general, the researchers deal with concept called data science that combination among three domain Data domain, Intelligence domain and Statistics domain [7, 12]. There are many types of data science (i.e., small, normal and big\hug data); small data organize in uniform structure such as table or list contain also that size of that data not exceed 30 samples, therefore, it do not surrender of the normal distribution and can't be used to take any decision, at the other side, normal data it also structure data surrender to the normal distribution and it useful to take different decisions such as (clustering, classification, prediction, optimization, etc.). Finally, big data is on the other hand, take different structure such as structure, semi structure, or unstructured, the size of this data lie in the rang 1 TB to 1 ZB. Extraction useful knowledge or pattern it by combination between two main concepts machine learning techniques with cloud computing [8, 11].

Intelligent data analysis (IDA) is new field of computer science search the ability of build new or pragmatic approach to discovery or recovery pattern. This term refer to find real problem need to solve, that problem have specific definition, design model to handle it, that model may be for clustering, classification, prediction, optimization or generated rules or recommendation, after that analysis the result to make it understandable by the users such as preparing reports, histogram or mathematical model or network of their result [8, 12, 13].

Optimization is a general term used in multi field such as computer science and mathematics, the aim of Optimization is to get values of the variables that minimize or maximize the objective function while satisfying the constraints. This thesis will used multi objective optimization function to satisfy the objective of it [9, 14].

## 2 Related Works

To generate Renewable Energy from one of energy resource (the wind). Therefore, in this part of the thesis we will try to review the works of the researchers in the same area of our problem and compare these works in the terms of five basic points such as Name of authors, Name of data base/data set, methodology, Evaluation measures, and the advantage of the method.

Adam Chehouri, et al. 2015 [1] This paper presents a review of the optimization techniques and strategies applied to wind turbine performance optimization. The topic is addressed by identifying the most significant objectives, targets and issues, as well as the optimization formulations, schemes and models available in the published literature and offers the performance optimization of horizontal wind turbines by highlighting the main aspects when tackling the wind turbine optimization, our work similarity of it by using the same design constraints while differ it by applying the different type of tools and algorithms.

Qiong Wu, et al. 2016 [2] showed a multi-objective optimization method is promoted for the design of an energy system integrating biomass CHP, PV and heat storage. The results give a trade-off between economic and environmental performance while providing feasible generation settlements that take in to account of CHP, PV, together utility grid, we will be using the same multi-objective optimization method but different algorithm.

Jose Garcia, et al. 2018 [3] presented a binary cuckoo search big data algorithm applied to different instances of crew scheduling problem. We used an unsupervised learning method based on the K-means technique to perform binarization. The quality, convergence, and scalability of the results were evaluated in terms of the number of solutions used by the algorithm. It was found that quality, convergence, and scalability are affected by the number of solutions; however, these depend additionally on the problem that is being solved. In particular, it is observed that for medium size problems, the effects are not very relevant as opposed the large problems such as G and H, where the effect of the number of solutions is much more significant, we will use the same algorithm to different instances of wind power to generate renewable energy.

Regina Lamedica, et al. 2018 [4] proposes a methodology that is based on mixed-integer linear programming (MILP) to calculate the optimal sizing of a hybrid wind-photovoltaic power plant in an industrial area. The proposed methodology considers the: (i) load requirements; (ii) physical and geometric constraints for the renewable plants installation; (iii) operating and maintenance costs of both wind and PV power plants, and the (iv) electric energy absorbed by the public network, our work similarity of it by using the same constraints while differ it by applying the different type of evaluation measures.

E. L. V. Eriksson, E. MacA Gray 2019 [5] This paper has provided an outline of a new method for simulation and optimization of hybrid renewable energy systems using a Normalized Weighted Constrained Multi Objective Meta heuristic search technique which takes into account technical, economic, environmental and socio-political objectives concurrently in a weighted fashion with a constraint capability, our work similarity of it by Calculates the power output of a wind turbine while we differ from it using other measurements.

After, we survey the main article related to the problem statement that attempting to solved it in this thesis, we found our work is near from the article discuss in [5] from the step of handle the same problem but it differ in determined the constraints and develop the search algorithm through using multi objectives optimization.

Table 1: compare among the previous works from the six points include the name of the authors (s) and daley name of database/dataset using in that article, methodology suggest by author(s)

To solved that problem, main measures used to evaluation results, the main advantages of that methodology and finally the main challenges and disadvantage.

**Table 1.** Compare among the previous works

| Name of authors | Name of data base/data set | Methodology | Evaluation measures | Advantage |
|---|---|---|---|---|
| Adam Chehouri, et al. 2015 [1] | Cost energy | Optimization by using Meta-heuristic algorithms | Accuracy by using The Cost of Energy (CoF) | Have Advantages of 3D composite (Woven, Braided, Stitched, Knitted) textiles |
| Qiong Wu, et al. 2016 [2] | – Primary energy<br>– Solar<br>– Bio-Energy | Muilt-objective optimization problem by using two objective functions:<br>– Economic objective<br>– Environmental objective | Precision using:<br>– Photo-voltaics (PV)<br>– Combined heat and power (CHP)<br>– Grid | Minimized annual energy cost and environmental impact |
| Jose Garcia, et al. 2018 [3] | Beasley's OR-Library | Optimization by:<br>– Cuckoo algorithm<br>– K-means | Accuracy by using K-means | (1) The evaluation of the average value through the variation of the solutions number<br>(2) The evaluation of iteration number through the solution number used to solve problems<br>(3) The evaluation of algorithm scalability through executor Number |
| Regina et al. 2018 [4] | Http://Capasso R Lamedica, Podest_a L, Ruvio A, Sangiovanni S, Lazaroiu GC, Maranzano GA. A measurement campaign in a metro-train deposit/maintenance and repair site for PV production optimal sizing. In: 15th conference on environment and electrical engineering. IEEE EEEIC; 2015 | Optimization by using mixed-integer linear programming (MILP) methodology | Precision by using Values of amortization (A), and Vr and Values of revenues, amortization, Ci; tot, C O&M; cash flows, and NPV | – Generates industrial energy and identifies the problem and sets its limits in an understandable way<br>– It is useful for correctly defining the investments according to the different Company's objectives |
| Eriksson and Gray, 2019 [5] | http://M. Sharafi, T. Elmekkawy, Multi-objective optimal design of hybrid renewable energy systems using PSO-simulation based approach, | Optimization by using Normalized Weighted Constrained Multi-Objective (NWCMO) meta-heuristic function | Accuracy by using Capital recovery factor, Levelized cost of energy, Loss of power supply probability, Total present value | Calculates the power generated by each turbine |

(*continued*)

# 3 Main Concept

## 3.1 Big Data

Big data is a collection of data comes from various sources these data such as words, images, audio messages and others [8]. Three main factors must be available to say that data is big data:

- Size: The number of terabytes of data that required long time to process it.
- Diversity: the diversity of this data between structured, unstructured and half structured.
- Speed: how fast the frequency of data occurs, for example, the speed of deployment of tweets differs from the speed of remote sensing of climate change.

## 3.2 Search Method

A search method is a way of efficiently and effectively finding the information you need to answer your research question, it very useful because This increases your chances of finding relevant information because the search is focused, you will spend less time reading irrelevant material.

## 3.3 Integrated System

It is defined in engineering as the process of bringing together the component subsystems into one system (an aggregation of subsystems cooperating so that the system is able to deliver the overarching functionality) and ensuring that the subsystems function together as a system, and in information technology as the process of linking together different computing systems and software applications physically or functionally to act as a coordinated whole.

The system integrator integrates discrete systems utilizing a variety of techniques such as computer networking, enterprise application integration, business process management or manual programming.

System integration involves integrating existing, often disparate systems in such a way "that focuses on increasing value to the customer" e.g., improved product quality and performance) while at the same time providing value to the company (e.g., reducing operational costs and improving response time). In the modern world connected by Internet, the role of system integration engineers is important: more and more systems are designed to connect, both within the system under construction and to systems that are already deployed (Fig. 1).

"A disciplined, unified and iterative approach to the management and technical activities necessary to:

- integrate support considerations into system and equipment design;
- develop support requirements that are related consistently to readiness objectives, to design, and to each other;
- acquire the required support; and
- provide the required support during the operational phase at minimum cost".

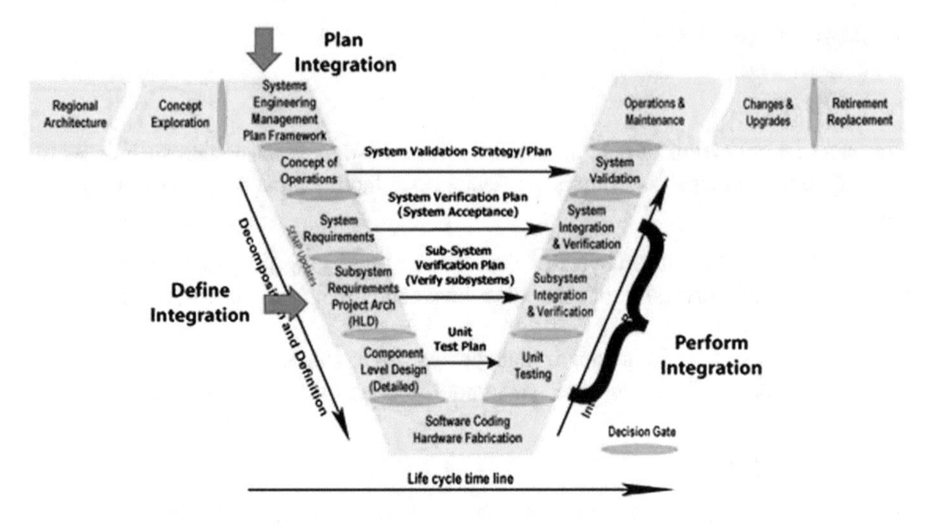


**Fig. 1.** Main stages of Integration System

### 3.3.1 Objective Function

The objective function is to find the highest electrical energy (renewable energy) by using the linear combination function for each x 1, x 2, x 3, x 4, x 5 where:

- X1: min wind
- X2: min area
- X3: optimal team
- X4: optimal material
- X5: min cost

$$\text{Max energy} = (\text{x1} + \text{x2} + \text{x3} + \text{x4} + \text{x5})$$

### 3.3.2 Constraints

The decision variables, together with data, must satisfy some technological, economic, and geometric constraints. Mathematically, these constraints are expressed by linear equalities and inequalities involving the variables, Each dataset related to FED has constraints different from other dataset, as follow:

- ***Wind Dataset***

After collected the wind dataset in step number one of this model, we found that dataset contain multi features (i.e. date, time, barometric pressure (INHG), temperature (F), wind direction (DEG), wind speed (MPH), resultant wind speed (MPH), standard deviation of wind direction (DEG), standard deviation of wind speed (MPH), peak time, peak speed (MPH)). The features more important of the problem descripted in chapter one are speed of the wind and time of capture this wind. In general, we descripted that constraints as follow:

$$F = \sum (\Delta wt), t = 1..T \quad (1)$$

Where $\Delta wt$ is wind power curtailment at time t, T is total dispatch period.

- ***Cost Dataset***

After collected the cost dataset in step number one of this model, we found that dataset contain the cost of the features (i.e. area, turbine, material), and to find the minimum cost, we first find the sum of all features for each i-th according the flowing equation:

$$F = \sum ki \quad (2)$$

Where $i = 1..N$, N is the number of features

$$\text{Min cost} = \min(F) \quad (3)$$

- ***Area Dataset***

These constraints respect to the wind-energy production, a constraints arises owing the limited available ground area for the installation of the wind turbines, in order to resolve these obstacles, consideration should be given to the following constraints:

First constraint, should be the total number of turbines in the area taken by each turbine should be less or equal than the total area available:

$$X1 * Ab <= Amax \quad (4)$$

Where X is the number of turbine, Ab is the area of turbine, Amax is the max area.

For example, let number of turbines = 5, the area that takes each turbine = 4 m and the max area = 400 m:

$$X1 * Ab <= Amax$$
$$5 * 4 = 20$$
$$20 < 400 \rightarrow \text{The area is acceptable}$$

Second constraint, it is achieved if the first constraint is processed:

$$X2 <= ([(S1 - 3)(S1 - 3)]/[L2 * \cos \square(1 + \tan \square/\tan \square)]) \quad (5)$$

$$D = L * \cos \square(1 + \tan \square/\tan \square) \quad (6)$$

Third constraint, it is achieved if both the first and second constraints are processed:

$$\text{Max turbine} = (S1 - 3)/D \tag{7}$$

$$X3 <= ([(S1 - 3)/L] * \text{Maxturbine}) \tag{8}$$

Where Xi $> = 0$, $i = 1, 2\ldots N$, Xi is the energy generated in each turbine, X2 is the type of turbine, S1 is the large side of the available area for installation of turbine, L is the feather length related of turbine plus 50 cm, □ is Wind speed, □ is the Wind direction, X3 Total values of turbine types.

- ***Material Dataset***

After collected the material dataset in step number one of this model, we found that dataset contain some features (i.e. wire, type, stats, years, batteries), and to find the best tools used in the power generation process, we first find the sum of all features for each t-th according the flowing equation as follow:

$$F = \sum Mt \tag{9}$$

Where $t = 1..N$, N is the number of features

$$\text{best material} = \max(F) \tag{10}$$

- ***Team Dataset***

In step number one of this model collected the team dataset after that, we found that dataset contain some features (i.e. certificate, number of work clock, number of experience certificate, number of cycles, age), and to find the best team used in generation process, we first find the sum of all features for each r-th according the flowing equation as follow:

$$F = \sum Tr \tag{11}$$

Where $r = 1..N$, N is the number of features

$$\text{best team} = \max(F)$$

### 3.4 Optimization

Optimization is a general term in computer science and mathematics and is used in several contexts including:

In mathematics, astrophysics is the branch that is interested in finding the ultimate and minimum ends of a mathematical function, sometimes with determinants. One example of optimization is to find the maximum limit for industrial production without exceeding the permissible limits of resources. Alastmthal has many applications in logistics and design problems.

In informatics, optimization is the process of improving system performance so that it reduces the effective operating time, bandwidth, or memory requirements.

Optimization algorithms are iterative. They begin with an initial guess of the optimal values of the variables and generate as sequence of improved estimates until they reach a solution. The strategy used to move from one iterate to the next distinguishes one algorithm from another. Most strategies make use of the values of the objective function f, the constraints c, and possibly the first and second derivatives of these functions. Some algorithms accumulate information gathered at previous iterations, while others use only local information from the current point. Regardless of these specifics, all good algorithms should possess the following properties [10]:

Robustness: They should perform well on a wide variety of problems in their class, for all reasonable choices of the initial variables.

Efficiency: They should not require too much computer time or storage.

Accuracy: They should be able to identify a solution with precision, without being overly sensitive to errors in the data or to the arithmetic rounding errors that occur when the algorithm is implemented on a computer (Fig. 2).

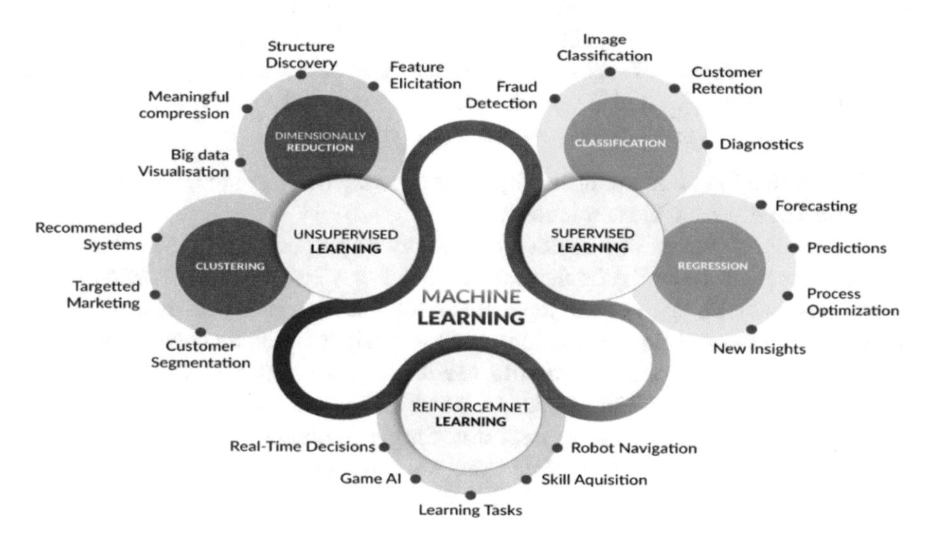


**Fig. 2.** Main techniques related to Machine Learning [9]

### 3.4.1 Cuckoo Optimization Algorithm

Cuckoos are a family of birds with unique reproductive strategy more aggressive compared to other bird's species. Some of cuckoo bird's species like Ani and Guira lay eggs in communal nests; however, they may remove others' eggs to increase the hatching probability of their own eggs. Other species use brood parasitism method of laying their eggs in the nests of other birds or host nests.

The parasitic cuckoos are good in sporting nests where eggs have just been laid and their timing of laying eggs is very precise.

They lay one egg in the host nest which will normally hatch quicker than the other eggs. When this happens, the foreign cuckoo would remove the non-hatched eggs from the nest by pushing the eggs out of the nest. This behavior is aimed at reducing the probability of the legitimate eggs from hatching. Furthermore, the foreign cuckoo chick can gain access to more food by mimicking the call of the host chicks. There are times when the host cuckoo discovers that one of the eggs is foreign. In that case the cuckoo either gets rid of the egg or abandon the nest altogether and moves to build a new nest somewhere else.

Cuckoo search algorithm is a nature-inspired algorithm developed based on reproduction of cuckoo birds. While working with CS algorithms, it is important to associate potential solutions with cuckoo eggs. Cuckoos normally lay their fertilized eggs in other cuckoos' nests with the hope of their off-springs being raised by proxy parents. There are times when the cuckoos discover that the eggs in their nests do not belong to them, in those cases the foreign eggs are either thrown out of the nests or the whole nests are abandoned. The CS optimization algorithm is basically based on the following three rules:

- Each cuckoo selects a nest randomly and lays one egg in it.
- The best nests with high quality of eggs will be carried over to the next generation.
- For a fixed number of nests, a host cuckoo can discover a foreign egg with a probability $pa\,\varepsilon[0, 1]$. In this case, the host cuckoo can either throw the egg away or abandon the nest and build a new one some where else.

The last rule can be approximated by replacing a fraction pa of the n host nests with new nests (with new random solutions).

The quality or fitness of a solution can simply be proportional to the value of the objective function. From the implementation point of view, the representation that is followed is that each egg in anest represents a solution, and each cuckoo can lay only one egg (thus representing one solution). We can safely make no deference between an egg, a nest or a cuckoo. The aim is to use the new and potentially better solution (cuckoo egg) to replace a bad solution in the nest.

Cuckoo search algorithm is very effective for global optimization problems since it maintains a balance between local random walk and the global random walk. The balance between local and global random walks is controlled by a switching parameter $p_a\,\varepsilon[0, 1]$. The local and global random walks are defined by Eqs. (1) and (2), respectively. Their parameters are defined in Table 2 (Fig. 3).

$$X_i^{t+1} = X_i^t + \alpha S \otimes H(p_a - \varepsilon) \otimes \left(X_j^t - X_k^t\right) \tag{12}$$

$$X_i^{t+1} = X_i^t + \alpha L(S, \lambda) \tag{13}$$

***Algorthim#3.1. MORE-G***

**Input**: Database called friendly environment database (FED) have wind, cost, area, material and team

**Output**: Generated maximum electrical energy from FED

**Initialization**: determined the weights for each dataset

w_wind=0.5,w_cost=0.4,w_area=0.3,w_material=0.2,w_team=0.1

**// Preprocessing Stage**

1: **For** each dataset i // i=1,2…..5

2: FEDi= Call Handle Missing value

3: NFEi= Call Handle Normalization (FEDi)

4: **End** for

5: **For** each dataset i // i=1,2…..5

6: Call Build constraints

7: **End** for

**// Building MORE-G Model**

8: **For** each dataset i

9: Di=Call DCuckoo

10: **End** for

11: **For** each dataset i in D

12: wi = Determined the weight for Di

13: sum=sum+(wi*Di)

14: **End** for

**// Evaluation Stage**

15: Call Evaluation DCuckoo

16: **End**

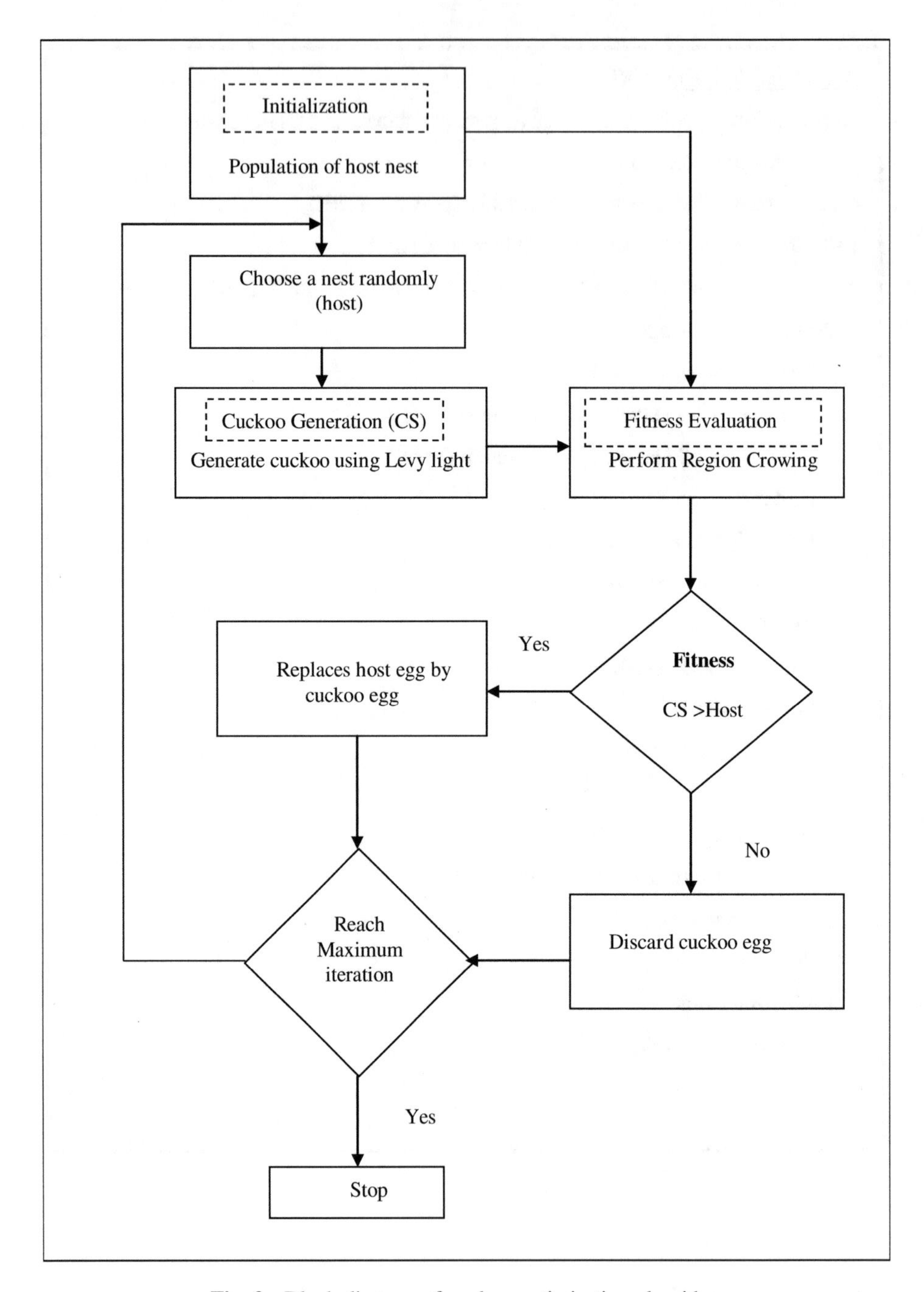


**Fig. 3.** Block diagram of cuckoo optimization algorithm

***Algorithm#2 . Lévy Based Cuckoo for A Global Optimization.***

```
Input: the database after pre-processing.
Output: Max Energy Generated FED.
Set of parameters: t: Number of Iteration, n: Number of nests host,
Fi: fitness function, Max Generation
1:  Determined the objective function f(x) , x=(x1,x2........xd)T
2:  Generate initial population of n host nests xi (i=1,2......n)
3:  While (t< Max Generation) or (stop criteria)
4:          Get a cuckoo (say i) randomly by Levy distribution
5:          Select two table equal size
6:          While (#feature<= # max feature)
7:                  Take first feature and put it in the determined location of  table1
8:                  IF (location (feature1) =location (feature2))
9:                          Replace from table1 to table2
10:                         Put first feature in the determined location in table2
11:                         Put the second feature in the determined location table1
12:                 Else
13:                         Put the second feature in the determined location table1
14:                 End if
15:         End While
16:         IF Fi < Fj then
17:                 Replace j by the new solution
18:         End if
19:         a fraction pa of the worse nests are abandoned and new ones are built
20:         Keep the best nests
21:         Rank the nests and find the current best
22:         Pass the current best solutions to the next generation
23:  End while
24: End
```

### 3.5 Multi Objective Optimization

Multi-objective optimization (also known as multi-objective programming, vector optimization, multicriteria optimization, multi attribute optimization or Pareto optimization) is an area of multiple criteria decision making that is concerned with mathematical optimization problems involving more than one objective function to be optimized simultaneously. Multi-objective optimization has been applied in many fields of science, including engineering, economics and logistics where optimal decisions need to be taken in the presence of trade-offs between two or more conflicting objectives. Minimizing cost while maximizing comfort while buying a car, and maximizing performance whilst minimizing fuel consumption and emission of pollutants of a vehicle are examples of multi-objective optimization problems involving two and three objectives, respectively.

Typically, there does not exist a single solution that simultaneously optimizes each objective. Instead, there exists a (possibly infinite) set of Pareto optimal solutions. A solution is called nondominated or Pareto optimal if none of the objective functions can be improved in value without degrading one or more of the other objective values. Without additional subjective preference information, all Pareto optimal solutions are considered equally good [9].

**Table 2.** Multi objective optimization functions

| Problem | $n$ | Variable bounds | Objective functions | Optimal solutions | Comments |
|---|---|---|---|---|---|
| SCH | 1 | $[-10^3, 10^3]$ | $f_1(x) = x^2$<br>$f_2(x) = (x-2)^2$ | $x \in [0, 2]$ | convex |
| FON | 3 | $[-4, 4]$ | $f_1(\mathbf{x}) = 1 - \exp\left(-\sum_{i=1}^{3}\left(x_i - \frac{1}{\sqrt{3}}\right)^2\right)$<br>$f_2(\mathbf{x}) = 1 - \exp\left(-\sum_{i=1}^{3}\left(x_i + \frac{1}{\sqrt{3}}\right)^2\right)$ | $x_1 = x_2 = x_3$<br>$\in [-1/\sqrt{3}, 1/\sqrt{3}]$ | nonconvex |
| POL | 2 | $[-\pi, \pi]$ | $f_1(\mathbf{x}) = \left[1 + (A_1 - B_1)^2 + (A_2 - B_2)^2\right]$<br>$f_2(\mathbf{x}) = \left[(x_1+3)^2 + (x_2+1)^2\right]$<br>$A_1 = 0.5\sin 1 - 2\cos 1 + \sin 2 - 1.5\cos 2$<br>$A_2 = 1.5\sin 1 - \cos 1 + 2\sin 2 - 0.5\cos 2$<br>$B_1 = 0.5\sin x_1 - 2\cos x_1 + \sin x_2 - 1.5\cos x_2$<br>$B_2 = 1.5\sin x_1 - \cos x_1 + 2\sin x_2 - 0.5\cos x_2$ | (refer [1]) | nonconvex, disconnected |
| KUR | 3 | $[-5, 5]$ | $f_1(\mathbf{x}) = \sum_{i=1}^{n-1}\left(-10\exp\left(-0.2\sqrt{x_i^2 + x_{i+1}^2}\right)\right)$<br>$f_2(\mathbf{x}) = \sum_{i=1}^{n}\left(\lvert x_i\rvert^{0.8} + 5\sin x_i^3\right)$ | (refer [1]) | nonconvex |
| ZDT1 | 30 | $[0, 1]$ | $f_1(\mathbf{x}) = x_1$<br>$f_2(\mathbf{x}) = g(\mathbf{x})\left[1 - \sqrt{x_1/g(\mathbf{x})}\right]$<br>$g(\mathbf{x}) = 1 + 9\left(\sum_{i=2}^{n} x_i\right)/(n-1)$ | $x_1 \in [0, 1]$<br>$x_i = 0,$<br>$i = 2, \ldots, n$ | convex |
| ZDT2 | 30 | $[0, 1]$ | $f_1(\mathbf{x}) = x_1$<br>$f_2(\mathbf{x}) = g(\mathbf{x})\left[1 - (x_1/g(\mathbf{x}))^2\right]$<br>$g(\mathbf{x}) = 1 + 9\left(\sum_{i=2}^{n} x_i\right)/(n-1)$ | $x_1 \in [0, 1]$<br>$x_i = 0,$<br>$i = 2, \ldots, n$ | nonconvex |
| ZDT3 | 30 | $[0, 1]$ | $f_1(\mathbf{x}) = x_1$<br>$f_2(\mathbf{x}) = g(\mathbf{x})\left[1 - \sqrt{x_1/g(\mathbf{x})} - \frac{x_1}{g(\mathbf{x})}\sin(10\pi x_1)\right]$<br>$g(\mathbf{x}) = 1 + 9\left(\sum_{i=2}^{n} x_i\right)/(n-1)$ | $x_1 \in [0, 1]$<br>$x_i = 0,$<br>$i = 2, \ldots, n$ | convex, disconnected |
| ZDT4 | 10 | $x_1 \in [0, 1]$<br>$x_i \in [-5, 5],$<br>$i = 2, \ldots, n$ | $f_1(\mathbf{x}) = x_1$<br>$f_2(\mathbf{x}) = g(\mathbf{x})\left[1 - \sqrt{x_1/g(\mathbf{x})}\right]$<br>$g(\mathbf{x}) = 1 + 10(n-1) + \sum_{i=2}^{n}\left[x_i^2 - 10\cos(4\pi x_i)\right]$ | $x_1 \in [0, 1]$<br>$x_i = 0,$<br>$i = 2, \ldots, n$ | nonconvex |
| ZDT6 | 10 | $[0, 1]$ | $f_1(\mathbf{x}) = 1 - \exp(-4x_1)\sin^6(6\pi x_1)$<br>$f_2(\mathbf{x}) = g(\mathbf{x})\left[1 - (f_1(\mathbf{x})/g(\mathbf{x}))^2\right]$<br>$g(\mathbf{x}) = 1 + 9\left[\left(\sum_{i=2}^{n} x_i\right)/(n-1)\right]^{0.25}$ | $x_1 \in [0, 1]$<br>$x_i = 0,$<br>$i = 2, \ldots, n$ | nonconvex, nonuniformly spaced |

### 3.6 Green Energy

Green energy comes from natural sources such as sunlight, wind, rain, tides, plants, algae and geothermal heat. These energy resources are renewable, meaning they're naturally replenished.

Renewable energy sources also have a much smaller impact on the environment than fossil fuels, which produce pollutants such as greenhouse gases as a by-product, contributing to climate change. Gaining access to fossil fuels typically requires either mining or drilling deep into the earth, often in ecologically sensitive locations [6].

Green energy, however, utilizes energy sources that are readily available all over the world, including in rural and remote areas that don't otherwise have access to electricity. Advances in renewable energy technologies have lowered the cost of solar panels, wind turbines and other sources of green energy, placing the ability to produce electricity in the hands of the people rather than those of oil, gas, coal and utility companies.

Since solar panel prices have fallen 80% since 2008, wind, geothermal and hydroelectric power are competing with fossil fuels (oil, coal and gas), prompting world countries and decision-makers to invest in clean energy to the benefit of the national economy. For all people, the sustainable development of all society.

### 3.7 Energy Returned on Energy Invested (EROER)

"It is the ratio of the amount of usable energy delivered from a particular energy resource to the amount of energy used to obtain that energy resource. It is a distinct measure from energy efficiency as it does not measure the primary energy inputs to the system, only usable energy. "EROI" is calculated from the following simple equation":

$$\text{"EROER"} = \frac{\text{"Energy Delivered"}}{\text{"Energy Required to Deliver That Energy"}}$$

## 4 Problem Statement

Most of the countries in the world, including Iraq, suffer from a clear lack of electricity production, which has led most countries to turn towards producing energy from natural sources or environmentally friendly sources that do not cause the emission of carbon dioxide gas while not causing pollution to the environment [add ref].

The problem of producing electrical energy from environmentally friendly sources with high efficiency and low cost is one of the most important challenges in this field.

This thesis will try to processing after the wind has been identified as the energy that will be used to generate electricity because of the compatibility of this kind of energy environment With the Iraq environment.

The main Objectives of that paper;

- To developing one of the search techniques (i.e., cuckoo search technique) through used multi objectives function satisfy optimization concept.
- To find the constrictions for each sub optimization problem of optimization problem.
- To minimize cost reduction and Increase the production of electric power generated from renewable energy method (REM) with determined the main limitations and Hypotheses of that problem.

To salsify the above goals(objectives), build Multi Objective Renewable Energy Generation Model (MORE-G) through develop the cuckoo algorithm through apply multi-objective optimization functions as core of that algorithm with a horizontal combination to generated energy.

Algorithm #1 explain MORE-G system stages in details

In horizontal combination, the candidates gain knowledge from various dimensions of other candidates to generate moderation solutions as given below: Assume that the parent individuals $X^{m1}$ and $X^{m2}$ (m1, m2) execute horizontal combination at $d^{th}$ dimension, the moderation solution are generated as follows:

$$MH_d^{m1} = r_1.X_d^{m2} + (1 - r_1).X_d^{m2} + c_1.(X_d^{m1} - X_d^{m2})$$
$$MH_d^{m2} = r_1.X_d^{m1} + (1 - r_1).X_d^{m1} + c_1.(X_d^{m2} - X_d^{m1})$$

here $MH^{m1}$, $MH^{m2}$ are two new generated moderation solutions. $r_1$, $r_2$ are uniformly distributed random values between [0, 1], $c_1$, $c_2$ are expansion coefficients which are uniformly distributed random values between [−1, 1]. After the moderation solutions obtained by horizontal combination, it needs to perform the competitive operation between $MH^m$ and $X^m(m = 1, 2, ..M)$. Only the individual with higher fitness can survive. Thus, X retains a set of personal best solutions called as dominant horizontal solutions ($DHS^m$). Below describes horizontal combination between two parent and their moderation solutions. Dominant solutions are those who are having better fitness values (Fig 4).

Parent 1 $X_1$ $X_2$ $X_3$ $X_4$ $X_5$
Parent 2 $Y_1$ $Y_2$ $Y_3$ $Y_4$ $Y_5$
Moderation Horizontal combination solutions

Moderation solutions ( $MH_1$) = $XY_1$ $XY_2$ $XY_3$ $XY_4$ $XY_5$
Moderation solutions ( $MH_2$) = $X_1Y$ $X_2Y$ $X_3Y$ $X_4Y$ $X_5Y$

Dominant Horizontal combination solutions
Dominant solutions ($DHS_1$) = $XY_1$ $XY_2$ $XY_3$ $XY_4$ $XY_5$
Dominant solutions ($DHS_2$) = $Y_1$ $Y_2$ $Y_3$ $Y_4$ $Y_5$

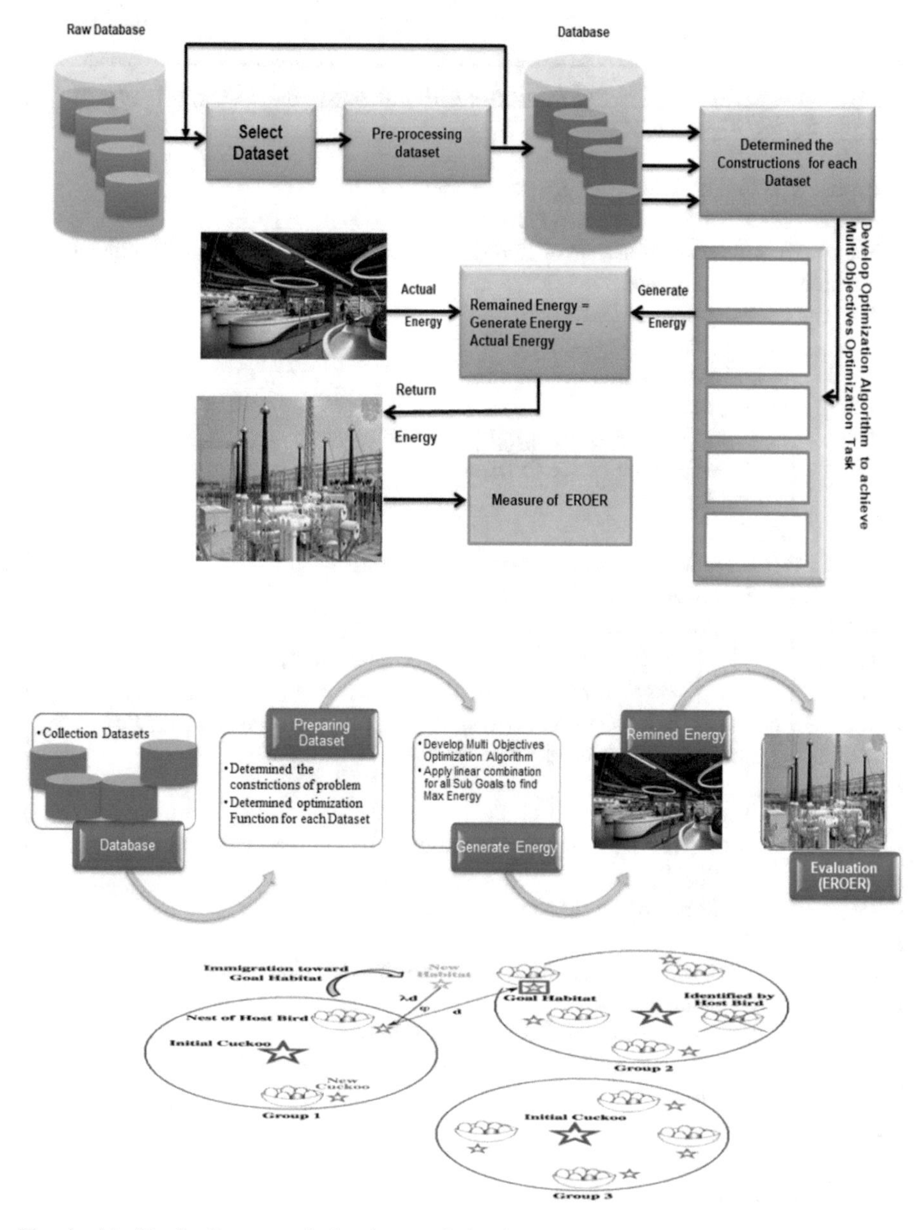


**Fig. 4.** (a) Block diagram of develop optimization algorithm (DOA) to Generated max renewable energy from wind energy, (b) Integrated of multi objectives optimization in single system

***Algorithm #3.Develop Optimization Algorithm to achieve Multi Objectives Optimization Task***

```
Input: a // the database after pre-processing
Output: Max Energy Generated FED
Set of parameters: t: number of iteration, n: number of nests host, Fi: objective
function, Max Generation
1:  For each dataset i  // i =1,2..5
2:      For each sample in dataset j
3:          For each column in sample k
// Generated population
4:              pop[i,j,k]=random a[i,j,k]
5:              compute objective function
6:          End for
7:      End for
8:      While (t< Max Generation) or (stop criteria)
9:          For each sample in dataset j
10:             For each column in sample k
11:                 select randomly(pop1[i,j,k],pop2[i,j,k])
12:                 child1=r1.pop2[i,j,k]+(1-r1). pop2[i,j,k]+c1.(pop1[i,j,k]- pop2[i,j,k])
13:                 child2=r1.pop1[i,j,k]+(1-r1). Pop1[i,j,k]+c2.( pop2[i,j,k]- pop1[i,j,k])
14:                 compute objective function for each child
15:             End for
16:         End for
17:         choice the Elite among population and children, whichever is better
18:         IF Fi < Fj then
19:             Replace j by the new solution
20:         End if
21          a fraction pa of the worse nests are abandoned and new ones are built
22:         Keep the best nests
23:         Rank the nests and find the current best
24:         Pass the current best solutions to the next generation
25:     End while
26: End for
27: End
```

# 5 Experiment

## 5.1 Data and Method

It is a collection of information (collections of raw dataset) organized in such a way as to easily access, modify and manage it. The databases are used by organizations to store, retrieve, and manage information, and each of these dataset contains a collections of constraints (features).

These dataset is wind, Area, Team, Material, and cost.

The wind consists of two features of wind speed and direction.

The area consists of five features size of turbine, the area of each turbine, type of turbine, number of turbine.

The team contains of five features: certificate, number of work clock, number of experience certificate, number of cycles, age.

The material consist of five features: wire, type, stats, years, batteries.

In this pager we use a database which consists of a set of databases related to wind, spaces, working team, tools and costs, where the lowest percentage of wind and less ground area is used by the best team and the lowest cost and highest efficiency (Table 3):

**Table 3.** Dataset of wind

| Direct | Speed |
|---|---|
| 200 | 1 |
| 220 | 2 |
| 250 | 3 |
| 300 | 4 |
| ….. | |
| ….. | |
| 350 | 8 |
| 400 | 5 |

The above table represents the wind dataset, which is two features of wind direction and wind speed (Table 4).

**Table 4.** Dataset of area

| Size of T | ab | Type | #T |
|---|---|---|---|
| Small | 50 | 0.7 | 20 |
| Normal | 70 | 0.3 | 30 |
| Large | 90 | 0.4 | 40 |
| …. | | | |
| …. | | | |
| Normal | 70 | 0.6 | 80 |
| Large | 90 | 0.1 | 90 |

The above table represents the area dataset, which contain five features size of turbine, the area of each turbine, type of turbine, number of turbine (Table 5).

**Table 5.** The dataset of team work

| Certif | #clock | #expert | #cycle | #age |
|---|---|---|---|---|
| 1 | 6 | 5 | 5 | 25 |
| 2 | 7 | 6 | 6 | 27 |
| ….. | | | | |
| ….. | | | | |
| 5 | 12 | 15 | 15 | 70 |
| 5 | 12 | 15 | 15 | 70 |

This table represents a dataset of the team work which consists of five features of the certificates, number of hours, number of years of experience, the number of training courses and age (Table 6).

**Table 6.** Dataset of material

| Wire | Type | Stats | Years | Batteries |
|---|---|---|---|---|
| 2 | 2 | 3 | 1 | 1000 |
| 1 | 2 | 1 | 2 | 2000 |
| ….. | | | | |
| ….. | | | | |
| 9 | 4 | 3 | 3 | 3500 |
| 5 | 4 | 3 | 3 | 3000 |

This table represents the dataset of material used in terms of wire, turbine type, origin, duration of warranty and batteries (Table 7).

**Table 7.** Dataset of cost

| C_A | C_T | C_M |
|---|---|---|
| 10 | 0 | 9 |
| 6 | 15 | 0 |
| ….. | | |
| ….. | | |
| 5 | 7 | 0 |
| 6 | 99 | 7 |

This table represents the cost database in terms of cost **of space, cost of turbines and** cost of tools used

### 5.2 Select Dataset

In here Dataset is selected from a raw database and then passed to the initial processing process per-processing

### 5.3 Pre-processing

In this stage Determined the constructions of problem and Determined optimization Function for each Dataset

### 5.4 Determined the Constructions for Each Dataset

#### 5.4.1 Objective Function

The objective function is to find the highest electrical energy (renewable energy) by using the linear combination function for each x 1, x 2, x 3, x 4, x 5 where:

- X1: min wind
- X2: min area
- X3: optimal team
- X4: optimal material
- X5: min cost

$$\text{Max energy} = (x1 + x2 + x3 + x4 + x5)$$

#### 5.4.2 Constraints

The decision variables, together with data, must satisfy some technological, economic, and geometric constraints. Mathematically, these constraints are expressed by linear equalities and inequalities involving the variables.

**A - Economic Constraints**

To consider the cost limitations of wind turbines, team works, area, return renewable energy, constraints have been considered, assuming a maximum available budget that should not be exceeded. Mathematically, the expression is a linear inequality involving only x1, x2, x3, x4, which ensures that the initial installation cost of wind turbines and PV panels does not exceed the fixed budget

$$C_{wind} * x1 + C_{team} * x2 + C_{area} * x3 + C_{return} * x4 \leq \text{budget}$$

Where

- $C_t$: is the unit installation cost of a wind turbine.
- $C_{team}$ is the cost of team work.
- $C_{area}$ is the cost of area.
- $C_{return}$ is the cost of return renewable energy.

**B - Geometric Constraints**

These constraints refer to some geometrical details. With respect to the wind-energy production, a constraint arises owing to the limited available ground area for the installation of the wind turbines:

$$x1 * A_b \leq Amax$$

where

- $A_b$ is the basic ground area occupied by a wind turbine.
- $A_{max}$ is the available area.

## 5.5 Develop Optimization Algorithm to Achieve Multi Objectives Optimization Task

Results after applying the development of optimization algorithm to give the lowest proportion of wind and less ground area and the best team work and the best tools and less expensive

- The results of wind dataset:

```
('sum for each sample=', [201, 222, 253, 304, 325, 336,
347, 358, 405])
('index, min sum, sample=', 0, 201, array([200,   1],
dtype=int64))
```

- The results of cost dataset:

```
('sum for each sample=', [19, 21, 62, 22, 21, 274, 12,
12, 112])
 ('index, min sum, sample=', 6, 12, array([4, 8, 0],
dtype=int64))
```

- The results of area dataset:

```
('sum for each sample=', [70.7, 100.3, 130.4, 120.8,
150.5, 120.2, 125.9, 150.6, 180.1])
('index, min sum, sample = ', 0, 70.7,  array([50. ,  0.7, 20. ]))
```

- The results of team work dataset:

```
('sum for each sample=', [42, 48, 56, 68, 78, 84, 93,
99, 117])
('index, max sum, sample=', 8, 117, array([ 5, 12, 15,
15, 70], dtype=int64))
```

- The results of material dataset:

```
('sum for each sample=', [1008, 2006, 3012, 4009, 4515,
5017, 2518, 3519, 3015])
('index, max sum, sample=', 5, 5017, array([   6,    3,
5,    3, 5000], dtype=int64))
```

Finally we apply the linear combination function:

$$F(x) = F(x_i) + F(x_j) + F(x_k) + F(x_L) + F(x_m)$$

Where

$X_i$: minimum wind
$X_j$: minimum area
$X_k$: best team work
$X_L$: optimal material
$X_m$: minimum cost

These datasets take different weights based on their important to generation energy, where:

The wind dataset take the weight (ww) = 0.4
The cost dataset take the weight (wc) = 0.1
The area dataset take the weight (wa) = 0.1
The team work dataset take the weight (wt) = 0.1
The material dataset take the weight (wm) = 0.1

$$\text{Max Energy} = ww * xi + wc * x_m + wa * x_j + wm * x_L + wt * x_k$$

```
('Max Energy=', 602.07)
```

# 6 Discussions and Conclusions

This paper has provided an outline of a new method for generation and optimization of renewable energy systems using Multi Objective optimization search technique through the use of cuckoo search optimization algorithm which Taking into consideration the natural determinants of the environment.

Cuckoo search optimization was used, to remedy this problem, because CS is a meta-heuristic optimization algorithm, which is used for solving optimization problems, and gives the highest amount of electrical energy through the lowest proportion of wind, less available ground areas, the best team work and the best tools used at the lowest cost and highest efficiency.

This is a nature-inspired meta-heuristic algorithm, which is based on the brood parasitism of some cuckoo species, along with Levy flights random walks.

To assess the results, the Energy Returned Energy Invested (EROER) is used.

# Appendix

See Table 8.

**Table 8.** Terms and their meaning

| Terms | Description |
|---|---|
| CoF | Cost of Energy |
| PV | Photo-voltaics |
| CHP | Combined heat and power |
| MILP | Mixed-integer linear programming |
| NWCMO | Normalized Weighted Constrained Multi-Objective |
| SEO | Search engine optimization |
| IS | Integrated system |
| EROER | Energy returned on energy invested |
| IMRNG | Integration Model of Renewable Energy Generation |
| $X_i^t$ and $X_k^t$ | Current positions selected by random permutation |
| A | Positive step size scaling factor |
| $X_i^t + 1$ | Next position |
| S | Step size |
| $\otimes$ | Entry-wise product of two vectors |
| H | Heavy-side function |
| Pa | Used to switch between local and global random walks |
| $\varepsilon$ | Random number from uniform distribution |
| L(S, $\lambda$) | Lévy distribution, used to define the step size of random walk |

## References


1. Chehouri, A., Younes, R., Ilinca, A., Perron, J.: Review of performance optimization techniques applied to wind turbines. Appl. Energy, **142**, 361–388 (2015).http://dx.doi.org/10.1016/j.apenergy.2014.12.043
2. Wu, Q., Zhou, J., Liu, S., Yang, X., Ren, H.: Multi-objective optimization of integrated renewable energy system considering economics $CO_2$ emissions. Energy Procedia **104**, 15–20 (2016).https://doi.org/10.1016/j.egypro.2016.12.004
3. García, J., Altimiras, F., Peña, A., Astorga, G., Peredo, O.: A binary cuckoo search big data algorithm applied to large-scale crew scheduling problems. Complexity **2018**, 15 pages (2018). https://doi.org/10.1155/2018/8395193. Article ID 8395193
4. Lamedica, E., Santini, E., Ruvio, A., Palagi, L., Rossetta, I.: A MILP methodology to optimize sizing of PV – Wind renewable energy systems. Energy **165**, 385–398 (2018). https://doi.org/10.1016/j.energy.2018.09.087
5. Eriksson, E., Gray, E.: Optimization of renewable hybrid energy systems e A multi-objective approach. Renew. Energy **133**, 971–999 (2019). https://doi.org/10.1016/j.renene.2018.10.053
6. Xavier, M.V.E., Bassi, A.M., de Souza, C.M., Barbosa Filho, W.P., Schleiss, K., Nunes, F.: Energy scenarios for the Minas Gerais State in Brazil: an integrated modeling exercise using System Dynamics. Energy Sustain. Soc. **3**, 17 (2013). http://www.energsustainsoc.com/content/3/1/17
7. Al-Janabi, S. Alkaim, A.F.: A nifty collaborative analysis to predicting a novel tool (DRFLLS) for missing values estimation. J. Soft Comput. (2019). https://doi.org/10.1007/s00500-019-03972-x
8. Vrochidis, S., Huet, B., Chang, E.Y., Kompatsiaris, I.: Big Data Analytics for Large-Scale Multimedia Search, 15 March 2019. https://doi.org/10.1002/9781119376996
9. Al_Janabi, S., Mahdi, M.A.: Evaluation prediction techniques to achievement an optimal biomedical analysis. Int. J. Grid Util. Comput. (2019)
10. Al_Janabi, S.: Smart system to create optimal higher education environment using IDA and IOTs. Int. J. Comput. Appl. (2018). https://doi.org/10.1080/1206212X.2018.1512460
11. Al-Janabi, S., Alwan, E.: Soft mathematical system to solve black box problem through development the FARB based on hyperbolic and polynomial functions. In: IEEE, 2017 10th International Conference on Developments in eSystems Engineering (DeSE), Paris, pp. 37–42 (2017). https://doi.org/10.1109/DESE.2017.23
12. Ali, S.H.: A novel tool (FP-KC) for handle the three main dimensions reduction and association rule mining. In: IEEE, 2012 6th International Conference on Sciences of Electronics, Technologies of Information and Telecommunications (SETIT), Sousse, pp. 951–961 (2012). https://doi.org/10.1109/SETIT.2012.6482042
13. Ali, S.H.: Miner for OACCR: case of medical data analysis in knowledge discovery. In: IEEE, 2012 6th International Conference on Sciences of Electronics, Technologies of Information and Telecommunications (SETIT), Sousse, pp. 962–975 (2012). https://doi.org/10.1109/SETIT.2012.6482043
14. Al-Janabi, S.: Pragmatic miner to risk analysis for intrusion detection (PMRA-ID). In: Mohamed, A., Berry, M., Yap, B. (eds.) Soft Computing in Data Science, SCDS 2017. Communications in Computer and Information Science, vol. 788. Springer, Singapore (2017). https://doi.org/10.1007/978-981-10-7242-0_23

# Towards a Rich and Dynamic Human Digital Memory in Egocentric Dataset

Khalid El Ansaoui[1(✉)], Youness Chawki[2], and Mohammed Ouhda[2]

[1] Mohammed VI Polytechnic University, Lot 660, Hay Moulay Rachid, 43150 Ben Guerir, Morocco
khalid.elasnaoui@ump6.com
[2] Faculty of Science and Techniques, M2I Laboratory, ASIA Team, Moulay Ismail University, BP 509 Boutalamine, Errachidia, Morocco
youness.chawki@gmail.com, ouhda.med@gmail.com

**Abstract.** Memories have always been a considerable importance of a person's life and experiences. A digital human memory as a field of study focuses on encapsulating this phenomenon, in digital form, during the thread of a lifetime. By spreading hardware everywhere, massive amount of data is being generated together by people and the surrounding environment. With all this demountable information available, successfully exploring, researching and collating, together, to form a human digital memory, is a new challenge and requires novel and efficient algorithmic solutions. The main goal of this work is going to propose a new method to automatically create rich and dynamic human digital memory in egocentric dataset from the lifelogging images of a person. For this purpose, we will propose a technique using Convolutional Neural Network (CNN) model. For validation, we will apply the proposed method on the Egocentric Dataset of University of Barcelona (EDUB) of 4912 daily images acquired by four persons.

**Keywords:** Lifelogging · Visual lifelogs · Human digital memory · CNN · EDUB

## 1 Introduction

Images in multimedia systems and on the Internet are successively growing. There are several researches works on visual information and automatic analysis of images. The image memorability has been appeared recently in the research community. The human brain processes simultaneously millions of images and other information from multiple sources. Among these different images and information, some of them are more memorable than others. The study of visual memorability requires knowing the probability with what a human can remember an image. A specific annotation tool for manually annotate image visual memorability with human interaction is needed.

Humans can remember thousands of images and a remarkable amount of their visual details [1]. While some images are forgotten or ignored quickly [2], others are more memorable, i.e. they are engraved in the human memory from the first exposure. Various fields are concerned with images memorability, for example, the field of communication (television programs, presentations, etc.) and the field of advertising (TV,


Y. Farhaoui (Ed.): BDNT 2019, LNNS 81, pp. 296–309, 2020.
https://doi.org/10.1007/978-3-030-23672-4_21

magazine, etc.) since they seek to attract the attention of viewers to convey their ideas. In order to understand the notion of image memorability, various studies [3–5] were carried out along two axes: a psychological/physiological axis because the researchers found a relation between the visual perception of the human being, Its psychic state and images memorability, and a technical axis dealing with the descriptors (color, gradient, shape, texture, etc.) characterizing these images, the movement of the eye and several other properties. Thus, the authors of [3] carried out a study in order to find memorability scores for a set of images. Thus, a memorability score represents the percentage of correct detections by participants. Visualizations are tools used for creating images, diagrams, or data representations to communicate a message. Visualization through visual imagery is an effective way to communicate both abstract and concrete ideas [6]. Different visual designs offer significantly distinct reading accuracy [7]. While memorability and visualizations memorability remain a new axis of study. The memorability of visualization as an image constitutes an intrinsic property of this visualization [8]. Thus, the visual characteristics extracted from the visualizations can explain their memorabilities. A representative model of visualization memorability can be realized based on a set of features inherent to this visualization.

In addition, MaiteGarolera, from the neuropsychology unit of Terrassa Hospital (Spain) said: The brain is designed to forget, we need to forget to survive, because we can't live remembering in each moment all that we have lived. For that reason, humans developed several tools and alternatives for persistent memory, like writings, drawings or photographs. These tools have artificially extended our capabilities for knowledge discovery and transference. Wearable cameras are new tools that take one step further, by allowing a much finer visual memory [9].

Our days have an average of 16 h awake or 960 min. A large amount of data is accumulated in that 960 min, on average of 1.400–2.000 images per day, because the lifelogging cameras usually take a photo every 30 s. This image storage equals about two gigabytes per day, with most imaging blurred from rapid movements or fast lighting changes. In addition, to this big data problem, most of these images are unintentionally taken, consequently, many of them can be blurred (because a camera wearer is moving quickly), with little information, fast illumination changes. To summarize, wearable cameras define a problem of big, noisy, unlabeled and unstructured visual data that present major challenges and require automatic and efficient algorithmic solutions for finding, indexing and retrieval specific images in this huge amount of data.

To sum up, wearable cameras define a problem of big and noisy visual data that require automatic tools for the indexing and retrieval. This work presents new tools in this field, by introducing the concept of visual memorability and the contextual data captured from physiological signals.

The rest of the paper is organized as follows: Sect. 2 introduces the definition of lifelogging, Sect. 3 deals with some related works. In Sect. 4, we will describe our contribution. Section 5 presents the data used on which we will apply our method. The conclusion is given in the last section.

## 2 Introducing Lifelogging

The earliest motivation behind automatic generation of personal digital archives can be traced back to 1945 when Bush expressed his vision that our lives can be recorded with the help of the technology and the access can be made easier to these digital memories. This new way of autobiography generation has become more and more realistic recently, with the advances of lightweight computing devices and highly accurate sensors. Mobile devices are approaching a more capable computing ability, dwarfing the most powerful computers in the past. The low price and the embedded nature of smaller and lightweight sensors (cameras, GPS, Bluetooth, accelerometers, etc.) make computing devices portable or even wearable to enable life recording to be done unobtrusively. The large volume of data storage and high-speed wireless networks needed for this help the mobile platform to turn into people-centric sensors capturing multidimensional sensory inputs besides spatial and temporal data. Lifelogging is the term describing this notion of digitally recording aspects of our lives, where the recorded multimedia content is the reflection of activities which we subsequently use to obtain the meaning of daily events by browsing, searching, or querying.

To build a mapping between the real world and the digital world, Various context scan be recorded for the capture of the true meaning of daily activities. Here, contexts refer to the information which can be used to characterize a situation. A large variety of contexts can be used in lifelogging such as textual information, photos, audio and video clips, environment information (light, temperature, pressure, etc.), bioinformation (heart rate, galvanic response, etc.) and spacial information (location, acceleration, co-presence, etc.). These contexts are changing dynamically and if captured then they can be used as cues to our activities and thus help with accessing information in our personal digital libraries.

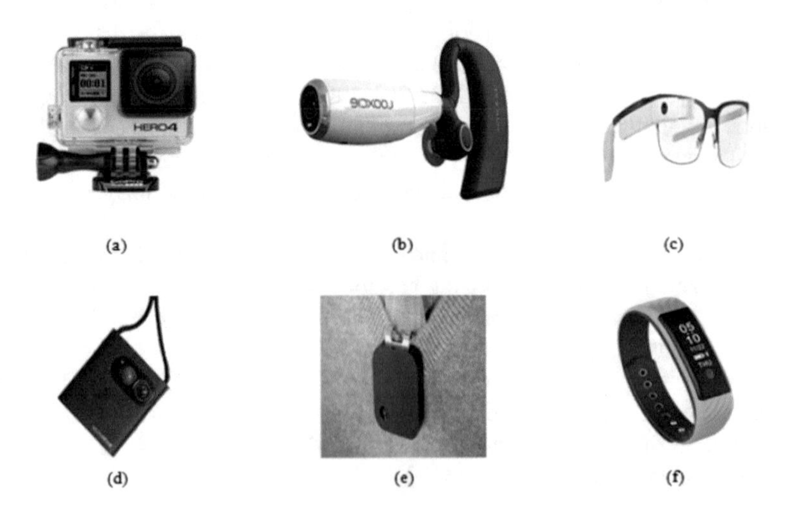


**Fig. 1.** A variety of life-logging wearable devices: (a). GoPro (2002). (b). Looxcie (2011). (c). Google glasses. (d). SenseCam (2005). (e). Narrative (2013). (f). Bracelet

To develop this further, a large number of digital devices with sensors can be applied to capture the above-mentioned contexts. Among all the devices emerging, the digital camera is the most widely-used lifelogging device (see Fig. 1) [10, 11].

The lifelogs formed by the data collected, over long periods of time, by continuously recording the user life, provide a large potential of mining or inferring knowledge about how people live [1], hence enabling a large amount of applications. Indeed, a collection of studies published in a special issue of the American Journal of Preventive Medicine [12] has proved the potential of visual lifelogs captured through a SenseCam from several viewpoints. In particular, it has been demonstrated that, used as a tool to understand and track lifestyle behavior, visual lifelogs would enable the prevention of non-communicable diseases associated to unhealthy trends and risky profiles (such as obesity or depression, among others). In addition, the lifelogs can be used as a tool for re-memory cognitive training; visual lifelogs would enable the prevention of cognitive and functional decline in elderly people [13–15].

When analyzing several days of a person (lifelogger) trying to characterize his behavior, habits or lifestyle, a natural question arises how to automatically create Richs human digital memory. Such information can be much interesting for different health applications: for example, days when the wearer of the camera is less active could predict beginning of depression or physical pain, while days that are too busy could lead to stress and fatigue.

Our main goal in this work is going to present new tools by introducing the concept of visual memorability and the contextual data captured from physiological signals. Toward this end, we will develop and test our algorithm to create rich and dynamic human digital memory based on 4912 daily images acquired by four persons using a wearable camera.

## 3 Related Work

In recent years, many interesting applications for lifelogging and human behavior have appeared and are being actively researched. Visual lifelogging data have been used to address different computer vision problems: informative image detection [16, 17], egocentric summarization [18, 19], content-based search and retrieval [20, 21], interaction analysis [22, 23], scene understanding [24], concept recognition [25], day similarity [10], this work is extended in [26], body movements [27], object-hand recognition [28, 29], Content Based Image Retrieval [30–41]. Fathi et al. [42] presented a new model for human activities recognition in short egocentric videos. This work is extended in [43] by a generative model incorporating the gaze features. Pirsiavash et al. [44] presented a temporal pyramid to encode spatio-temporal features along with detected active objects knowledge. Moreover, Ma et al. [45] proposed a twin stream Convolutional Neural Network (CNN) architecture for activity recognition from videos.

The memorability of an image is a topic that has been appeared recently in the research community. These works [3, 4, 46–48] study what makes an image memorable and explore visual features and composition of the image.

Decide when an image is memorable or not, might be interesting for advertising, photography, etc. Generally, memorability is useful because from a day, we want to select those images that are relevant, and we will assume that an image is relevant if we can remember it. Indeed, the work presented in [3] defines that an image is memorable if we just saw it for a second, we can detect that it is a repetition. This assumption seems to be coherent. The research's authors did an experiment with many users. The task for the users was a visual memory game. The game consists in an application that shows images for 5 min. They define two types of images: the targets are the images that they want to annotate. There are also fillers, the images that they put between targets. Targets are shown only twice, while fillers can be repeated many times. Some fillers are considered for vigilance, to ensure that a user is paying attention in the game and the results of the game can be used in the research. With the data collected during the game, they compute a memorability score for each image.

Dobbins et al. [49] suggested the DigMem system, which used distributed mobile services, machine learning and linked data in order to create such memories. Along with the design of the system, a prototype had been developed, and two case studies have been undertaken, which successfully created memories.

A DigMem has been presented in [50]. Indeed, it was a platform for creating human digital memories, based on device-specific services and the user's current environment. Therefore, information was semantically structured to create temporal "memory boxes" for human experiences. A working prototype has been successfully developed, which demonstrated this approach.

Borkin et al. in [51], carry out a visualization study using 2,070 single-panel visualizations, categorized with visualization type (e.g., bar chart, line graph, etc.), collected from different sources: news media sites, government reports, scientific journals, and infographics. They assign memorability scores for hundreds of these visualizations, and they find that these scores are consistent between observers. Thus, memorability is an intrinsic property of visualizations. Also, they annotate visualization with different attributes like ratings for data-ink ratios and visual densities. These attributes are used to analyze visualizations memorability.

Kim et al. [52] presented a crowd-sourced study, in which they investigate the utility of using mouse clicks as an alternative for eye fixations in the context of understanding data visualizations. Participants were presented with a series of images containing graphs and diagrams and asked to describe them. Each image was blurred so that the participant needed to click to reveal bubbles - small, circular areas of the image at normal resolution. Then, they compare the bubble click data with the fixation data from a complementary eye-tracking experiment by calculating the similarity between the resulting heat maps. Thus, a high similarity score suggests that the proposed methodology is a practical crowd-sourced alternative to eye tracking experiments. Actually, they design their approach to measure which information people consciously choose to examine for understanding visualizations.

The work presented in [53], examines how visualizations are recognized and recalled. They annotate a dataset of 393 visualizations and analyze the eye movements of 33 participants. Also, they gather text descriptions of the visualizations generated by thousands of participants. Thus, they determine what components of visualizations attract people's attention, and what information is encoded into memory.

Following the same objective of human digital memory, the question is how to use visual life logs for it, since lifelogging images give much richer information about human behavior since visual lifelogging images contain information about the environment of the person, the events he/she is involved, interactions and daily activities, etc. To the best of our knowledge, this is the first work suggesting an automatically creating rich human digital memory using lifelogging images.

## 4 Description of the Proposed Contribution

As we mentioned, memories are an important aspect of a person's life and experiences. The digital domain of human memories focuses on the encapsulation of this phenomenon, in digital form, during the thread of a lifetime. By spreading hardware everywhere, people and their environments generate an enormous collection of data. With all of this demountable information available, successfully exploring, researching and collating, to form a human digital memory, is a challenge. This is especially true when the age of the data is verified. The linked data provides an ideal and new solution to overcome this challenge, where a variety of their sources can be extracted for detailed information about a given event. The digital human memory, created with the data from the lifelogging devices, produces a dynamic and rich memory. Memories, created in this way, contain living structures and various data sources, which result from the semantic compilation of content and other memories. Information can be created as how you feel, where you are, and the context of the environment.

The contribution of this paper consists to introduce the concept of visual lifelog to create a dynamic and rich human digital memory using visual image content. In this section a method to compute memorability maps will be presented. A memorability map is an image where each pixel has a normalized value between 0 and 1 related with the contribution of this region to the visual memorability of the image. By this way, we can know what parts or regions of an image make it memorable.

Furthermore, there is a map called saliency map that describe with an image where the human fixes his eye gaze where he observes a scene in the firsts moments. From memorability and saliency maps, it is interesting to relate both maps to know what parts of the image make it memorable. Toward this goal, we need to compute saliency and memorability maps for each image. Saliency maps have been explored before and there is one convolutional neural network model that obtains these maps from image. SalNet is a CNN model for saliency map prediction that will be used due to the good results obtained in a challenge for this purpose. The result of the algorithm is a grey map where saliency points are labeled with a value near to 1 while points out of the eye gaze have a label near 0.

To our best knowledge, there is no algorithm or CNN model to compute memorability maps using EDUB. For this reason, our contribution will be the first one that will use CNN model to create dynamic and rich human digital memory using EDUB [10, 11, 26, 53].

### 4.1 Machine Learning and Deep Learning

Machine learning can be defined as sets of techniques automatically learn a model of the relationship between a set of descriptive features (or attributes, the input) and a target feature (or attributes, the output, what we want to predict) from a set of historical examples.

We use a machine learning to induce a prediction model from a training dataset. These algorithms learn from image features, and there are a lot of different features we can select from images and different ways to manage this data in order to construct model. But for all these techniques, features are handcrafted, and we define a set of rules that the algorithm must apply from data, defining arbitrary threshold and operations.

A recent technique called deep learning outperforms all existing techniques. The implementation of deep learning known as convolutional neural network (CNN) have been extensively applied for image recognition, detection and segmentation problems. CNNs are capable of automatically learning complex features that cannot be designed for the programmer of the algorithm, features learned by the own structure, which tries to simulate brain architecture, achieve superior results performance to hand-crafted features.

### 4.2 Convolutional Neural Network

Convolutional neural networks are biologically inspired variants, as we know the brain contains a complex arrangement of cells. These cells are sensitive to small sub-regions of visual field, called receptive field. These cells act as local filters and are connected between them, structured in many layers. As we know that each cell is a filter, all cells that belong to a same layer shared the same weights, weights that will be applied to the input. In computer vision the input is an image and each cell correspond to a convolutional filter that act to a certain part of the image. Many layers are connected between them, where the output of a layer is the input of the next layer, as we can see in the next figure:

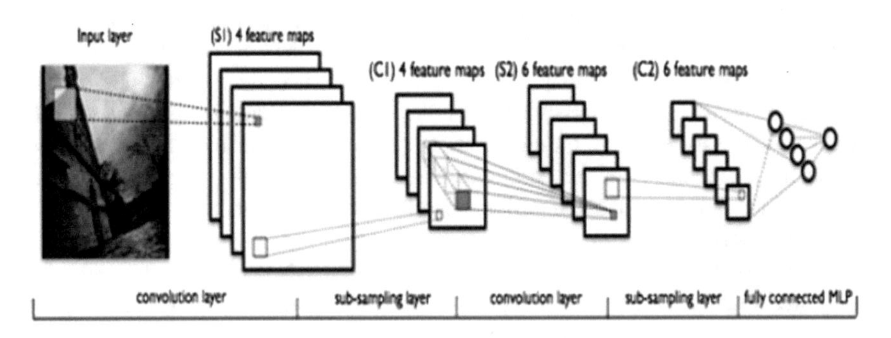


**Fig. 2.** Convolutional neural network design

How the network understands an image or process the image is difficult, but it is known that features obtained in different layers of a network works better than human-built features. As a need of understanding what a CNN see, some works have tried to represent the output of each layer, showing that if we go deeper in the network, features are more complex, and more details of image are understand, started understanding textures, shapes and finally objects [34].

# 5 Experimental Results

Our work is aimed at creating rich human digital memory, usually captured with lifelogging wearable camera.

## 5.1 Data

For this purpose, our experiments will be performed on the EDUB public dataset of images acquired with a Narrative wearable camera. This device is typically clipped around the chest area or on the user's clothes under the neck. For this purpose, we will use the public image dataset EDUB http://www.ub.edu/cvub/dataset/ on which we will validate our results (Fig. 2). This dataset is a set composed of 4912 images, their sizes are 384 $\times$ 512, and is acquired by four persons using Narrative camera (see Fig. 1(e)). This set is acquired during 8 different days, two days per person. Figure 3 shows an example of the images in the EDUB dataset [10, 11, 26, 53].

**Fig. 3.** Example of images of the EDUB dataset acquired

### 5.2 Memorability Maps with a Fully Convolutional Network

With that size (384 × 512) each image is divided in regions of 227 × 227 pixels, size equal to the input size of the CNN model. The result of this method is an image with the same size than the input. Despite that, every patch has a different value and the memorability score between two connected pixels can be very different. For this reason, a filtering of resultant memorability map is proposed. As a filtering technique a Gaussian filter will be applied to the memorability map.

Convolutional neural networks have the capability to keep localization in images. This capability is due to the convolutional layers because there are built with convolutional filters that have local impact.

First approach to compute memorability maps is not use this capability because localization is done making a grid of the image. In this case memorability map will be directly obtained without grid the image (Fig. 4).

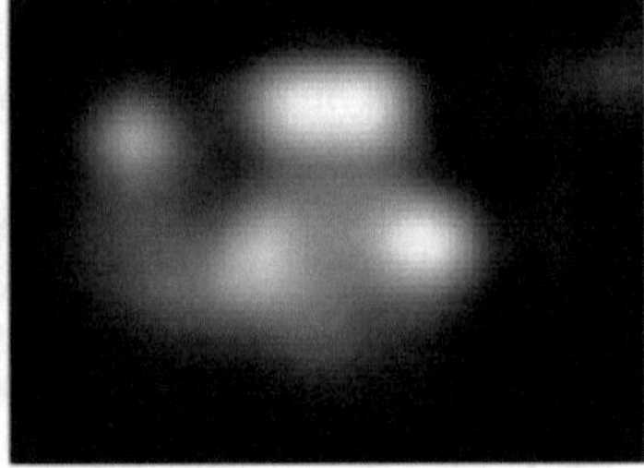

**Fig. 4.** Saliency map example

To explain the results of this method we consider three images that will be used in all section to explain the results and compare between them (Fig. 5):

**Fig. 5.** Images to explain results

After applying the method to these images correspondent memorability maps were obtained observing is a low-resolution memorability map because a single value is

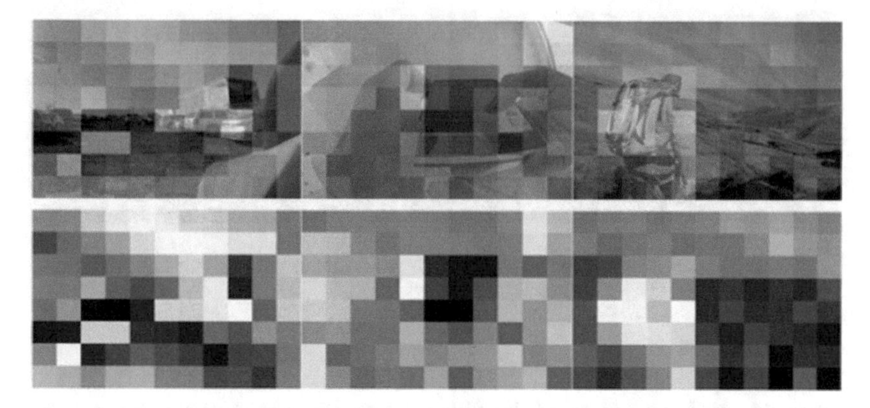

**Fig. 6.** On top, overlap between original image and the memorability map computed with a transparency of fifty percent. On the bottom, raw memorability map for these images.

assigned to a large number of pixels, as result of apply the convolutional neural network model to them (Fig. 6):

Before binarize the map is needed to filter it. To do that a Gaussian filter using MATLAB was applied to each map and the values binarized. This filtering requires selecting some parameters for the Gaussian function. For this work a kernel size of $256 \times 256$ pixels was selected (Fig. 7).

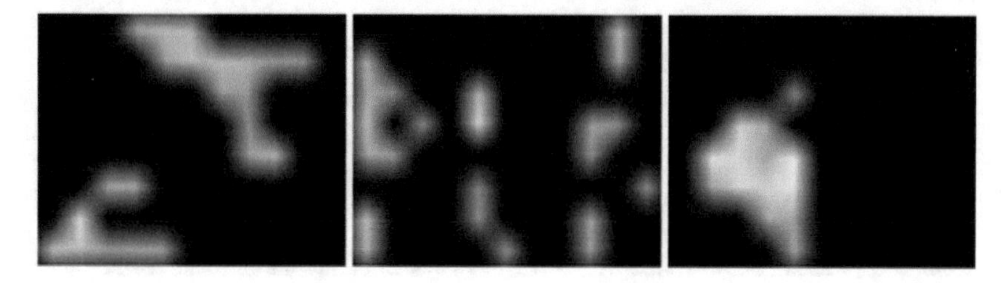

**Fig. 7.** Filtered memorability maps obtained with the test images presented before.

Following the approach, saliency and memorability maps were binarized and multiplied, obtaining a binary mask. Selected threshold for binarize was 0.2 in the range between 0 and 1. Once this mask was obtained for each image, the mask arithmetically multiplied original image. Intersection pixels have a value equal to 1 and the other pixels have a value equal to 0. When multiply only these pixels where have value equal to 1 in the mask keep the pixel value of the original image. The rest of pixels acquire 0 value. The results for these three example images are (Fig. 8):

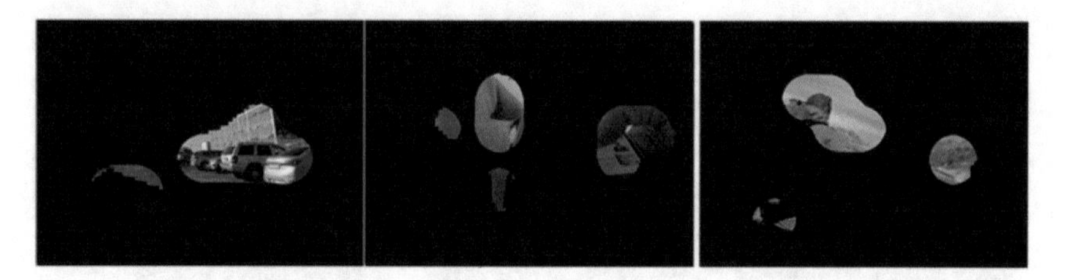

**Fig. 8.** Visual intersection between saliency and memorability binarized maps.

## 6 Conclusions

In this paper, we have addressed the following problem: how to create a rich human digital memory captured by a wearable camera. We presented new approach that will use convolutional neural networks which have the capability to keep localization in images. This capability is due to the convolutional layers because there are built with convolutional filters that have local impact.

The proposed approach will be able to automatically create rich human digital memory. Although the work presented a preliminary validation, we believe it demonstrates the potential of lifelogging techniques to create dynamic and rich memory.

## References


1. Standing, L.: Learning 10,000 pictures. Q. J. Exp. Psychol. **25**, 207–222 (1972)
2. Konkle, T., Brady, T.F., Alvarez, G.A., Oliva, A.: Scene memory is more detailed than you think: the role of categories in visual long-term memory. Psychol. Sci. **21**(11), 1551–1556 (2010)
3. Isola, P., Xiao, A., Torralba, A., Oliva, A.: What makes an image memorable? In: IEEE Conference on Computer Vision and Pattern Recognition (CVPR), pp. 145–152 (2011)
4. Konkle, P., Parikh, A., Torralba, A., Oliva, A.: Understanding the intrinsic memorability of images. In: Advances in Neural Information Processing Systems, pp. 2429–2437 (2011)
5. Mancas, M., Le Meur, O.: Memorability of natural scene: the role of attention. In: ICIP, pp. 196–200 (2013)
6. Cairo, A.: The Functional Art: An Introduction to Information Graphics and Visualization. New Riders, Berkeley (2013)
7. Cleveland, W.S., Mcgill, R.: Graphical perception: theory, experimentation, and application to the development of graphical methods. J. Am. Stat. Assoc. **79**, 531–554 (1984)
8. Borkin, M.A., Vo, A.A., Bylinskiin, Z., Isola, A., Sunkavalli, S., Oliva, A., Pfister, H.: What makes a visualization memorable? IEEE TVCG **19**(12), 2306–2315 (2013)
9. Herrera, M.C.: Visual Memorability for Egocentric Cameras. Universitat Politcnicade Catalunya Escola Superior denginyeria Industrial, Aeroespacial I Audiovisualde Terrassa (2006)
10. El Asnaoui, K., Petia, P., Aksasse, B., Ouanan, M.: Using content-based image retrieval to automatically assess day similarity in visual lifelogs. In: The International Conference on Intelligent Systems and Computer Vision. IEEE Conference Publications (2017)

11. El Asnaoui, K., Aksasse, H., Aksasse, B., Ouanan, M.: A survey of activity recognition in egocentric life-logging datasets. In: International Conference on Wireless Technologies Embedded and Intelligent Systems. IEEE Conference Publications (2017)
12. Doherty, A.R., Hodges, E.S., King, A.C., Smeaton, A.F., Berry, E., Moulin, J.C., Lindley, A., Kelly, P., Foster, C.: Wearable cameras in health: the state of the art and future possibilities. Am. J. Prev. Med. **44**(3), 320–323 (2013)
13. Hodges, S., Williams, E., Berry, E., Izadi, J., Srinivasan, J., Butler, A., Smyth, G., Kapur, N., Wood, K.: SenseCam: a retrospective memory aid. In: UbiComp: Ubiquitous Computing, pp. 177–193. Springer (2006)
14. Doherty, A.R., Pauly-Takacs, K., Caprani, N., Gurrin, C., Moulin, C.J., OConnor, N.E., Smeaton, A.F.: Experiences of aiding autobiographical memory using the sensecam. Hum. Comput. Interact. **27**(1–2), 151–174 (2012)
15. Lee, M.L., Dey, D.K.: Lifelogging memory appliance for people with episodic memory impairment. In: Proceedings of the 10th International Conference on Ubiquitous Computing, pp. 44–53. ACM (2008)
16. Xiong, B., Grauman, K.: Detecting snap points in egocentric video with a webphoto prior. In: European Conference on computer Vision, pp. 282–298 (2014)
17. Lidon, A., Bolaos, M., Dimiccoli, M., Radeva, P., Garolera, M., Giro-iNieto, X.: Semantic summarization of egocentric photo stream events. arXiv preprint arXiv:1511.00438 (2015)
18. Smeaton, A.F., Over, P., Doherty, A.R.: Video shot boundary detection: seven years of trecvid activity. Comput. Vis. Image Underst. **114**(4), 411–418 (2010)
19. Jinda-Apiraksa, A., Machajdik, J., Sablatnig, R.: A key frame selection of lifelog image sequences. Erasmus Mundus M.Sc. In: Visions and Robotics thesis, Vienna University of technology (2012)
20. Wang, Z., Hoffman, M.D., Cook, P.R., Li, K.: Video shot boundary detection: seven years of trecvid activity. Vferret: content-based similarity search tool for continuous archived video. In: ACM Workshop on Continuous Archival and Retrieval of Personal Experiences, pp. 19–26 (2006)
21. Chandrasekhar, V., Tan, C., Min, W., Liyuan, L., Xiaoli, L., Hwee, L.J.: Incremental graph clustering for efficient retrieval from streaming egocentric video data. In: IEEE International Conference on Pattern Recognition, pp. 2631–2636 (2014)
22. Doherty, A.R., Smeaton, A.M.: Combining face detection and novelty to identify important events in a visual lifelog. In: IEEE International Conference on Computer and Information Technology Workshops, pp. 348–353 (2008)
23. Alletto, S., Serra, G., Calderara, S., Cucchiara, R.: Head pose estimation in first person camera views. In: 22nd International Conference on Pattern Recognition (ICPR). IEEE (2014)
24. Kikhia, A., Boytsov, A.Y., Hallberg, J., Jonsson, H., Synnes, K.: Structuring and presenting lifelogs based on location data. In: Pervasive Computing Paradigms for Mental Health, pp. 133–144. Springer (2014)
25. Byrne, D., Doherty, A.R., Snoek, C.G.M., Jones, G.J.F., Smeaton, A.F: Everyday concept detection in visual lifelogs: validation, relationships and trends. Multimedia Tools Appl. **49** (1), 119–144 (2010)
26. El Asnaoui, K., Petia, R.: Automatically assess day similarity using visual lifelogs. J. Intell. Syst. (2018)
27. Kitani, M.K., Okabe, T., Sato, Y., Sugimoto, A.: Fast unsupervised ego-action learning for first-person sports videos. In: IEEE Conference on Computer Vision and Pattern Recognition, pp. 3241–3248 (2011)
28. Fathi, A., Farhadi, A., Rehg, J.M.: Understanding egocentric activities. In: IEEE International Conference on Computer Vision and Pattern Recognition, pp. 407–414 (2011)

29. Sundaram, S., Mayol-Cuevas, W.W.: Egocentric visual event classification with location-based priors. In: Advances in Visual Computing, pp. 596–605. Springer (2010)
30. El Asnaoui, K., Aksasse, B., Ouanan, M.: Content-based color image retrieval based on the 2D histogram and statistical moments. World Acad. Sci. Eng. Technol. Comput. Inf. Eng. **2**, 603–607 (2015)
31. El Asnaoui, K., Aksasse, B., Ouanan, M.: Color image retrieval based on a two-dimensional histogram. Int. J. Math. Comput. **26**, 10–18 (2015)
32. El Asnaoui, K., Chawki, Y., Aksasse, B., Ouanan, M.: A content-based image retrieval approach based on color and shape. Int. J. Math. Comput. **29**, 37–49 (2016)
33. El Asnaoui, K., Chawki, Y., Aksasse, B., Ouanan, M.: Efficient use of texture and color features in content-based image retrieval (CBIR). Int. J. Appl. Math. Stat **54**, 54–65 (2016)
34. Ouhda, M., El Asnaoui, K., Aksasse, B., Ouanan, M.: Content-based image retrieval using convolutional neural networks. In: Lecture Notes in Real-Time Intelligent Systems. Advances in Intelligent Systems and Computing, pp. 463–476. Springer (2019)
35. Ouhda, M., El Asnaoui, K., Aksasse, B., Ouanan, M.: A content based image retrieval method based on K-means clustering technique. J. Electron. Commer. Organ. **16**(1), 82–96 (2018)
36. El Asnaoui, K., Chawki, Y., Aksasse, B., Ouanan, M.: A new color descriptor for content-based image retrieval: application to COIL-100. J. Digit. Inf. Manag. **13**, 472–479 (2015)
37. El Asnaoui, K., Aksasse, B., Ouanan, M.: Content-based color image retrieval based on the 2-D histogram and statistical moments. In: Second World Conference on Complex Systems (WCCS), Agadir, Morocco. IEEE Conference Publications, pp. 653–656 (2014)
38. Chawki, Y., Ouanan, M., Aksasse, B.: CBIR using the 2-D ESPRIT method: application to coil 100 database. Int. J. Imaging Robot. **16**(2), 66–77 (2016)
39. Chawki, Y., El Asnaoui, K., Ouanan, M., Aksasse, B.: Content-based image retrieval using Gabor filters and 2-D ESPRIT method. In: Lecture Notes in Networks and System, vol. 25, pp. 95–102. Springer (2018)
40. Chawki, Y., El Asnaoui, K., Ouanan, M., Aksasse, B.: Content frequency and shape features based on CBIR: application to color images. Int. J. Dyn. Syst. Differ. Eqn. **8**(1–2), 123–135 (2018)
41. Chawki, Y., El Asnaoui, K., Ouanan, M., Aksasse, B.: New method of content based image retrieval based on 2-D ESPRIT method and the Gabor filters. TELKOMNIKA Indones. J. Electr. Eng. Comput. Sci. **15**(12), 313–320 (2015)
42. Fathi, A., Farhadi, A., Rehg, J.M.: Understanding egocentric activities. In: 2011 International Conference on Computer Vision, pp. 407–414. IEEE (2011)
43. Fathi, A., Li, Y., Rehg, J.M.: Learning to recognize daily actions using gaze. In: European Conference on Computer Vision, pp. 314–327. Springer (2012)
44. Pirsiavash, H., Ramanan, D.: Parsing videos of actions with segmental grammars. In: Computer Vision and Pattern Recognition (CVPR) (2014)
45. Ma, M., Fan, H., Kitani, K.M.: Going deeper into first-person activity recognition. In: The IEEE Conference on Computer Vision and Pattern Recognition (CVPR) (2016)
46. Khosla, A., Xiao, J., Isola, P., Torralba, A., Oliva, O.: Image memorability and visual inception. In: SIGGRAPH Asia 2012 Technical Briefs, p. 35. ACM (2012)
47. Khosla, A., Xiao, J., Torralba, A., Oliva, O.: Memorability of image regions. In: NIPS 2 4 (2012)
48. Bainbridge, W.A., Isola, P., Oliva, A.: The intrinsic memorability of face photographs. J. Exp. Psychol. Gen. **142**(4), 13–23 (2013)
49. Dobbins, C., Merabti, M., Fergus, P., Jones, D.L.: Creating human digital memories with the aid of pervasive mobile devices. Pervasive Mob. Comput. **12**, 160–178 (2014)

50. Dobbins, C., Merabti, M., Fergus, P., Llewellyn-Jones, D., Bouhafs, F.: Exploiting linked data to create rich human digital memories. Comput. Commun. **36**, 1639–1656 (2013)
51. Borkin, M., Bylinskii, Z., Kim, N., Bainbridge, C., Yeh, C., Borkin, D., Pfister, H., Oliva, A.: Beyond memorability: visualization recognition and recall. IEEE Trans. Vis. Comput. Graph. **22**(1), 519–528 (2016)
52. Kim, N., Bylinskii, Z., Borkin, M., Oliva, A., Gajos, K.Z., Pfister, H.: A crowd sourced alternative to eye-tracking for visualization understanding. In: CHI15 Extended Abstracts, pp. 1349–1354. ACM, Seoul (2015)
53. Bolanos, M., Dimiccoli, M., Radeva, P.: Towards storytelling from visual lifelogging: an overview. J. Trans. Hum. Mach. Syst. **47**, 77–90 (2015)

# Predictive Analytics and Optimization of Wastewater Treatment Efficiency Using Statistic Approach

I. Bencheikh[1(✉)], J. Mabrouki[1], K. Azoulay[1], A. Moufti[2,3], and S. El Hajjaji[1]

[1] Laboratory of Spectroscopy, Molecular Modeling, Materials, Nanomaterial, Water and Environment, Faculty of Science, CERNE2D, Mohammed V University in Rabat, Avenue Ibn Battouta, BP1014, Agdal, Rabat, Morocco
Imaneben98@gmail.com
[2] Laboratoire des Sciences des Matériaux, des Milieux et de la Modélisation (LS3M), FPK, University Hassan 1, BP145, 25000 Khouribga, Morocco
[3] Regional Center for Careers in Education and Training, Casablanca-Settat, Morocco

**Abstract.** The high charge, as well as the complicity of the effluents, make that the assimilative capacity of water is ineffective in the majority of the cases. Wastewaters treatment plants are a suitable alternative to solve the problem of this inefficiency of the natural assimilative capacity of water. The control and the determination of the optimal values of parameters that influence the function and the effectiveness of the treatment is a necessity. A classical study can be done to determine the good conditions of the treatment, this way is effective but it requires more time and reactive products that increase the cost invested in the study and design of the treatment. In our case, we will try to apply predictive analytics based on a statistic approach to optimize the efficiency of manganese dioxide extracted from a mining waste to treat synthesized textile effluent. In this study, we will use the response surface method using the JMP[11] software to visualize the effect of the presence of salts in textile effluents on the removal of dyes using manganese dioxide. According to the p-Value of each studied parameter, both manganese dioxide dose and contact time have a significant effect on dye elimination. Also, the best retention rate (70.5%) is observed with a value of manganese dioxide dose equal to 6 mg/L and contact time of 180 min with an initial charge of 1.5 g/L for dye and 20 g/L for NaCl.

**Keywords:** Wastewater's treatment · Predictive analytics · Statistic approach · Manganese dioxide ($MnO_2$)

## 1 Introduction

The textile industries, like all human activities, produce various effluents and wastes, with or without treatment, are discharged into the sea, rivers or soils [1]. The high chemical charge generated from effluents is the main concern of these industries. The most important environmental impacts are due to the presence of salts, detergents and


Y. Farhaoui (Ed.): BDNT 2019, LNNS 81, pp. 310–319, 2020.
https://doi.org/10.1007/978-3-030-23672-4_22

organic acids in these effluents. Which contaminates directly or indirectly the superficial and internal water resources. These contaminated water can be subject to different types of pollution-related illnesses, skin wounds, respiratory tract irritation [2], typhoid, cholera, viral hepatitis, etc., and, in severe cases, death [3]. For that, many techniques have been developed to find an economical and efficient way to treat textile wastewater, processes and other technologies. These methods are generally highly effective for dyeing wastewater's treatment.

The use of IT tools is part of the desire to limit as much as possible the costs (financial and temporal). Practical and cost constraints strongly encourage the use of predictive analytics studies, for optimized design of new electro-Technical devices. The conceptual objective being to obtain the variation rate of the effects of parameters, as well as the real optimal conditions of the different factors, controlling the satisfactory functioning of the wastewater treatment methods [4].

The use of the methodology of experimental design favored the use of simulations, and more broadly that of numerical modeling based on statistic approaches allowing the prediction of the variation of wastewater treatment conditions and their optimum.

An excess or minimum value of one of the reagents involved in wastewater treatments can have a negative influence on the treatment efficiency or even can disrupt the operation of the polluted water treatment plant, so determining the optimal values for each parameter of the method is the key to successful treatment. A classical study can be done to determine the good conditions of the treatment, this way is effective but it requires more time and reactive products that increase the cost invested in the study and design of the treatment. In our case, we will try to apply this method to test the efficiency of manganese dioxide extracted from a mining waste to treat synthesized textile effluent at the laboratory scale. In this study, we will use the response surface method to visualize the effect of salts in the effluents on the removal of dyes using manganese dioxide.

## 2 Application of JMP[11] Software in the Optimization of Wastewater Treatment by Adsorption Technique

To understand the evolutions of the characteristic variables of the adsorption, their interactions and their influence on the efficiency of the treatment, we applied in this work the formalization conferred by the experimental design method.

This term ultimately refers to a complete methodology for the behavioral characterization of a system. It is based on the predictive analytical study of the treatment's method and the measurement of the variables specific to the technique under consideration; this mainly includes the factors influencing the adsorption as well as their optimums. This statistical approach makes it possible to determine the relations existing between the different parameters of the treatment, by establishing the analytical relations linking them.

In order to follow the study by the methodology of experimental design, many software was used for the study of water treatment by adsorption as well as the study of the efficiency of certain new adsorbents to eliminate the organic and mineral pollution in wastewater [4].

Aydın and Aksoy [5] used the Minitab software to study the Optimization of the adsorption of chromium on chitosan: using the Surface response methodology. For the study of removal of Pb(II)from water solution by Pistacia vera. L Yetilmezsoy et al. [6] used STATISTICA software package (Trial version 8.0. StatSoft Inc.. the USA) using box Behnken experimental design. Asfaram et al. [7] studied the Optimization of parameters affecting the removal of basic dye Auramine-O by ZnS: Cu nanoparticles loaded on activated carbon using a composite design using STATISTICA 10.0 software. Bencheikh et al. conducted a study aimed at testing the efficiency of a low-cost adsorbent to remove an azo dye by the methodology of experimental research using Nemrodow software [8]. In order to study the adsorption of metallic ions onto fly ash Ricou-Hoeffer et al. used (Nemrod 3.0 software) [9].

Depending on the type of these mathematical models, the experimenter can deduce more or less precise information. Qualitative or quantitative.

The theoretical bases of this approach are algebro-statistics. They thus give the MEP the ability to manage the error terms inherent in the existence of experimental variability.

The JMP[11]software used in this study allows the follow-up and the study by different design including the definitive screening design, screening design, full factorial design, response surface designs and mixture design [10].

# 3 Case Study: Predictive Analytics and Optimization of Synthesized Textile Wastewater Treatment by Adsorption Technique Using Manganese Dioxide Extracted from Mining Waste

## 3.1 Preparation of Synthesized Textile Effluent

A solution of methylene blue (1.5 g/L) was prepared by dissolving 1.5 g of Methylene blue 1 L of a solution of NaCl (20 g/L). In a beaker of 100 mL, we placed 50 ml of the solution with the amount of manganese dioxide presented in (Table 1) at pH = 8. The mixture is placed in a magnetic stirrer at room temperature. Absorbance is measured by using the UV-visible spectrophotometer with at methylene blue wavelength (664 nm).

The retention rate is calculated by the following equation:

$$\%R = \frac{(C_0 - C_e)}{C_0} x100 \quad (1)$$

Where $C_0$ is the initial concentration of dye (mg. $L^{-1}$);
$C_e$ is the equilibrium concentration of dye (mg. $L^{-1}$).

### 3.2 The Methodology of Statistic Approach

We chose the methodology of surface response to visualizing the effect of the time and the dose of manganese dioxide in the adsorption treatment of synthesized textile wastewaters as well as the determination, of the optimal conditions of treatment, for that, we have opted for the methodology of experimental design.

- **First step: Choice of influencing parameters on the treatment and determine the interval of the study**

The use of the methodology of response surface for the study the effectiveness of a synthetic textile effluent's treatment by the method of adsorption into the manganese dioxide extracted from mining waste. This effluent (containing a concentration of 1.5 mg/L of Methylene Blue and 20 mg/L of NaCl). First of all, we fixed the factors affecting the phenomenon. as well as their upper and lower limit. The adsorbent dose and the contact time are the parameters targeted in this work. (Table 1) represents the values of upper and lower limits of each parameter.

**Table 1.** Limits for studied parameters

| Parameters | Level (-1) | Level (+1) |
|---|---|---|
| Adsorbent dose (mg/L) $X_1$ | 1 | 6 |
| Contact Time (min) $X_2$ | 15 | 180 |

- **Second step: Choosing the software. choosing the surface design and obtaining the experimental matrix**

The design used for our study is the Central Composite Design (CCD). This design is the most used surface response design which can fit a complete quadratic model. Central Composite Design includes a factorial design or a fractional factorial design with central points. Plus a group of axial points (or stars) that estimate the curvature [11].

$$\text{Number of experiments} = 2^f + 2*f + N_0 \tag{2}$$

With: f is the number of the studied factors;

$N_0$ is the number of center 'points.

In order to follow this study, we choose the JMP[11] software based on exploratory data analysis and visualization. They are designed to allow users to search for data to learn something unexpected. as opposed to confirming a hypothesis. JMP[11] links statistical data to the graphs that represent it. So users can drill forward and backward with data and their different visual representations [12, 13]. After choosing the design and the entrance of the upper and lower limits of each parameter the following experimental matrix is obtained (Table 2).

**Table 2.** Experimental matrix for dye removal by $MnO_2$

| Experience | The dose of $MnO_2$ (g/L) | Time (min) |
|---|---|---|
| 1 | 0.8 | 97.5 |
| 2 | 1 | 15 |
| 3 | 1 | 180 |
| 4 | 3.5 | 8.5 |
| 5 | 3.5 | 97.5 |
| 6 | 3.5 | 97.5 |
| 7 | 3.5 | 186.44 |
| 8 | 6 | 15 |
| 9 | 6 | 180 |
| 10 | 6.2 | 97.5 |

The generalized polynomial equation containing six coefficients is as follows:

$$Y = b_0 + b_1X_1 + b_2X_2 + b_{12}X_1X_2 + b_{11}X_1^2 + b_{22}X_2^2 \quad (3)$$

With:
**Y**: experimental studied response;
$\mathbf{X_i}$: coded value of the variable i;
$\mathbf{b_0}$: medium effect;
$\mathbf{b_i}$: principle effect of the variable i;
$\mathbf{X_ib_{ij}}$: interaction effect between $X_i$ et $X_j$.

- **Third part: performing the requested tests and data processing by JMP[11] software**

After carrying out the experiences listed in the experimental matrix (Table 2), we obtain the response presented in the (Table 3).

**Table 3.** Experimental matrix for dye removal by $MnO_2$ with the obtained retention rates.

| Experience | Dose of $MnO_2$ (g/L) | Time (min) | Retention rate (%) |
|---|---|---|---|
| 1 | 0.8 | 97.5 | 28.34 |
| 2 | 1 | 15 | 22.57 |
| 3 | 1 | 180 | 28.44 |
| 4 | 3.5 | 8.5 | 38.58 |
| 5 | 3.5 | 97.5 | 31.31 |
| 6 | 3.5 | 97.5 | 33.42 |
| 7 | 3.5 | 186.44 | 62.48 |
| 8 | 6 | 15 | 33.22 |
| 9 | 6 | 180 | 68.21 |
| 10 | 6.2 | 97.5 | 49.2 |

## 4 Results of Analysis and Discussion

**Statistic validation of the model**

This step allows us to check the validation of the used model via several analyses like predicted plot residual plot and analyze of variance.

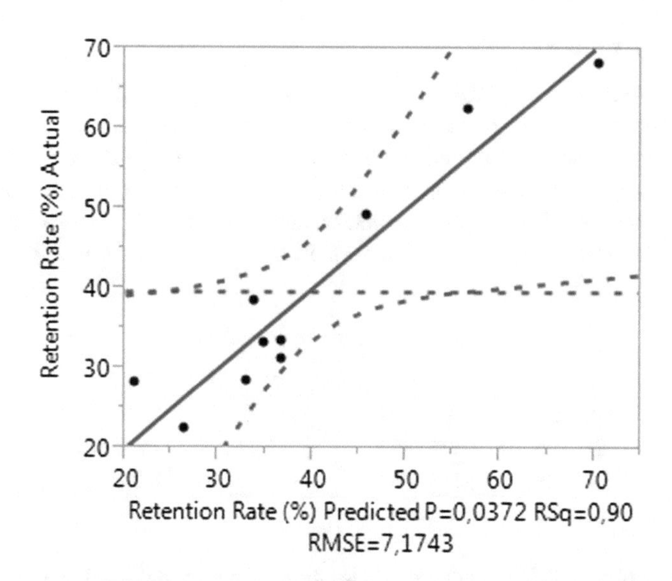


**Fig. 1.** Experimental and predicted response for MB onto $MnO_2$.

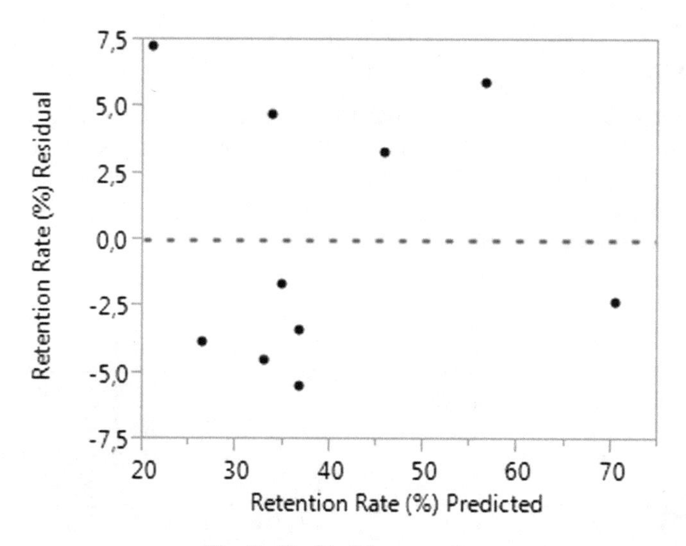


**Fig. 2.** Residual by row-plot.

- **Predicted plot**

Determination factor $R^2$ is a measure reflecting the correlation of the experimental studies and predicted one [14]. The Fig. 1 indicated a good alignment of the points. So we can confirm the correlation between experimental and predicted studies. Also, the aleatory distribution of the experimental point present in Fig. 2 indicates the validation of the used model.

- **Analyze of variance**

Variance Analysis (ANOVA) is a method for analyzing the influence of qualitative factors on one or more response variables. If all factors (variables) are quantitative, regression analysis is the recommended method for analyzing the data. The goal of regression analysis is to develop prediction equations and optimize the response variable. If there is a mix of qualitative and quantitative (variable) factors, the method is known as covariance or generalized regression analysis (Tables 4 and 5).

**Table 4.** Model's fit parameters.

| Source | DF | Sum of Squares | Mean Square | F Ratio |
|---|---|---|---|---|
| Model | 5 | 1918.6044 | 383.721 | 7.4551 |
| **Error** | 4 | 205.8844 | 51.471 | **Prob > F** |
| **C. Total** | 9 | 2124.4887 | – | 0.0372* |

The value of C.Total indicates the validation of the model used in the study if the value is under 0.05 the model is valid if it is higher than 0.05 the model is not valid. In our case, the chosen model for this study is valid because the C.Total is equal to $0.0372 < 0.05$.

- **Effect tests**

The estimate is a parameter that indicates the positive and negative effect of each parameter basing on the sign of the parameter's coefficient and also the p-value is the main parameters that indicated the most influencing parameters. if the value of p-value is under 0.05 the parameter is significant if it is higher than 0.05 the parameter is considered as insignificant [10].

**Table 5.** CCD's parameters for the system (Dye/$MnO_2$).

| Term | Estimate | Std Error | t Ratio | Prob > \|t\| |
|---|---|---|---|---|
| Intercept | 36.793959 | 4.518481 | 8.14 | 0.0012* |
| Dose of $MnO_2$ (mg/L)(1.6) | 11.520332 | 2.850912 | 4.04 | 0.0156* |
| Time (min)(15.180) | 10.535696 | 2.852445 | 3.69 | 0.0210* |
| Dose of $MnO_2$ (mg/L)*Time (min) | 7.2806419 | 3.587168 | 2.03 | 0.1123 |
| Dose of $MnO_2$ (mg/L)*Dose of $MnO_2$ (mg/L) | −2.84574 | 4.357674 | −0.65 | 0.5494 |
| Time (min)*Time (min) | 7.248824 | 4.3635 | 1.66 | 0.1720 |

So the efficiency of the treatment depends mainly on dose of manganese dioxide and time and p-value (dose of $MnO_2$) = 0.156 < 0.05; p-value (Time) = 0.0210 < 0.05. And negligee the interaction between these parameters (p-value (dose of $MnO_2$-Time) = 0.1123 > 0.05; p-value (dose of $MnO_2$- dose of $MnO_2$-Time) = 0.5494 > 0.05; p-value (Time-Time) = 0.1720 > 0.05).Thus the optimized polynomial equation is as follows:

$$\text{Retention rate}\,(\%) = 36.793959 + 11.520332X_1 + 10.535696X_2 \tag{4}$$

For interaction between parameters to have an effect, prediction traces must change their slope and curvature when modifying the current values of other terms. If there is no interaction effect, the traces only change in height this last case illustrates and explains our obtained results Fig. 3. This confirms the results obtained in the analysis of the effect test.

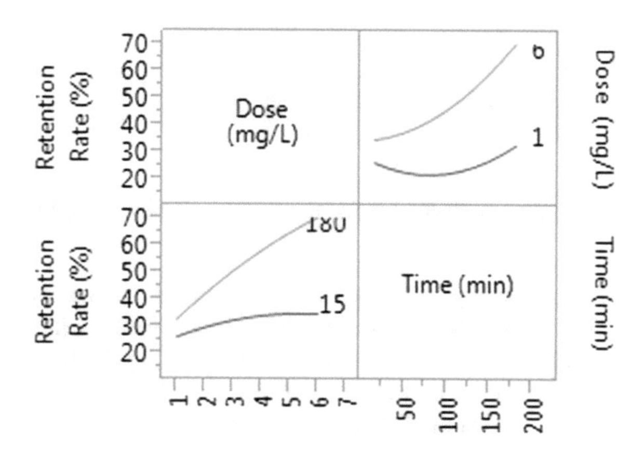


**Fig. 3.** Interaction profile of the adsorption treatment using manganese dioxide.

- **Prediction profiler**

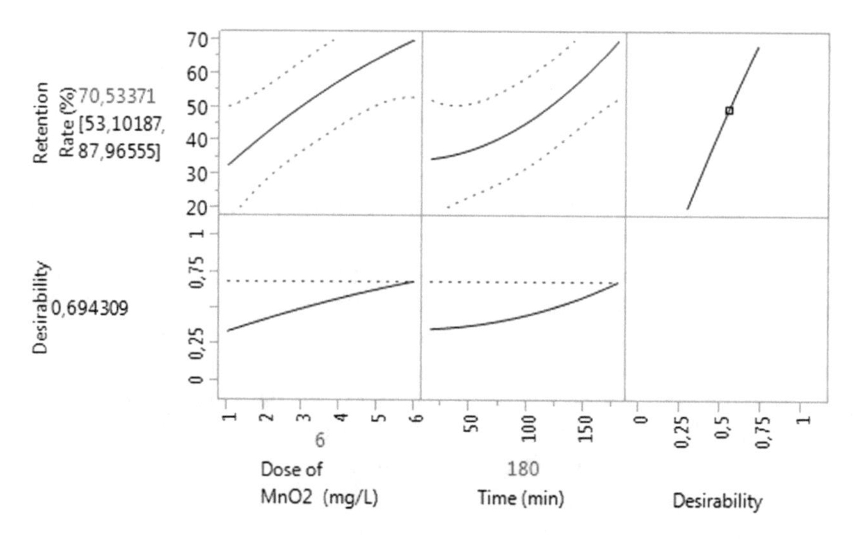


**Fig. 4.** Profile of predicted of the optimal conditions of adsorption of the MB on $MnO_2$.

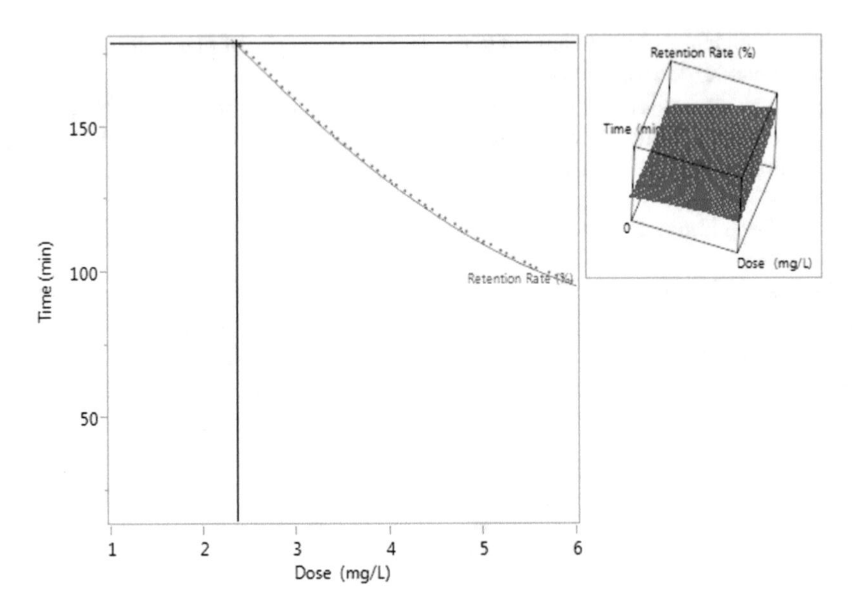


**Fig. 5.** Contour's profile for the treatment by adsorption on $MnO_2$.

From the profiler, we can get the optimizing value of the parameters depending on the desirability. Also, we can see and follow the variation of the treatment's efficiency depending on the different studied parameters. In our case, by fixing the high value of desirability we can see that the best retention rate (70.5%) is observed with a value of manganese dioxide equal to 6 mg/L and contact time of 180 min.

The contour profiler displays the response contours of the $MnO_2$ dose and the contact time at a time. The interactive contour profiling feature is useful for graphically optimizing response surfaces. Figure 5 confirms the same condition obtained in Fig. 4.

## 5 Conclusion

The methodology of experimental design used the study of the efficiency of extracted dioxide manganese to treat textile synthetic effluents was effective.

The Determination factor $R^2$ confirm the correlation between experimental and predicted studies. Also, the aleatory distribution of the experimental point observed in the residual plot indicates the validation of the used model.

Also, the study of effect indicates that the efficiency of the treatment depends mainly on the dose of manganese dioxide and time because their p-value is under 0.05.

The study shows it gave us also the optimal condition of treatment using our adsorbent with minimum experiences (10 experiences) and in a short time. From the study, we can conclude that both adsorbent dose and time affect the efficiency of treatment. Also, the best retention rate (70.5%) is observed with a value of manganese dioxide equal to 6 mg/L and contact time of 180 min.

**Acknowledgment.** We thank the "National Center for Scientific and Technical Research" in Rabat (Morocco) for their support by the Excellence Scholarship.

## References

1. Bawa, M.L., Djaneye-Boundjou, G., Boukari, Y.: Caractérisation de deux effluents industriels au togo: étude d'impact sur l'environnement. Afrique Sci. **02**(1), 57–68 (2006)
2. Sultana, M.S., Islam, M.S., Saha, R., Al-Mansur, M.: Impact of the effluents of textile dyeing industries on the surface water quality inside DND embankment, narayanganj. Bangladesh J. Sci. Ind. Res. **44**, 65–80 (2009)
3. Islam, J.B., Sarka, M., Rahman, A.L., Ahmed, K.S.: Quantitative assessment of toxicity in the Shitalakkhya River. Bangladesh. Egypt. J. Aquat. Res. **41**, 25–30 (2015)
4. Vivier, S.: Stratégies d' optimisation par la méthode des Plans d' Expériences, et Application aux dispositifs électrotechniques modélisés par Eléments Finis Stéphane Vivier To cite this version : HAL Id : tel-00005822, p. 83 (2004)
5. Aydin, Y.A., Aksoy, N.D.: Adsorption of chromium on chitosan: optimization, kinetics and thermodynamics. Chem. Eng. J. **151**(1–3), 188–194 (2009). https://doi.org/10.1016/j.cej.2009.02.010
6. Yetilmezsoy, K., Demirel, S., Vanderbei, R.J.: Response surface modeling of Pb(II) removal from aqueous solution by Pistacia vera L.: Box-Behnken experimental design. J. Hazard. Mater. **171**(1–3), 551–562 (2009). https://doi.org/10.1016/j.jhazmat.2009.06.035
7. Asfaram, A., Ghaedi, M., Agarwal, S., Tyagi, I., Gupta, V.K.: Removal of basic dye Auramine-O by Zns: Cu nanoparticles loaded on activated carbon: optimization of parameters using response surface methodology with central composite design. RSC Adv. **5** (24), 18438–18450 (2015). https://doi.org/10.1039/c4ra15637d
8. Bencheikh, I., Abourouh, I., Kitane, S., Dahchour, A., El Mrabet, M., et El hajjaji, S.: Adsoptive removal of an azo dye (Methyle Orange) onto a low cost adsorbent using the methodology of experimental research, Recent Advances in Environmental Science from the Euro-Mediterranean and Surrounding Regions book, pp. 235–236 (2018). DOI: https://doi.org/10.1007/978-3-319-70548-4_75
9. Ricou-Hoeffer, P., Lecuyer, I., Le Cloirec, P.: Experimental design methodology applied to adsorption of metallic ions onto fly ash. Water Res. **35**(4), 965–976 (2016). https://doi.org/10.1016/S0043-1354(00)00341-9
10. Jacquez, J.A.: Design of experiments. J. Franklin Inst. **335**, 131–238 (2002). https://doi.org/10.1016/s0016-0032(97)00004-5
11. https://support.minitab.com
12. Altman, M.: A review of JMP 4.03 with special attention to its numerical accuracy. Am. Stat. **56**(1), 72–75 (2002). https://doi.org/10.1198/000313002753631402
13. Carver, R.H.: Practical data analysis with JMP, p. 61. SAS Institute, Cary (2010)
14. Elmoubarki, R., Taoufik, M., Moufti, A., Tounsadi, H., Mahjoubi, F.Z., Bouabi, Y., Qourzal, S., Abdennouri, M., Barka, N.: Box-Behnken experimental design for the optimization of methylene blue adsorption onto Aleppo pine cones. J. Mater. Environ. Sci. **8**(6), 2184–2191 (2017)

# A Novel Software to Improve Healthcare Base on Predictive Analytics and Mobile Services for Cloud Data Centers

Muhammed Abaid Mahdi and Samaher Al_Janabi(✉)

Department of Computer Science, Faculty of Science for Women (WSCI), University of Babylon, Babylon, Iraq
wsci.muhammed.a@uobabylon.edu.iq,
samaher@itnet.uobabylon.edu.iq

**Abstract.** The main idea of this work is generated from studying the main challenges in the healthcare field. The goal is to identify patients who will be admitted to the hospital within the next year by using historical claims data including Information about patient and analysis it to solve the problem. But this work is dealing with very huge databases. Therefore, the new challenge is occurred the time required for executing the prediction methods and analysis, it bases on five error predicate measures, it including (Maximum error, RMSE, MSE, MAE and MAPE). As a result, cloud is suggested as a tool to solve the problem of analysing the predictions of a huge health care database. A new predictor is proposed to determine how many days in the next year a patient will spend in the hospital. In this work, we attempt to satisfy the idea that explains in Fig. 1. The main challenge here, how we can build the predictor that satisfies the gold triangle. The traversal of the T-Graph) cloud computing "i.e., Speed execution", data mining algorithm "i.e., Abilities to deal with very huge databases" and Predication techniques" i.e., Ability to plan of the next years.

**Keywords:** Predictive techniques · Data mining · Deep analysis · Mobile services · Healthcare · T-Graph

## 1 Introduction

Today, more than 71 million individuals in the United States have been admitted to hospitals each year, according to the latest survey from the American Hospital Association. However, a considerable number of hospital admissions turn out to be unnecessary, which causes huge wastes of money as a result; understanding how long a patient will spend in a hospital each year can be beneficial. Based on such knowledge, health care providers can develop new care plans and strategies to reach patients before emergencies occur, thereby reducing the number of unnecessary hospitalizations. Recently, the Heritage Provider Network (HPN) has organized an online competition (HPN 2011) for this predictive problem. Although the competition is still ongoing, the best result of the top teams on the Leaderboard is around 45% Root Mean Squared Logarithmic Error, which is not very promising. As we will discuss in 2011, Association for the Advancement of Artificial Intelligence, all rights reserved. This work, the


Y. Farhaoui (Ed.): BDNT 2019, LNNS 81, pp. 320–339, 2020.
https://doi.org/10.1007/978-3-030-23672-4_23

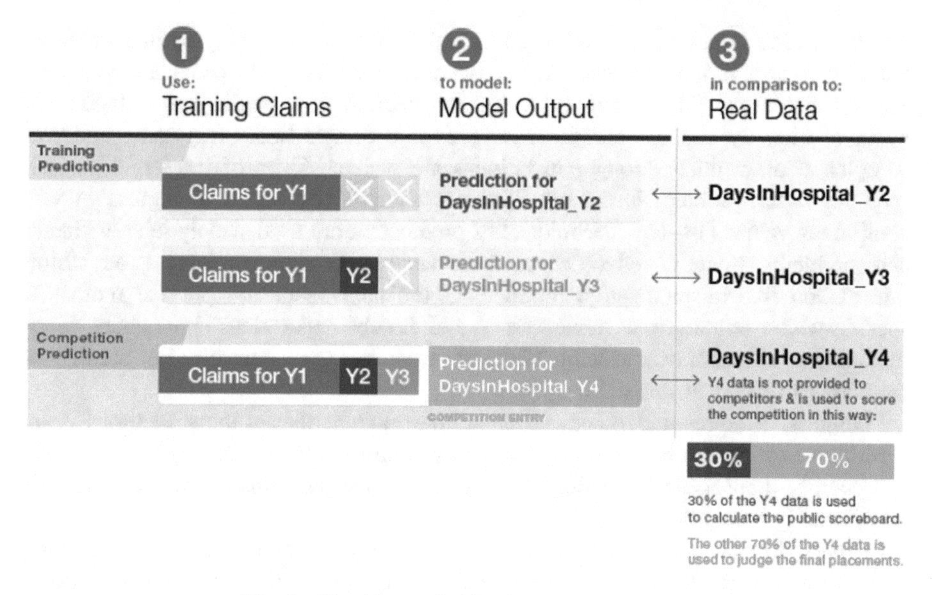


**Fig. 1.** The idea and objective of this proposal

problem turns out to be challenging in practice, for reasons from both the data we have and the problem itself.

Unlike most of the machine learning problems for which training data are scarce, the dataset used in this problem consists plenty of records. However, the set of features used to characterize each record is limited, much less than the number of records we have. On the other hand, due to the highly skewed distribution of the target attribute (i.e., The number of days a patient will stay in a hospital) values it is difficult to find indicative features that are highly correlated with the target attribute, even after applying some feature selection algorithms. • Instead of predicting whether a patient will be in hospital next year, the problem requires to predict the exact number of days he/she will stay. Basically, this is a regression problem since we need to predict some continuous attribute. It is generally believed in machine learning re- search community that regression problems are more difficult than classification problems, for which we are only supposed to answer yes or no.

## 2 Literature Reviews

The health care problem is one of the costliest problems in the modern society that has a significant effect on a person's life and on the financial resources of the countries. High quality health care system contributes with other factors to increase life expectancy, the average death age in US increased from 68.2 years in 1950 to 78.7 years in 2010 [1] but the cost of this care is growing rapidly to reach $2.6 trillion in 2010 [2]. Some major factors control those costs such as number of patients that attend two hospitals and the number of days that spend in hospital, unnecessary hospital admissions cause a waste of resources amount to $30 billion in 2006 [3].

Predication is one of data mining techniques that has an ability to find unknown value of a target variable based on values of some other variables. The relation between target variable and other interested variables specifies how predication is hard, with linear relation we can use simple techniques like Simple linear regression and with complex relation other stronger techniques are needed. Type of the target variable specifies if the problem is the classification with binary values or regression with continuous values [14–16]. Techniques of predication are used widely to solve health care problems, thereby, reduce costs, these techniques allow to forecast any future information like hospitalization numbers in the next years and provide models to specify which admission is unnecessary and finally make clear sight about future requirement in health care system. Many of researches were introduced in predication methods and use it in this field.

Dimitris, Margrét and others, 2008 utilize modern data mining methods as an algorithmic prediction for health care costs by Aggregate many sources of claims data and classify them using Classification Tree with some performance measures like hit ratio, penalty error and $R^2$ [4].

Duana, Streeta and Xu, 2010 created a recommended nursing clinical system which helps to make right decision and improve clinical quality control by using Association Rules to find patterns in itemsets of community hospitals in the Midwest Dataset and use support, confidence, lift and IV as utility measurements [5].

Santhanam and Shyam, 2010 used a classification technique to classify blood donor then detect safe donor from unsafe by using CART algorithm that has ability to provide easy understanding for classification rules with the blood transfusion dataset (UCI ML repository) then evaluate the result by Precision and Recall measures [6].

Xiang and Wentao and Jia, 2011 used Machine Learning algorithms to reduce unnecessary hospitalizations by predicate how long the patient will stay in hospital in the next year, according to his record in last year, researchers used SVM, Random forest, Regression tree and Boosting Ensemble with HPN 2011 Dataset [7].

Jyoti, Ujma, Dipesh and Sunita, 2011 compare the performance of predictive data mining techniques such as KNN, Neural Networks, Bayesian and classification Decision Tree on Cleveland Heart Disease database, then applied genetic algorithm to reduce the actual data size and get the optimal subset of attribute sufficient for heart disease prediction [8].

Necdet and Osman, 2011 Applied six regression tree methods CART, CHAID, E-CHAID, QUEST, RFRC and BTCR to predict risk of death from head injury with 10-fold cross validation on each method based on Real clinical data set of 1603 patients with head injuries. The result was that BTCR is the best one depends on accuracy 93% and ROC (AUC = 0.954) [9].

Rashedur and Fazle, 2011 Used and compared different decision tree classification techniques to classify admitted patients according to their critical condition and develop an application to diagnose and measure the criticality of the newly arrived patient to mining Hospital surveillance unit of ICDDR, B using TDIDT algorithm (C4.5 Decision Tree Classifier) and evaluated result using False positive rate (FP), Recall, Precision [10].

Sung, Suk-Young and others, 2012 developed Decision tree model to find patterns in smoking behaviour among elders and find factors to prevent smoking behaviour or to decrease heavy smoking habits using CART method based on the National Survey on Drug Use and Health (NSDUH, 2006) then compared it with Logistic regression model using accuracy performance measure [11].

Jufen, Kevin and Alan Rigby, 2013 constructed CHAID classification tree with 10-fold cross-validation to predict probability of death or hospitalization for heart failure and compared the result with logistic regression (LR) models using ROC curve analysis based on TEN-HMS Dataset, they found CHAID tree performed better than the LR-model for predicting the composite outcome [12].

Dengju, Jing And Xiaojuan Zhan, 2013 introduced a novel method to predicate diseases by the combination of Random Forest and Multivariate Adaptive Regression Splines on The Wisconsin Diagnostic Breast Cancer Dataset (WDBC), researchers found that RF&MARS method has a higher classification accuracy than an RF model, but lower classification accuracy than MARS model, accuracy, sensitivity, specificity and confusion matrixes used as a Measures for Performance Evaluation [13].

**Research Questions**

The proposed work sets out to answer the following research questions:

- ***Which of Predication tools is best in healthcare problem and why*?** This work attempts to answer of that equation by determining and analysis the main parameters used in predicate in each of the above tools to building new tools of the predicate. But, before building that a tool, you might want to look at another equation,
- ***What the user needs from predication techniques high Quality of Recognition and Prediction or speed predication?*** In this work, we hope to satisfy the both goals (i.e., Quality and speed). But we know the fact that say when the accuracy and speed of predication these leads to increase the cost.
- How we can combine among these three mains permeates (highest accuracy, highest speed and less cost) in the new suggest predicate technique used to solve the health care problem?
- ***Why health care problem?*** To identify patients who will be admitted to a hospital within the next year, using historical claims data includes Information about patient and analysis it to the solution of the problem. Add to that, healthcare providers can take proactive steps to prevent hospitalizations for high-risk patients. This will not only save money, but will also improve patient quality-of-life.
- ***How we can satisfy the goal of this project?*** This can satisfy by taking the advantage of each the following concept (Cloud Computing, Data Mining algorithm and Predication Techniques). We hope to building the gold triangle (cloud computing "i.e., speed execution", data mining algorithm "i.e., abilities to deal with very huge databases" and Predication techniques" i.e., Ability to planning of the next years").

**Table 1.** Characteristic of each concept

| Term | Advantage | Disadvantage |
|---|---|---|
| Cloud computing | Speed execution | Cost |
| Data mining Algorithm | Abilities to deal with very huge databases | Time |
| Predication techniques | Ability to planning of the next years | Accuracy |

## 3 Predication Techniques

- **CART** is one of Decision tree techniques that used to classify data with an easier and more understandable way. For more detail see [4]. CART works with one variable (x) in every split operation, so it needs to decided which variable will be the best choice according to split criterion. In Regression model, target variable Y has continuous values, mean of values in terminal nodes is one way to set a value to (Y) but it has a problem with noise so median method can be used [20].
- **ECHAID**: Like CHAID, it allows for each parent to have more than two child nodes [17]. For more detail, see [4].
- **RFRC**. "Has an accuracy as good as Adaboost and sometimes better. It's relatively robust to outliers and noise and faster than bagging or boosting. It gives useful internal estimates of error, strength, correlation and variable importance. It's simple and easily parallelized [18]. When using Random Forest importance, its training time required from hours to even days of computation, especially for larger sets [19]. RFRC depends on randomness that make it suffer from problems of randomness".
- **MARS**: "is a data driven regression procedure that builds a model based on "divide and conquer" concept of the number of equations (basis function) and coefficients, each equation for a region in input space. It can handle high dimensional data (from 2 to 20 variables) that represent the main problem in other techniques. [28] ". It's automatically extended to cover the nonlinear relation between dependent variable (Y) and independent variables (X). Basis functions used specify the relation between Y and X for each equation, every two-basis function have not (shared point in decision boundary) that specified for data, when terms of basic functions are complete backward steps is needed as a post pruning to avoid overfitting [13, 17]. Here, we don't have a tree like CART or CHAID but series of equations performs regression task, so it depends totally on mathematical functions lead to have a strength of mathematic for finding optimal solutions. MARS has many interesting features, there is no user specific parameters are needed, thereby to be more flexible because it adapts by date. Also, variables don't need transformation that eliminates preprocessing steps leads to less computational time, variables are automatically selected by MARS. Other interesting features that MARS has the ability to handle more than one target variable (Y). Despite of slower model building compare with recursive partitioning [13, 17], 20].
- ***BTCR:*** is a finite loop of CART's operations that produce sequences of binary trees with forward strategy. Like RFRC, BTCR builds number of trees to reach optimal accuracy by cover all possible variable choices but BTCR doesn't depend on bootstrap sampling or selection variable by random but in each iteration, tree is built to predicate residual of previous tree. It prevents tree from grow without control, in each iteration the size of current tree was detected to fixed value. [17, 21–24].
- **BNNC** is one of classification methods that used both probability theory and graph theory to solve the problem of non-deterministic relation between variables set X and target class Y. [25, 26].

- ***SVM*** is one of prediction and classification manners analysis with all parameters in [25, 27].

### 3.1 Objectives of Proposed Research

The goal of the suggest new predicate tool, we hope that tool can achieve the following equation (less cost, less time and highest accuracy) with providing the report for both sides: the first for hospital about the number of days that the patient spend on it in the next years to take all the necessary requirements, and the another report send to patient (him\her) visit the hospital in previous years, these report, the patient receive it by the your' Mobil.

This study embarks on the following objective:

- To analysis the parameters of eight data mining forecasting techniques and determine each of that parameters is considered effective in the take the decision. Then establishing new predication tool called HCP based on the previous analyzed, through building fundamental modeling to prove the theory and establish its viability.
- To prove a novel tool, satisfy gold triangle or T-Graph by combination among the main three sides [ cloud computing "i.e., speed execution", data mining algorithm "i.e., abilities to deal with very huge databases" and Predication techniques" i.e., ability to planning of the next years")].
- To assess the algorithm and framework analysis of the proposed strategy.
- To make recommendations for both sides: the first for hospital about the number of days that the patient spend in it in the next years to take all the necessary requirements, and the another report send to patient (him\her) visit the hospital in previous years, these report, the patient receive it by the your' Mobil.
- Using web applications make healthcare providers can take proactive steps to prevent hospitalizations for high-risk patients. This will not only save money but will also improve patient quality-of-life.
- To patent the resulting algorithmes if HCP hit rates exceed those of curent.

### 3.2 Research Methodology

The following points represent the main steps of research methodology while the Fig. 3 shows block diagram of the proposal system while Fig. 4 system architecture

- Determine the main permeates that effect in the predication in each predication techniques (we used in this study nine predication techniques)
- We attempt to design new tool called HealthCare Predicator (HCP) base on the steps in *Pseudo1*
- Reduce all the irrelevant features base on calling the rough set procedure as explained in Pseudo2
- Compare the Result of HCP with all the other predicate tools (CART, MARS, RFRC, BTCR, CHAIR, E-CHAID, Bayesian Network, SVM, Self-Organization Network)

- Evaluate HCP with all the other predicate tools (CART, MARS, RFRC, BTCR, CHAIR, E-CHAID, Bayesian Network, SVM, Self-Organization Network) base on the error predicate measures
- Finally, Determine the best Predicate tool for healthcare database that give "number of days which the patient spend in the hospital in next year with **two reports**
- The first for hospital to take all the necessary requirements and
- The another send to patient (him\her) by the your' Mobil.

## 4 Description of Methodology

*Stage 1:* Problem Reformualtion against latest Research Questions, Definition and Requirements Capture

- Update study on T-Graphs, Healthcare Web applications, cloud computing fesilities, Predication Techniques, the main architectures of cloud computing/technology issues and regulatory norms & new perspectives.
- Identify requirements and select modelling tools.
- Produce review report and two paper.

*Stage 2:* Preprocessing the Healthcare Database.

- Study the natural of the Healthcare DB
- Apply the normalization on the description features of that DB
- Remve the irrelvent features based on Rough set

*Stage 3:* Specify and Design the new predicator and Cloud Computing Archtechure.

- Analysis previous predication techniqiues to determine which of parameters are mainly effect in given the predication result and which parmaters are less effect in given the predication result.
- Specify and Design new predicator base on mainly effect parmaters get by above step.
- Specify and Design prototype cloud computing architecture to examine the new predicator with all the other neine predication techniques by working on the same database (i.e., Heathcare DB).
- Produce report and two papers on proposed framework, new predicator and archtecture respectively.

*Stage 4:* Analysis and Evaluation

- Test the new framework, models against evaluation crietria.
- Evaluate the framework

Determine *Best Predicate tool for healthcare database that give "number of days which the patient spend in the hospital in next year with two reports the* ***first for hospital*** *to take all the necessary requirements* ***and the another*** *send to patient (him \her) by the your' Mobil. And* produce paper on final results of total system (Fig. 2) and (Tables 2, 3, 4, 5 and 6).

**Pseudo code of Healthcare Predicator (HCP)**
**Input:** Healthcare Database.

**Output:** Best Predicate tool for healthcare database that give "number of days which the patient spends in the hospital in next year with two reports the first for hospital to take all the necessary requirements and the another send to patient (him\her) by the your' Mobil.

**1:** Preprocessing by Convert all the description features to digital base on the normalization

**2:** Reduce all the irrelevant features base on calling the rough set procedure.

**3:** Set the main permeates for HCP tool, Max Number of epochs.

**4:** Pass all the important features to HCP

**5:** Do for each training pattern

**6:** Begin the process stage

**7:** While the process stage completes

- Compute number of epochs,
- Time require to complete training process,
- Accurate of predicate
- The cost requires using the cloud computing services.

**8:** Call the results of apply another predication tool on the same database that work at the same time "in parallel" using different servers based on the cloud computing

**9:** Compare the Result of HCP with all the other predicate tools (CART, MARS, RFRC, BTCR, CHAIR, E-CHAID, Bayesian Network, SVM, Self-Organization Network)

**10:** Evaluate HCP with all the other predicate tools (CART, MARS, RFRC, BTCR, CHAIR, E-CHAID, Bayesian Network, SVM, Self-Organization Network) base on the error predicate measures

**11:** Draw the plot of results step9 and another for results of step 10.

**Pseudo code of Rough Set to Reduce Irrelevant Features**

**Input:** All the Features (F) of Healthcare Database, F=[F1,….,Fn], Set of predefined patterns "# days the patient sped in hospital in years "Y1,Y2" Input for each feature FT, Fa=1…….,m. y in one of patterns

**Output:** All FT that are marked with no match removed from pattern

**1:** **For** all selected predefined patterns FT
**2:** **For** all properties used in F
**3:** Machin between FT and F in terms of the domain
**4:** Compare based on the (exact, plugin, Subsumed, uncertain, and no match) the machine values
**5:** **End**
**6:** Label F as non-match if FT not associate with any F
**7:** **End**
**8:** **Repeat** step (2-5) until no more patterns to be compared
**9:** Stop if all FT label as non-match

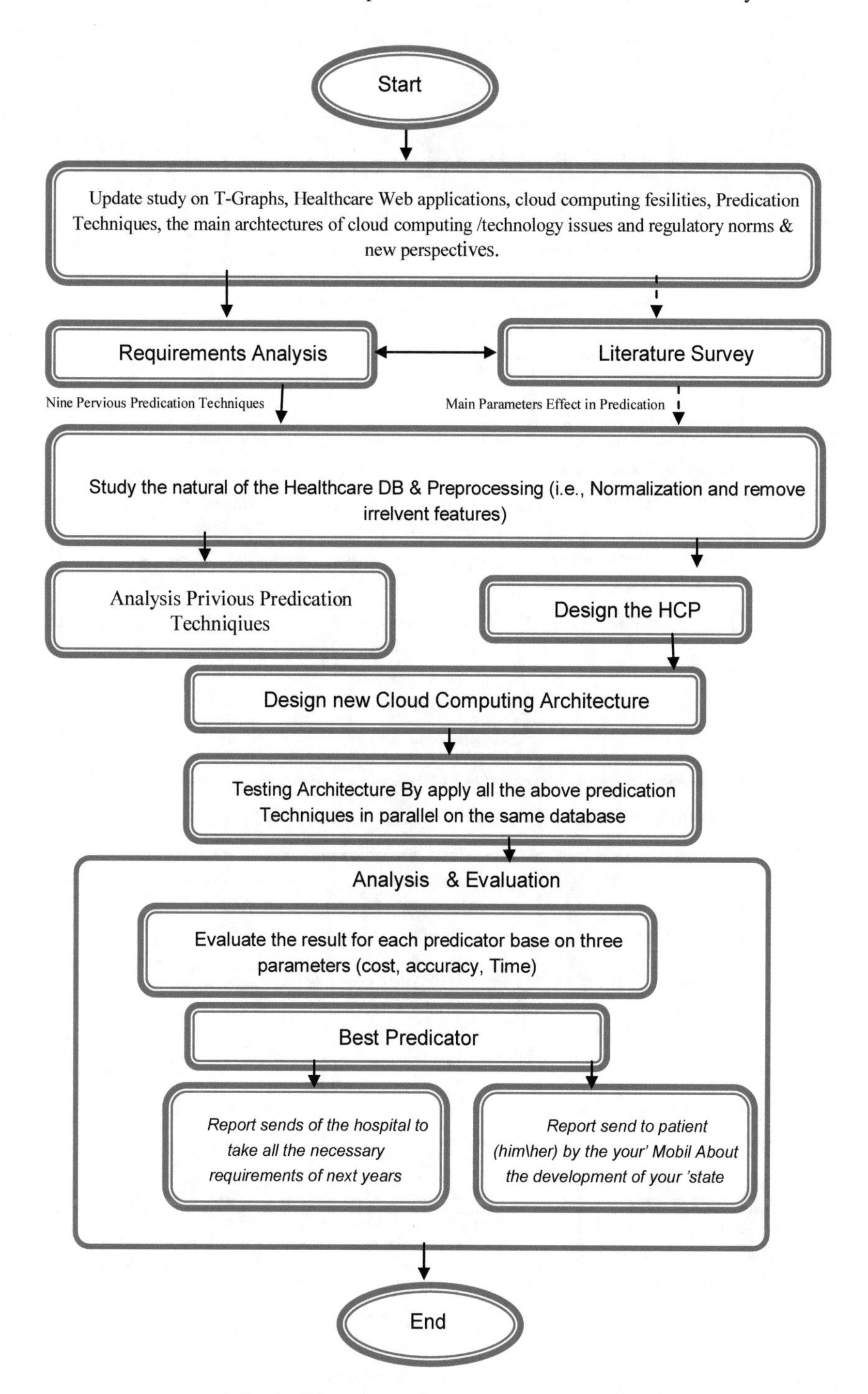


**Fig. 2.** Flow chart of research work activities

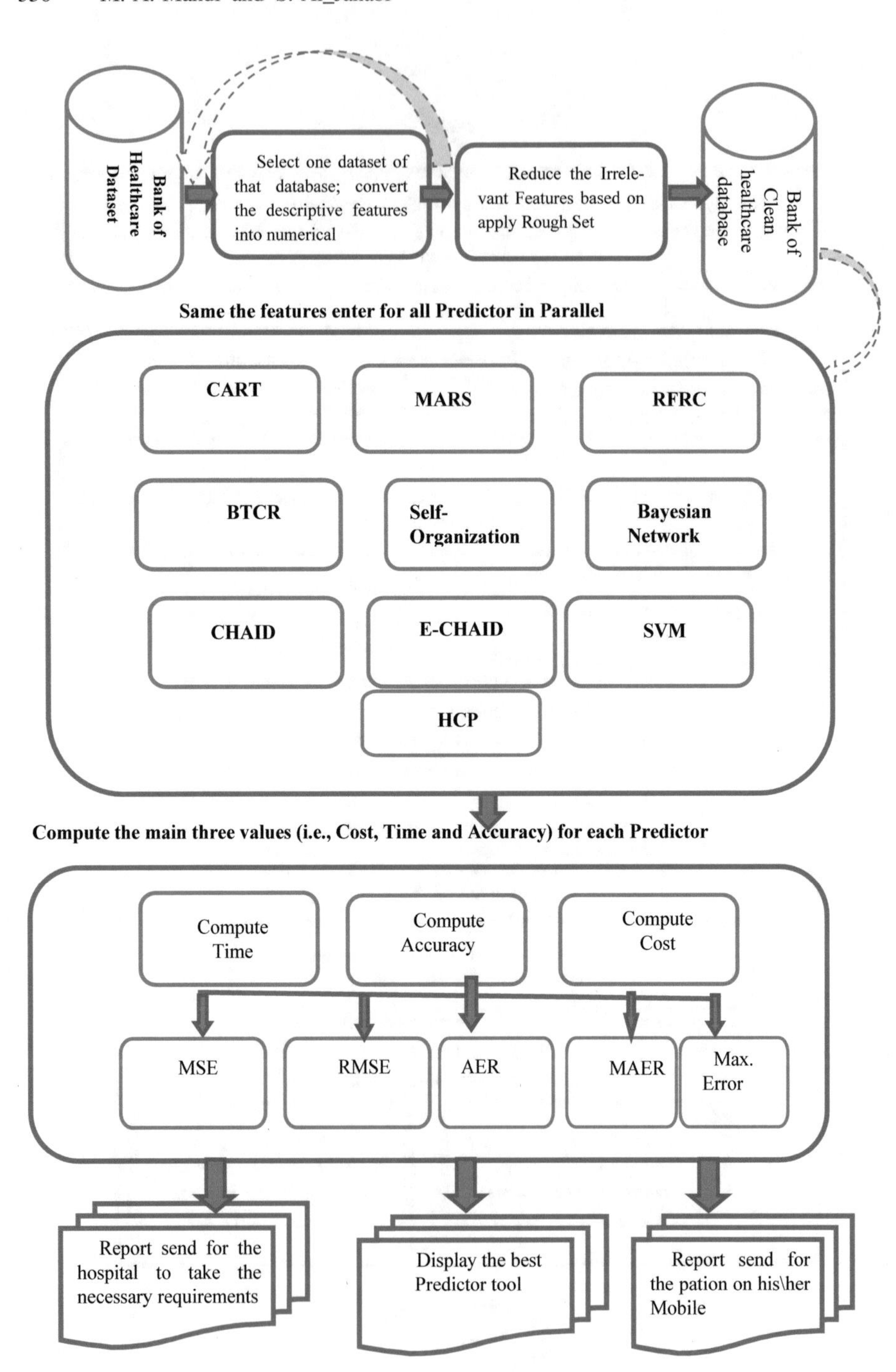


**Fig. 3.** System architecture

## 5 Experiment

Data set

F1: A wound that splits open after surgery on the abdomen or pelvis
F2: Accidental cuts and tears from medical treatment
F3: Blood stream infection after surgery
F4: Broken hip from a fall after surgery
F5: Collapsed lung due to medical treatment
F6: Deaths among Patients with Serious Treatable Complications after Surgery
F7: Infections from a large venous catheter
F8: Pressure sores
F9: Rate of complications for hip/knee replacement patients
F10: Serious blood clots after surgery
F11: Serious complications
F12: Overall Percentage

**Table 2.** Summary of main parameters of model

| Model summary | | |
|---|---|---|
| Specifications | Growing method | QUEST |
| | Dependent variable | MeasureName |
| | Independent variables | ProviderID, HospitalName, Address, City, State, ZIPCode, CountyName, PhoneNumber, MeasureID, ComparedtoNational, Denominator, Score, LowerEstimate, HigherEstimate, Footnote, MeasureStartDate, MeasureEndDate |
| | Validation | None |
| | Maximum tree depth | 5 |
| | Minimum cases in parent node | 100 |
| | Minimum cases in child node | 50 |
| Results | Independent variables included | MeasureID, MeasureStartDate, MeasureEndDate, Footnote, ComparedtoNational |
| | Number of nodes | 17 |
| | Number of terminal nodes | 9 |
| | Depth | 5 |

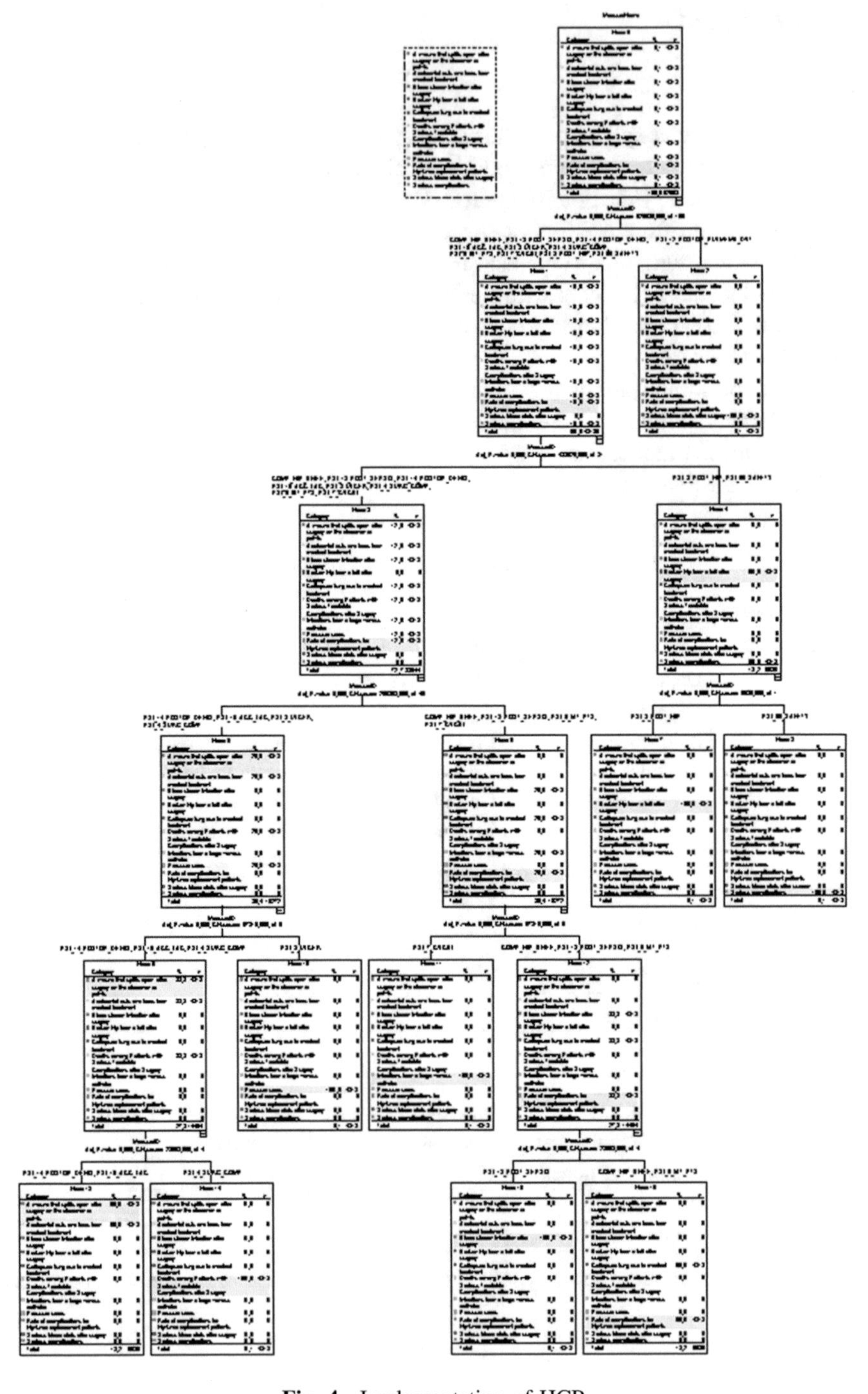

**Fig. 4.** Implementation of HCP

**Table 3.** Relation among observations and predictor

| | Predicted | | | | | | | | | | | |
|---|---|---|---|---|---|---|---|---|---|---|---|---|
| Observed | A wound that splits open after surgery on the abdomen or pelvis | Accidental cuts and tears from medical treatment | Blood stream infection after surgery | Broken hip from a fall after surgery | Collapsed lung due to medical treatment | Deaths among Patients with Serious Treatable Complications after Surgery | Infections from a large venous catheter | Pressure sores | Rate of complications for hip/knee replacement patients | Serious blood clots after surgery | Serious complications | Percent Correct |
| F1 | 4818 | 0 | 0 | 0 | 0 | 0 | 0 | 0 | 0 | 0 | 0 | 100.0% |
| F2 | 4818 | 0 | 0 | 0 | 0 | 0 | 0 | 0 | 0 | 0 | 0 | 0.0% |
| F3 | 0 | 0 | 4818 | 0 | 0 | 0 | 0 | 0 | 0 | 0 | 0 | 100.0% |
| F4 | 0 | 0 | 0 | 4818 | 0 | 0 | 0 | 0 | 0 | 0 | 0 | 100.0% |
| F5 | 0 | 0 | 0 | 0 | 0 | 0 | 0 | 0 | 4818 | 0 | 0 | 0.0% |
| F6 | 0 | 0 | 0 | 0 | 0 | 4818 | 0 | 0 | 0 | 0 | 0 | 100.0% |
| F7 | 0 | 0 | 0 | 0 | 0 | 0 | 4818 | 0 | 0 | 0 | 0 | 100.0% |
| F8 | 0 | 0 | 0 | 0 | 0 | 0 | 0 | 4818 | 0 | 0 | 0 | 100.0% |
| F9 | 0 | 0 | 0 | 0 | 0 | 0 | 0 | 0 | 4818 | 0 | 0 | 100.0% |
| F10 | 0 | 0 | 0 | 0 | 0 | 0 | 0 | 0 | 0 | 4818 | 0 | 100.0% |
| F11 | 0 | 0 | 0 | 0 | 0 | 0 | 0 | 0 | 0 | 0 | 4818 | 100.0% |
| F12 | 18.2% | 0.0% | 9.1% | 9.1% | 0.0% | 9.1% | 9.1% | 9.1% | 18.2% | 9.1% | 9.1% | 81.8% |

**Table 4.** Multivariate tests[a] a. design: intercept

| Multivariate Tests[a] | | | | | | |
|---|---|---|---|---|---|---|
| Effect | | Value | F | Hypothesis df | Error df | Sig. |
| Intercept | Pillai's Trace | 1.000 | 19069437602.318[b] | 4.000 | 52994.000 | .000 |
| | Wilks' Lambda | .000 | 19069437603.271[b] | 4.000 | 52994.000 | .000 |
| | Hotelling's Trace | 1439365.785 | 19069437603.271[b] | 4.000 | 52994.000 | .000 |
| | Roy's Largest Root | 1439365.785 | 19069437603.271[b] | 4.000 | 52994.000 | .000 |

a. Design: Intercept
b. Exact statistic

**Table 5.** Tests of between-subjects effects

| Tests of between-subjects effects | | | | | | |
|---|---|---|---|---|---|---|
| Source | Dependent variable | Type III sum of squares | Df | Mean square | F | Sig. |
| Corrected model | ProviderID | .000[a] | 0 | . | . | . |
| | ZIPCode | .000[b] | 0 | . | . | . |
| | PhoneNumber | .000[c] | 0 | . | . | . |
| | MeasureStartDate | .000[d] | 0 | . | . | . |
| | MeasureEndDate | .000[a] | 0 | . | . | . |
| Intercept | ProviderID | 3806060972739508.500 | 1 | 3806060972739508.500 | 155138.154 | .000 |
| | ZIPCode | 154291503122102.530 | 1 | 154291503122102.530 | 213013.523 | .000 |

**Table 6.** Tests of between-subjects effects

| Tests of between-subjects effects | | | | | | |
|---|---|---|---|---|---|---|
| Source | Dependent variable | Type III sum of squares | df | Mean square | F | Sig. |
| Corrected model | ProviderID | .000[a] | 0 | . | . | . |
| | ZIPCode | .000[b] | 0 | . | . | . |
| | PhoneNumber | .000[c] | 0 | . | . | . |
| | MeasureStartDate | .000[d] | 0 | . | . | . |
| | MeasureEndDate | .000[a] | 0 | . | . | . |
| Intercept | ProviderID | 3806060972739508.500 | 1 | 3806060972739508.500 | 155138.154 | .000 |
| | ZIPCode | 154291503122102.530 | 1 | 154291503122102.530 | 213013.523 | .000 |
| | PhoneNumber | 18134056760498860000000000.000 | 1 | 18134056760498860000000000.000 | 334157.678 | .000 |
| | MeasureStartDate | 9785844724635072000000000.000 | 1 | 9785844724635072000000000.000 | 76281342158.630 | .000 |
| | MeasureEndDate | 9880923218415728000000000.000 | 1 | 9880923218415728000000000.000 | 1934035451745.220 | .000 |
| Error | ProviderID | 1300194753918060.500 | 52997 | 24533365170.067 | | |
| | ZIPCode | 38387172244667.914 | 52997 | 724327268.424 | | |
| | PhoneNumber | 2876039272666098800000000.000 | 52997 | 54267963708626883000.000 | | |
| | MeasureStartDate | 6798784580808712200.000 | 52997 | 128286215838796.770 | | |
| | MeasureEndDate | 270759921868982592.000 | 52997 | 5108966957921.818 | | |
| Total | ProviderID | 5106255726657569.000 | 52998 | | | |
| | ZIPCode | 192678675366771.000 | 52998 | | | |
| | PhoneNumber | 21010096033165518000000000.000 | 52998 | | | |
| | MeasureStartDate | 9785851523424612000000000.000 | 52998 | | | |
| | MeasureEndDate | 9880923489167507000000000.000 | 52998 | | | |
| Corrected total | ProviderID | 1300194753918060.200 | 52997 | | | |
| | ZIPCode | 38387172244668.660 | 52997 | | | |
| | PhoneNumber | 2876039272666141400000000.000 | 52997 | | | |
| | MeasureStartDate | 6798784580812865500.000 | 52997 | | | |
| | MeasureEndDate | 270759921868798240.000 | 52997 | | | |

**Table 7.** Compare among all the Predictor Techniques based on time, accuracy and cost

| Techniques | Evaluation measures | | Values of each measures |
|---|---|---|---|
| CART | Times | Training | 134 (Time in Seconds) |
| | | Testing | 43 (Time in Mminutes) |
| | Accuracy | MSE | 0.8177557 |
| | | AMSE | 0.8839217 |
| | | AER | 0.9401711 |
| | | MAER | 2.3548387 |
| | | Max_Error | 33.762315 |
| | Space | | O(d + N)//d the number of input dimensions, N number of tree |
| MARS | Times | Training | 128 |
| | | Testing | 35 |
| | Accuracy | MSE | 1.6216028 |
| | | AMSE | 0.5568312 |
| | | AER | 0.3100609 |
| | | MAER | 0.4650237 |
| | | Max_Error | 23.475241 |
| | Space | | O(d*k)//d the number of input dimensions, K power of equations |
| RFRC | Times | Training | 187 |
| | | Testing | 56 |
| | Accuracy | MSE | 2.3175829 |
| | | AMSE | 0.9429556 |
| | | AER | 0.8891653 |
| | | MAER | 0.8303335 |
| | | Max_Error | 40.599391 |
| | Space | | O(d^2*k)//d the number of input dimensions, K max of voting in classification, and average of voting in regression |
| BTCR | Times | Training | 170 |
| | | Testing | 32 |
| | Accuracy | MSE | 1.4516129 |
| | | AMSE | 0.441175 |
| | | AER | 0.1946357 |
| | | MAER | 9.0180131 |
| | | Max_Error | 31.268403 |
| | Space | | O(d^2*k) |
| Self-organization | Times | Training | 120 |
| | | Testing | 30 |
| | Accuracy | MSE | 1.9556348 |
| | | AMSE | 0.5401644 |
| | | AER | 0.2839092 |
| | | MAER | 0.4177195 |
| | | Max_Error | 20.26639 |
| | Space | | O(d^2*k)//d the number of input dimensions, k number of layers |

*(continued)*

**Table 7.** (*continued*)

| Techniques | Evaluation measures | | Values of each measures |
|---|---|---|---|
| Bayesian network | Times | Training | 64 |
| | | Testing | 43 |
| | Accuracy | MSE | 0.5459787 |
| | | AMSE | 0.3982987 |
| | | AER | 0.1586418 |
| | | MAER | 0.3575301 |
| | | Max_Error | 27.087073 |
| | Space | | O(d^2*P)//d the number of input dimensions: Probability for each input |
| E-CHAID | Times | Training | 189 |
| | | Testing | 68 |
| | Accuracy | MSE | 0.6254390 |
| | | AMSE | 0.4192831 |
| | | AER | 0.1655549 |
| | | MAER | 0.4356112 |
| | | Max_Error | 30.98220 |
| | Space | | O[(d*P)^2//d the number of input dimensions: probability for each input |
| SVM | Times | Training | 85 |
| | | Testing | 34 |
| | Accuracy | MSE | 2.9878169 |
| | | AMSE | 1.2383184 |
| | | AER | 1.5334325 |
| | | MAER | 0.8632506 |
| | | Max_Error | 39.07786 |
| | Space | | O(d^2*k)//d the number of input dimensions number of layers |
| HCP | Times | Training | 55 |
| | | Testing | 28 |
| | Accuracy | MSE | 0.7938039 |
| | | AMSE | 0.3651354 |
| | | AER | 0.1333238 |
| | | MAER | 0.286571 |
| | | Max_Error | 21,00037 |
| | Space | | O((d + k)^2)//d the number of terminal nodes, k number of layers |

# 6 Conclusion

We can summarize the main benefits of this paper through comparing among all the predictive techniques discussed in this work as explained in Table 7 besides (i) Developing and evaluation decision support system by present deep and accuracy framework with architecture for satisfy the goal (ii) Evaluate functions and features as well as reduce the execution time. (iii) Design integrated and useful prototype (iv) Implementation of prototype on hardware. (v) Risk factors to be resolved. (vi) The predictor proves it ‘ability from two sides speed and accuracy in solve the healthcare problem. While it achieved: (i) New model & supporting framework with algorithm of exploring solution of the healthcare problem based on Treasure Graph (T-Graph). (ii) Prototype system and test-bed to experiment with more advanced models and algorithms.

## References


1. National Center for Health Statistics, Health, United States, 2012: With Special Feature on Emergency Care, p. 342, Table 113
2. National Vital Statistics Reports, vol. 61(4) 8 May 2013, p. 81, Table 8
3. HPN. 2011. The heritage health prize competition. http://www.heritagehealthprize.com
4. Al_Janabi, S., Abaid Mahdi, M.: Evaluation Prediction Techniques to Achievement an Optimal Biomedical Analysis, International Journal of Grid and Utility Computing (2019)
5. Duana, L., Streeta, W.N., Xu, E.: Healthcare information systems: data mining methods in the creation of a clinical recommender system. Enterp. Inf. Syst. **5**(2), 169–181 (2011)
6. Santhanam, T., Sundaram, S.: Application of CART algorithm in blood donors classification. J. Comput. Sci. **6**(5), 548–552 (2011)
7. Peng, X., Wu, W., Xu, J.: Leveraging machine learning in improving healthcare, Association for the Advancement of Artificial Intelligence (2011)
8. Soni, J., Ansari, U., Sharma, D., Soni, S.: Predictive data mining for medical diagnosis: an overview of heart disease prediction. Int. J. Comput. Appl. **17**(8), 975–8887 (2011)
9. Sut, N., Simsek, O.: Comparison of regression tree data mining methods for prediction of mortality inhead injury. Expert Syst. Appl. **38**(12), 15534–15539 (2011). Elsevier
10. Rahman, R.M., Hasan, FRMd: Using and comparing different decision tree classification techniques for mining ICDDRB hospital surveillance data. Expert Syst. Appl. **38**(9), 11421–11436 (2011). Elsevier
11. Moon, S.S., Kang, S.-Y., Jitpitaklert, W., Kim, S.B.: Decision tree models for characterizing smoking patterns of older adults. Expert Syst. Appl. **39**(1), 445–451 (2012)
12. Zhang, J., Goode, K.M., Rigby, A., Balk, H.M.M., Cleland, J.G.: Identifying patients at risk of death or hospitalization due to worsening heart failure using decision tree analysis. Int. J. Cardiol. **163**(2), 149–156 (2013). Elsevier
13. Al-Janabi, S., Alkaim, A.F.: Springer, Soft Computing Journal (2019). https://doi.org/10.1007/s00500-019-03972-x
14. Wu, X., Kumar, V. and others, Top 10 algorithms in data mining, Knowl. Inf. Syst. (2008)
15. Timofeev, R.: Classification and Regression Trees (CART) Theory and Applications Master thesis, Humboldt University, Berlin (2004)
16. Al-Janabi, S.: Pragmatic miner to risk analysis for intrusion detection (PMRA-ID). In: Mohamed, A., Berry, M.W., Yap, B.W. (eds.) SCDS 2017. CCIS, vol. 788, pp. 263–277. Springer, Singapore (2017). https://doi.org/10.1007/978-981-10-7242-0_23
17. Ali, S.H.: Novel approach for generating the key of stream cipher system using random forest data mining algorithm, IEEE, 2013 Sixth International Conference on Developments in eSystems Engineering, Abu Dhabi, 2013, pp. 259–269. https://doi.org/10.1109/dese.2013.54
18. Kalajdzic, K., Hussein Ali, S., Patel, A.: Rapid lossless compression of short text messages. Comput. Stan. Interfaces **37**, 53–59 (2015). https://doi.org/10.1016/j.csi.2014.05.005. ISSN 0920-5489
19. Kursa, M.B.: Robustness of the Random Forest-based gene selection methods (2013)
20. Friedman, J.H.: Multivariate Adaptive Regression Splines, Tech Report (1990)

21. Al-Janabi, S., Patel, A., Fatlawi, H., Kalajdzic, K., Al Shourbaji, I.: Empirical rapid and accurate prediction model for data mining tasks in cloud computing environments, IEEE, 2014 International Congress on Technology, Communication and Knowledge (ICTCK), pp. 1–8, Mashhad (2014). https://doi.org/10.1109/ictck.2014.7033495
22. Jun, S.-H.: Boosted regression trees and random forests, Statistical Consulting Report (2013)
23. Schonlau Rand, M.: Boosted regression (boosting): an introductory tutorial and a Stata plugin, The Stata Journal (2005)
24. Ali, S.H.: A novel tool (FP-KC) for handle the three main dimensions reduction and association rule mining, IEEE, 2012 6th International Conference on Sciences of Electronics, Technologies of Information and Telecommunications (SETIT), Sousse, pp. 951–961 (2012). https://doi.org/10.1109/setit.2012.6482042
25. Al_Janabi, S., Al_Shourbaji, I., Salman, M.A.: Assessing the suitability of soft computing approaches for forest fires prediction. Appl. Comput. Inform. **14**(2), 214–224 (2018). https://doi.org/10.1016/j.aci.2017.09.006
26. Friedman, N., Geiger, D., Goldszmidt, M.: Bayesian Network Classifiers, Machine Learning, vol. 29. Kluwer Academic Publishers, The Netherlands (1997)
27. Xiong, W., Wang, C.: A hybrid improved ant colony optimization and random forests feature selection method for 56 v/' microarray data. IEEE Computer Society, Fifth International Joint Conference on INC, IMS and IDC (2009)
28. Al_Janabi, S.: Smart system to create optimal higher education environment using IDA and IOTs, International Journal of Computers and Applications, Taylor & Francis (2018). https://doi.org/10.1080/1206212x.2018.1512460

# The Scientific and Technical Information System: An Engine for Capitalizing and Protecting the Results of Scientific Research Work of Higher Education Institutions with Limited Access. Case of the Faculty of Sciences and Techniques of Errachidia

Khalid Lali[1(✉)], Abdellatif Chakor[1(✉)], and Yousef Farhaoui[2(✉)]

[1] Faculty of Legal, Economic and Social Sciences,
Mohammed V University, Souissi/Rabat, Morocco
khalid.lali1979@yahoo.fr,
abdellatif.chakor@um5s.net.ma

[2] Faculty of Sciences and Techniques, Department of Computer Science,
IDMS Team, Moulay Ismail University of Meknes, Meknes, Morocco
youseffarhaoui@gmail.com

**Abstract.** Decision-makers within the Faculty of Science and Technique of Errachidia are now convinced that the use of computerized knowledge is a stimulator of productivity and excellence in terms of scientific production. In the same vein, it is important to note that international competitiveness in scientific research has also encouraged Moroccan higher education institutions to invest in scientific and technical information systems. It is therefore a flexible platform that integrates computer applications whose purpose is the processing, analysis, storage, extraction, dissemination, sharing and the capitalization of a voluminous mass of data resulting from scientific research work. Thus and in order to continue to be perceived as space abolishing all the existing boundaries in front of the scientific community, the bearers of such a project must emphasize the complementarity between technological determinism and organizational determinism. The leaders of this establishment are convinced that these digital infrastructures will ensure the transfer of expertise, knowledge and know-how. In addition, these automatisms will also offer opportunities for scientific exchanges and alliances, and will therefore promote the rapprochement of academic communities. Finally, these tool will enable the Faculty to publicize its own policy of promoting scientific research whose main objective is to provide credible and objective solutions to the various problems encountered in the socio-economic sector at local, regional and national levels.

**Keywords:** Scientific and technical information systems


Y. Farhaoui (Ed.): BDNT 2019, LNNS 81, pp. 340–348, 2020.
https://doi.org/10.1007/978-3-030-23672-4_24

# 1 Introduction

Stimulated by the need to respond of the specific needs or to cope with certain difficulties, scientific research at the Faculty of Science and Technology Errachidia has set itself as a catalyst and priority: the promoting an ecosystem of innovation, growth, sharing and discovery of new scientific and technical knowledge.

In order to achieve this goal, the Faculty with limited access has made it a priority to implement information systems dedicated to scientific and technical research. The importance of these scientific and technical information systems is concretely illustrated as a support for decision-making insofar as they make it possible to identify specialists in a given research theme as well as new researchers who will appear and therefore open the feasibility of scientific alliances.

The purpose of this paper is therefore to answer the following question: Can a scientific and technical information system be considered as a better tool for promoting the results of scientific research of public higher education institutions? Case of the Faculty of Sciences and Techniques of Errachidia. To answer this problematic we could retain the following two hypotheses:

H1: A successful implementation of an information system dedicated to the promotion of scientific research depends above all on a rational strategic planning that places the greatest importance on financial, material and human resources and the latter should be provided with specific skills.
H2: A robust scientific and technical research information system is first and foremost a system with secure access that opens the possibility for alliances to the scientific community.

In order to test the validity of our two hypotheses, it seemed interesting to draw up a questionnaire made up of several indicators.

- The answers to the survey questions led to the creation of a quantifiable database.
- The answers were coded, entered and processed by the SPSS software.
- Data processing was done using Factor Correspondence Analysis (FCA) since it is a qualitative variable followed by dynamic cloud analysis.

# 2 Methodology for Quantitative Research

In order to be able to answer the problematic of this article and also to test the validity of our two hypotheses initially established, we based ourselves on two analyzes (intra bloc and inter bloc) supplemented by an automatic classification:

## 2.1 Intra Bloc Analysis

The purpose of this analysis is to group people and variables related to the different indicators and to identify certain characteristics from this group. A descriptive analysis of the survey data gives an idea of the characteristics of the people surveyed. It is a question of checking whether respondents are young or old, whether there are more men than women, to identify the researchers by grade, the number and type of some organized scientific events, to illustrate their needs and means of communication, without forgetting to highlight the difficulties encountered in the execution of their missions daily.

## 2.2 Inter Block Analysis

In order to test the validity of each of our two hypotheses initially stated at the beginning of this work we proceed to the simultaneous analysis of the indicators constituted for each of the two variables taken separately. The objective is to group people and variables and to identify, from the grouping of the indicators (of the two variables), certain identities.

**Definition of Variables and Indicators**

Before starting our analyzes of the data extracted from our sample, we first made sure to identify each variable by indicators. Thus the variable "APSPI = Appropriate strategic planning to implement" was represented by the following indicators:

- Strengthening skills through continuous and sufficient training programs, especially in the area of decision computing "CSTPDC".
- Recruitment of new administrators in the field of network and computer development "RAFNCD".
- Evolution of the budget reserved for the organization of scientific events and scientific productions "BOSESP".
- Purchase in renewal of computer hardware and software "PRCHS".
- Pedagogical approach and credibility of the supplier of the scientific and technical information system "PASSIS".
- Quality of the technical assistance provided by the staff of the software solutions dedicated to the scientific research in the event of unexpected failures "QTAPSS".

The variable "IRCUSO = Improvement of research conditions and university scientific output following the establishment of the scientific and technical information system" can be represented by the following indicators:

- Number of publications in indexed journals "NPRI"
- Number of organized scientific congresses "NOSC".
- Number of PhD students enrolled "NPhDSE".
- Number of Sustained University habilitation "NSUH".
- Collaboration and communication between researchers having the same thematic "CBRST".
- Collaboration and communication between research teachers having different research problems "CBRDRP".
- Number of scientific agreements and partnerships signed "NSAPS".
- Daily frequency of use of the applications proposed by the new information system "DFUAIS".

And finally the variable "CRRSIS = Characteristics of a reliable and robust scientific and technical information system" was determined from the following indicators:

- Fast and secure access to updated data "FSAUD".
- Adaptation in a turbulent environment "ATENV".
- Automatic reorganization of work "RW".
- Flexibility of piloting staff "FLEX".
- Centralization and decentralization of decisions taken "CDDT".

### 2.3 Automatic Classification

The purpose of automatic classification is to group individuals into homogeneous classes based on the study of certain characteristics of teacher-researchers. It thus makes it possible to describe the data by reducing the number of individuals. By homogeneous classes, we mean grouping people who are alike and separating those who are distant. For our work we have opted for the non-hierarchical automatic classification based on the method of dynamic clouds according to which the number of classes of the partition is fixed in advance.

## 3 Analysis of Results and Implications of Results

After having collected and entered the data of our investigation, we took the care to analyze them thoroughly in such a way as to be able to invalidate or confirm our two hypotheses stemming from our problematic object of this article.

### 3.1 Analysis of Quantitative Results

In this analysis, a descriptive analysis (intra-bloc analysis) was first carried out followed by a simultaneous analysis of two large sets of indicators that were composed to test the validity of our two hypotheses applied to sample subject of our investigation.

**Intra Bloc Analysis**
This analysis allowed us to come out with the following results:

- 89.76% of the research professors of the Faculty of Sciences and Techniques of Errachidia are the men while 10.24% are the women
- 29.92% are professors of higher education, PES (37 of which are the men).
- 22.92% are teachers empowered, PH (25 of whom are the men and 3women).
- 48.03% are assistant professors, PA (52 of whom are men and 9women).
- 25.20% of FSTE researchers specialize in related topics in the field of physics and more specifically, mechanical engineering and the energy field (including 30 men and 2 women).
- 15.75% work on topics related to animal biology and plant biology (aromatic and medicinal plants), 15 of which are men and 5 women.

- 11.81% of the researchers of the establishment are specialized in the geological field and more precisely the geo-resource and environment but also the mining extraction and studies on the dams (among which 13 men and two women).
- 10.24% are researchers specialized in business intelligence, network and also in image processing and cryptography including 11 men and two women.
- 17.32% are researchers in the field of operations research and artificial intelligence, including 22 men.
- 19.69% work on topics related to organic chemistry.
- 18.90% of researchers from the Faculty of Science and Technology Errachidia and under the University Moulay Ismail Meknes are between 32 years and 41 years.
- 34, 65% of the researchers in the limited access institution are aged between 42 and 51 years old.
- 46.46% are aged between 52 and 61 years old.
- Concerning PhD theses supported from the academic year 2009/2010 to the academic year 2018/2019; the number of theses defended during this period was 39 theses including 6 theses in mathematics, 8 theses in computer science, 6 theses in physics, 9 theses in biology, 10 theses in chemistry and no thesis in geology. It is also important to note that, of the totality of the 39 theses supported, 4 theses were supported in 2009/2010, 2 theses in 2011/2012, only one thesis defended in 2012/2013, 5 theses defended in 2013/2014, 6 theses defended in 2014/2015, 5 theses defended in 2015/2016, 7 theses were supported in 2017/2018 and finally 4 theses were supported in 2018/2019.
- For habilitations supported between the 2013/2014 academic year and the 2017/2018 academic year; this number was 44, of which 10 were supported in mathematics, 8 in computer science, 9 in physics, 8 in biology, 5 in chemistry and 4 in geology. It should also be noted that, of all the 44 habilitations supported, 5 were supported in 2013/2014, 8 in 2014/2015, 13 in 2015/2016, 14 in 2016/2017 and 4 in 2017/2018.
- Until September 14, 2018, the number of doctoral students was 122 as follows: 26 are enrolled in the first year, 36 doctoral students enrolled in the second year, 26 enrolled in the third year, 19 doctoral students enrolled in the fourth year, 12 enrolled in fifth grade and finally 3 are enrolled in sixth grade.

### Inter Block Analysis

#### Simultaneous Analysis of "APSPI" and "IRCUSO"

In order to test the validity of our first hypothesis, we decided to apply the factorial analysis of the correspondences and the classification into dynamic clouds in two classes on the two variables "APSPI" and "IRCUSO" which allowed us to regroup the teaching staff of the FST Errachidia which is attached to the University Moulay Ismail of Meknes in two groups dispersed according to the indicators of the two variables. Hence the figure below, which shows the dispersion of the teachers questioned according to the indicators of the two variables (Fig. 1).

The two factors F1 and F2 resulting from the reduction of the indicators representing the two variables "APSPI" and "IRCUSO" explain 56% of the total variability (30% for F1 and 26% for F2). From the analysis it can be said that there are two categories of teacher-researchers. The first class occupies the North-West and South-West of the F1F2

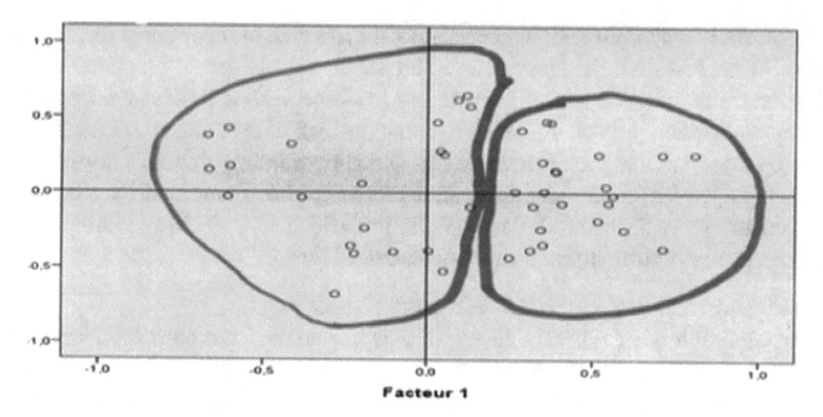


**Fig. 1.** Simultaneous analysis of "APSPI" and "IRCUSO" Source: Survey conducted among teacher researchers at the Faculty of Science and Technique of Errachidia 2018–2019

factorial plane. These are the teacher-researchers who have affirmed the existence of a great improvement in the conditions of university research and scientific production as a result of the setting up of the scientific and technical information system. According to respondents in this group, this new system has strengthened collaboration and communication between researchers with the same research themes as well as those with different research issues and who are interested in the current promising sectors in Morocco today. As a result, the number of signed scientific partnerships has increased especially with researchers from foreign universities. In addition, the arrival of this computer tool dedicated to scientific research made it possible to be informed of the new topics discussed in the field of scientific research and this explains the increase in the number of publications in indexed journals, the number of scientific congresses organized by the FSTE faculty, the number of the habilitations supported and the number of PhD students enrolled in the various doctoral training. It is very important to note that this favorable development was the result of an appropriate strategic planning prior to the implementation of this information system dedicated to scientific research and which is characterized by the increase in the budget reserved for the organization of scientific events as well as to support scientific productions, renewal of computer and software equipment, specific and sufficient continuing training programs, especially in the area of decision-making informatics, with the aim of strengthening skills and finally, the elected members of the Faculty council have also decided to reserve budget posts for the recruitment of new administrators in the field of network and computer development. According to this first group of teacher-researchers, the pedagogical approach and the credibility of the supplier of the scientific and technical information system as well as the quality of the technical assistance provided by the staff of the software solutions dedicated to the scientific research in case of Unexpected failures have also led to an increase in the daily usage frequency of the applications proposed by the new information system.

The second category is in the North-East and the South-East, which believes that the arrival of these new technologies has not led to significant changes, especially in the development of scientific research because of the low rate of recruitment of skills in the field of IT. In addition, the amount reserved for the acquisition of computer hardware and software remains very low compared to the expectations of teachers interested in this implementation project, without forgetting the lack of experience and the flexibility of some executives and the insufficiency high quality training programs. Moreover this category judges that these new applications have not been preceded by a good identification of the real needs of the organization and that they do not bring credible solutions to the problems previously encountered.

**Simultaneous Analysis of "IRCUSO" and "CRRSIS"**

To test the validity of our second hypothesis we applied the correspondence factor analysis and dynamic cloud analysis on the two variables "IRCUSO" and "CRRSIS" and this allowed us to see that the factor F1 explains 24% of the total variability and the factor F2 explains 20% of the total variability.

The Factor analysis of correspondence and automatic classification in 2 classes made it possible to identify the 2 groups represented in the figure below (Fig. 2):

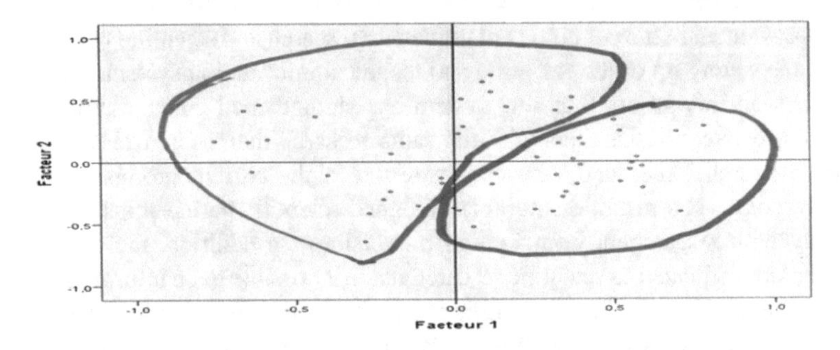


**Fig. 2.** Simultaneous analysis of "IRCUSO" and "CRRSIS" Source: Survey conducted among teacher researchers at the Faculty of Science and Technique of Errachidia 2018–2019

The first group is located in the North-East and South-West and is composed of teacher-researchers who declare the improvement of the conditions of university research and scientific production following the implementation of this scientific information system and technical, leading to the conclusion of scientific alliances with national and foreign partners, as well as an increase in the number of publications in indexed journals and an increase in the number of organized scientific events. According to its respondents, all of its favorable results were the result of a robust, reliable and secure information system providing better collaboration and communication between geographically dispersed researchers. Moreover this system gives a secure and updated access to the data of each researcher. One can even add according to this first group - that in spite of the limited number of persons in charge of the piloting of this system of scientific and technical information - the latter are flexible and arrive through these automatisms to adapt to an environment in continuous mobility. The robustness of this information system can also be seen through its ability to automatically rearrange the various tasks performed and this is what leads us to argue that it undeniably presents aspects of an intellectual and intelligent work.

The second group consists of research professors who have stated that there have been no favorable changes in the valorization of scientific production following the setting up of a scientific and technical information system and which justified it by the fact that it is a system suffering from several deficiencies, especially in terms of data security.

According to this group, the planning preceding the implementation of this system has failed at all levels, especially in terms of equipment acquired and recruited profiles that do not meet the characteristics required to make this system a real success.

## 4 Analysis and Discussion of the Results

This analysis allowed us to come out with the following results:

- Of 186 accredited research teams at the Moulay Ismail University of Meknes we have 28 accredited research teams belonging to the Faculty of Science and Technology of Errachidia.
- Among the 37 accredited research laboratories at the Moulay Ismail University of Meknes we have two accredited research laboratories from the Faculty of Science and Technology of Errachidia.
- The evolution of the number of publications submitted in the context of the support of the University Moulay Ismail of Meknes to the scientific production was as follows: 105 publications in 2010, 153 publications in 2011, 182 publications in 2012, 244 publications in 2013 and 535 publications in 2014.
- As part of the support for relevant and structuring research projects, the University has allocated a special allocation that rose from 2,000,000 dirhams in 2011 to 4,500,000 dirhams in 2013 and finally to 5,000,000 dirhams in 2016.
- With regard to support for the mobility of teacher-researchers giving them the opportunity to participate in international scientific events and continuing training at national and international level, with a view to enabling beneficiaries to acquire new qualifications and skills to adapt to the evolutions of the scientific research we find that 14,50% of these beneficiaries (during the year 2017/2018) fall within the Faculty of Sciences and Techniques of Errachidia (more precisely on a total of 69 beneficiaries, 10 teachers beneficiaries are under the responsibility of the FSTE).
- 80% of the researchers of the FSTE are convinced that a new scientific and technical information system will enhance all the scientific production that is the result of the research work of the teachers of the Faculty with limited polarization, as well as other researchers from the other recognized universities at the national and international levels. In addition and according to these same teachers investigated, a climate of networking and transparency will prevail, which will allow exchanges with other researchers working on the same themes but also working on priority projects in Morocco today. We can say that this scientific and technical information system will open the possibility to alliances between researchers, against only 20% who do not share the same vision.

- 64.5% of respondents argue that the current human, material and financial resources are insufficient to guarantee the success of the new decision support systems to implement against 35.5% who do not share the same opinion.
- 6.8% state that the advent of these automated systems that are scientific information systems will enable researchers to develop a new culture of networking, to promote creativity and to be aware of new developments in the field scientific research compared to only 3.2% do not share the same opinion.

## 5 Conclusion

In order to highlight and capitalize on all the major research carried out by research professors at the FST Errachidia, decision makers have been called for several years to invest in scientific and technical information systems, being aware of the great their role in the performance, competitiveness, attractiveness and protection of tangible and intangible heritage, which have become major challenges. These systems are intended to convey scientific information to solve specific problems and this after appropriate treatment to users in a timely manner. Research professors are very optimistic and are convinced of the great added value that these automatisms will bring in the promotion of scientific research. However, it can be concluded that there is still a lack of competent financial and human resources to make this project a success.

## References

Chaudhary, P., Hyde, M., Rodger, J.A.: Exploring the benefits of an agile information system. Intell. Inf. Manag. **9**(5), 133–155 (2017)

Salvador, A.B., Ikeda, A.A.: Big data usage in the marketing information system. J. Data Anal. Inf. Process. **2**(3), 77–85 (2014)

Harrizi, D., Dafi, N.: Towards a new managerial culture: methodology for setting up a management control system in a public institution - case of a regional education and training academy. ScienceLib Editions Mersenne, vol. 5, no. 130105 (2013). ISSN 2111-4706

Fabre, I., Gardies, C.: University of Avignon, The open archives: new information practices for teachers-researchers? In: Proceedings of the International Colloquium, Scientific Edition and Publication in the Humanities and Social Sciences: Forms and Issues, pp. 81–93, 17–19 March 2010

Fabre, I., Gardies, C.: Access to digital scientific information: organization of knowledge and stake of power in a scientific community. Sciences of society, University Press of midday, pp. 84–99 (2008)

Desq, S., Fallery, B., Reix, R., Rodhain, F.: The specificity of French-language research in information systems. RFG French Review of Management, vol. 33, pp. 63–80 (2007)

Rostaing, H., Toledo, E.G., Kister, J.: University of Toulouse, France, Information system for assistance in the strategic management of research in a public research establishment. Strategic, Scientific and Technological Watch, VSST'2004, IRIT, pp. 157–168. October 2004

Verry-Jolivet, C.: Practices and expectations of researchers: the scientific media library of the Pasteur Institute. Bull. Libr. Fr. (BBF) **46**(4), 26–30 (2001)

Niosi, J., Bellon, B., Saviotti, P., Crow, M.: National systems of innovation: in search of a usable concept. Fr. Rev. Econ. **7**(1), 215–250 (1992)

# Trust in Cloud Computing Challenges: A Recent Survey

Zakariae Tbatou, Ahmed Asimi(✉), and Chawki El Balmany

Departments of Mathematics and Computer Sciences, Faculty of Sciences, Ibn Zohr University, B.P. 8106 City Dakhla, Agadir, Morocco
tbatou.zakariae@gmail.com, asimiahmed2008@gmail.com

**Abstract.** With the widespread use of computer networks, the challenge of distributed systems is to support multiple security policies. Notably, the cloud computing that represents a revolution in computer system, with its hype today, the services providers migrate to the integration of the said systems in different areas. Despite his advantages such speed, Qos and performance, actually, the cloud does not present a solution to address the problems of computer systems which stay more complex, namely security. In practice, cloud computing defines several security limitations because of the problem of virtualization and segmentation of data, therefore, several approaches have been proposed to address security issues, including authentication. In this paper we implement cloud authentication issues while emphasizing the different principles namely the Trusted Third Party (TTP) and the Third Party Auditor (TPA) as well as the authentication techniques based on these two principles in order to define the specific security requirements for the cloud.

**Keywords:** Computer networks · Distributed systems · Security policies · Cloud computing · Security limitations · Authentication issues · TTP · TPA

## 1 Introduction

When the networks started using a client-server model and the terminals were replaced by the PCs, thus, to entrust the complex tasks to remote computers. This exploitation leads to the widespread use of distributed systems given their performance and efficiency. In this sense, cloud computing takes a large part of the current architecture being a specific case of distributed systems and which offers the advantage of virtualization [1]. Cloud computing presents the current IT trend which offers processors and software as a service more powerful and cheapest use, the reason why almost all organizations are trying to get migrate into. It offers several advantages such the reduction of hardware and maintenance costs, accessibility regardless of geographical location, flexibility and highly automated process for its users and a higher QoS comparing

Supported by organization x.


Y. Farhaoui (Ed.): BDNT 2019, LNNS 81, pp. 349–359, 2020.
https://doi.org/10.1007/978-3-030-23672-4_25

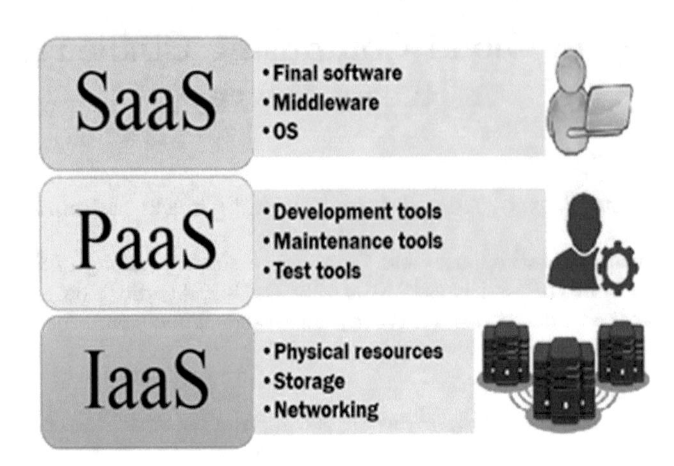


**Fig. 1.** Description of Cloud standard architecture.

with other systems (storage space, speed, availability ...). In terms of virtualization, the researchers [1–3] identified three layers to specify the different tasks for each layer (Fig. 1): Software as a Service (SAAS): Provides end-user software, enables funding and maintenance of security and collaboration Infrastructure as a Service (IAAS): Represents deployment development tools and maintenance tests. Platform as a Service (PAAS): Defines the physical structures and physical resources that enable data cloning, monitoring and storage and archiving software. In spite of all furnished advantages by cloud computing, it also brings new security challenges. In fact, the migration of software and data into the cloud makes the security and privacy priorities seen sensitive data on remote machines that are not owned or even managed by the customers themselves [13, 14]. In this sense, the trust becomes a dominant property that must be satisfied in such system. In this paper, our main goal is highlighting the authentication in the cloud computing. In the second section, we implement the authentication, in these categories, we illustrate the authentication challenges in the cloud and we describe some proposed principles to remedy the problem of authentication focusing on the principles of TTP and TPA. The third section, we analyze and we specify the security issues of different approaches based on TTP and TPA principles and we end with a conclusion.

## 2 Review of Authentication in Cloud

In the literature, there exist a few survey authentication policy in cloud computing to solve the problem of access control management in this section we treat the authentication in cloud computing in it categories and the two famous principles TTP and TPA.

### 2.1 Authentication in Cloud

Cloud computing seems to be the ideal solution for achieving robustness and continuity of services [28,32]. This solution crosses its complexity which is a major disadvantage caused by the large number of requests exchanged between the entities. For this reason, the security policy must be defined with the evolution of the architecture [9,11]. In terms of confidentiality and authentication, two major problems arise: data and software migration makes the user unaware of the way which his own data stored and distributed, consequently, the number of requests becomes important. This increases the chance of capturing so much data, therefore it weakens the robustness of the set of protocols, including authentication protocols. As the cloud computing is achieving popularity, the concept of trust become more crucial [9,12–14], in fact the external storage of owner data made this later doubted the reliability and policies of the cloud, at this level, Authentication plays an important role in protecting resources against unauthorized use, it must be defined and clear for cloud users [12,22], unfortunately, the majority of users remain unaware of the notion of security. As described in Fig. 2 TTP or TPA play a very important role to solve the problem of trust in the cloud. We distinguish that the different components are based on the use of keys for encryption/decryption operations and key generation at the level of launch or storage of the virtual machine. In this context, the keys used must meet NIST's Cloud infrastructure-as-a-Service delivery requirements, including authentication between different entities to ensure a reliable exchange between the different entities. In [2], security is adopted as a cooperated responsibility between cloud user cloud service provider. To redress this problem several approaches were proposed [32,36,37] which guarantees on one side, the scalability of the system and on the other side the optimization of the rate of the use of shared resources and the cryptographic calculation task by integrating a TTP or TPA. In term of security, the authentication models proposed so far remain based on traditional password authentication [22,25] either, which will be detailed in the next section, or on physical primitives (biometrics, smart card) [8,35]. In the rest of this paper we admit the following definitions:

*Definition 1.* Mutual authentication, also known as bi-directional authentication, is a process in which two entities of a communication link authenticate to one another based on the identity knowledge of the other [25].

*Definition 2.* Strong authentication or multi-factor authentication, in computer security, is identification evidence based on several independent factors or derived from distinct parameters in order to allow their access to resources.

*Definition 3.* Strong mutual authentication is a hybrid authentication process that is both strong and mutual.

### 2.2 TTP (Trusted Third Party)

Generally, securing an information system involves identifying threats and challenges that need to be addressed by implementing the appropriate countermea-

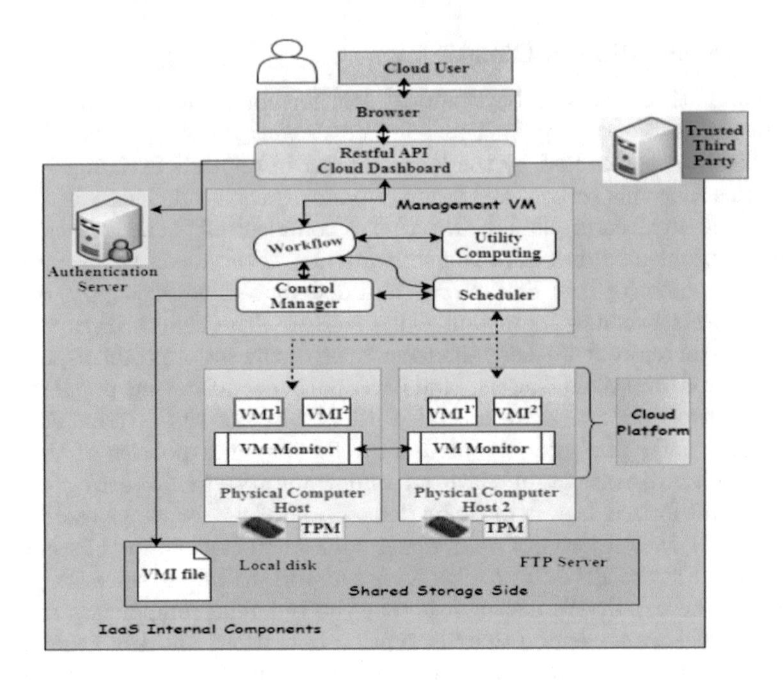


**Fig. 2.** Public IaaS layers and components.

sures. Due to architectural design and features, cloud computing imposes many security benefits (segmentation, availability ...). In term of authentication, most of the used mechanisms are still experiencing significant weakness in implementation and design due to the complexity of the architecture and the prevention of attacks [3,4,11,32]. In this sense, most of the standard authentication models based on password are demonstrably breakable due to weakness and memorization of passwords [24,25]. To solve these problems in cloud, several alternatives have been proposed as the addition of authentication factors, the use of human characteristic such as print finger [8]. According to many analyzes [9,21,37], although these new primitives are intended to replace the use of the password and offer more robustness to the authentication policy, it is still difficult to exploit and implement them in public infrastructures. In such situation, cryptanalyses impose on protocol designers who bypass the aforementioned problems to present "improved" schemes that satisfy more than optimal confidence for users without affecting the performance of the cloud, in this context many approaches were developed [1,8,11,15,18], we focus on the TTP (Trusted Third Party) [19,22] and TPA (Third Party Auditor) principles [20].

### 2.3 TPA (Third Party Auditor)

Recently, great interest has been shown in ensuring remotely stored data integrity under different system and security models the reason that made the use of a third-party auditor (TPA) efficiency [19,20], in fact, many company try to consult service organization that provides a variety of services to audit tasks as well as its serves as a bridge between the cloud service users (CSUs) and the cloud services provider. Which made TPA a potential solution for securing a cloud [29]. The Trusted Third Party (TTP) is an audit-based entity that facilitates secure interactions between two communicating parties, the cloud user and cloud provider, in which both trust this third party. The Third Party Auditor (TPA) being the third trusted party and auditor (TPA) allows the registration of security service provider assigned by the cloud service provider giving a strong authentication and authorization [19], it can perform multiple audits simultaneously. TPA is seen as a cost-effective approach in which CSUs outsource and entrust the management of encryption keys to an approved TPA. The TPA, in meanwhile, is responsible for managing encryption keys and verifying the integrity of outsourced data for a CSU account [20]. This would enable efficient key management and ensure data integrity, thus spreading the problems of data encryption due to virtualization and segmentation of user data in the cloud [24]. In addition, the use of TPA allows CSUs to reduce the relatively of high computing power on their hardware, which is necessary to perform data encryption, authentication of externalized data messages, and verify the convenience of data in the cloud on demand without retrieving a copy of all data or creating an additional online charge for cloud users [19,20]. Therefore, the use of TPA allows each CSU to easily access data and take advantage of the real cloud computing benefits. Unfortunately, there are still many problems with TPAs, like introducing another third party, authentication, authorization, and auditing for secure cloud storage are may needs external auditor to audit outsourced user data in the cloud [27]. Therefore, the user or organization is required to trust the new entity without acquiring knowledge about the content of the data. This would pose a bigger problem for the cloud, since the CSU and the CSP (Cloud Services Provider) should both trust the TPA [27,31]. Another problem is that the introduction of more entities, the risk of internal threat increases. The damage that can be caused by a malicious entity is the violation or the complete deletion of data [14], depending on the amount of access granted to the auditor by the CSU/CSP. When we compare all these approaches, we believe that the use of TPA can offer several benefits namely knowing, ensuring data integrity, managing encryption keys, easing the computational burden and reducing the trust relation between CSU and the CSPs. Many existing works assume that a TPA is a trusted entity [27,28]. However, the fact that the TPA can be transformed into a malicious entity cannot be ignored [27]. For a complete security solution, it is required that the use of the TPA is to achieve the benefits mentioned above and enable other cloud actors to audit the TPA, which is reflected in the requirement of mutual strong authentication.

### 2.4 Cryptographic Primitives

According to the description of the operating mechanisms of TTP and TPA, these two principles take several aspects in different proposed approaches [9,19,20], to address authentication issues in general, both actors play the role of a key management center that must be assured is trustworthy for entrusting him the task of regenerating keys within the cloud. For these reasons, several researchers have proposed schemes based on these principles with the integration of cryptographic primitives such as elliptic curve [6,17,21] or Diffie Hellman principle [7,25] to perform a robust key regeneration and create a reliable communication channel by requiring authentication between different actors.

## 3 Security Analysis

Generally, security is related to important aspects of confidentiality, availability and integrity. For cloud infrastructure [31], it suggests unique security issues that need to be examined in detail [2,32]. Regarding authentication, that implicitly affects computer security properties namely confidentiality, integrity, availability and access control, must establish an end-to-end relationship between the different entities while keeping the authenticity of each actor and implicitly verified the identity and the authorization attributed to this actor [12,13]. In this section we analyze and discuss some well-known and used authentication protocol in cloud infrastructure founded on TTP and TPA such as SSO, Oauth, Two factors authentication and Smart card authentication.

### 3.1 A. SSO (Single Sign On)

SSO (Single Sign On) is a mechanism based on TTP principle that allows to a user of a distributed system to access a variety of different application services distributed over different systems with only one authentication, it can be one or many applications that allows the authentication process [4,10]. The results of this authentication are automatically propagated to the extremities of systems as needed. In fact, this simple concept hides a whole range of complications and security implications [3], making this standard open, highly extensible and attractive as a basis for further development. It offers an optimal security and share mutuality for its users. There are different types of SSO architecture [23,30]. Some of them handle only one set of credentials, and others handle multiple sets of credentials. In the first case, there is a token-based system (Token) and a PKI-based system. In the latter case, there is a system based on credential synchronization, a secure system for caching client-side credentials, and a secure server-side credential caching system. The analysis of this protocol describes three major problems that define certain limitations for the principle of SSO: (1) it uses symmetric cryptography, in particular, the complexity of the processing of the authentication status for several Web sites [30], (2) more effort is needed to improve security on the authentication process in the event of a failure, in case of cloud the authority must be proved to remedy the problems of internal attacks [10,23], and (3) it centralizes access to the authentication server level. The protocol could be attacked with identity theft. In addition, the Man-in-the-middle attack is still effective on this process [30].

### 3.2 OAuth (Open Authorization)

The OAuth protocol (open standard for authorization) [5] is a framework for defining an ideal third party entered by all protocol participants and providing a generic framework for a resource owner to allow a third party to access the owner's resources without revealing third party the owner's identification parameters like username or password [16,23]. The OAuth specification defines the management of delegated permissions in a variety of situations, such as Web and desktop applications. This mechanism defines an authorization area based on token (Token) [33], the use of the certified Public Key Infrastructure (PKI) and the general authority certificate in order to assign several authorization flows to meet the specific needs of each user profile. Despite the success experienced of this protocol, for cloud computing, many analysis prove an uncovered weakness in the protocol or implementations [16,33]. In reality, an attacker could collect private data in front of the important number of users or steal their identities, in addition, Users on the Internet must have public keys to authenticate themselves, assuming that the service provider's public key is certified by a global Certificate Authority (CA), the client's public key must at least be prerecorded with this service globally certified which represents certain opacity on the way of storage of the keys [16], for the service provider side, it must have a public key certified worldwide [23], so the requests for authentication and authorization are often managed by the SSL which also represents certain limitations and weaknesses.

### 3.3 Two Factors Authentication

In traditional authentication based on a single factor, often the user password, the main goal is to provide the communicating parties with some assurance that one part model knows other's true identities [7,25], such suffer from several potential security vulnerabilities and several attacks as described in many analysis [29,34], we cite guessing attacks and dictionary attacks where the password can be compromised. The two factors authentication has been widely deployed for various kinds of daily applications and prove its robustness, it guaranties strong authentication [35], and as described in many approaches it can satisfy mutual authentication. In most cases this task is assigned to a TTP or TPA [18,29] which affects the calculation and storage of authentication factors. The security requirement in this technique indicates that the authentication parameters must be different and separated (in most case derived from a private parameter), to ensure a strong robustness it is noted that the factors must be per session in order to guarantee the non-traceability of the authentication parameters, this technique still be weak against the dictionary attack [35].

### 3.4 Smart Card Authentication

The smart card is a peripheral that allows the storage and the cloning of the client authentication parameters and reduce the compute load at the client level

[15], this technique has made it possible to reduce several attacks due to the storage of the authentication parameters in the client side [36]. With this technique, the chance that the password (generally authentication parameters) can be compromised becomes rare, the scientific advice of the use of such techniques but it still questionable face some attacks such as stolen attack (recovery session) [35], given its effectiveness in many applications such banking applications is require. Unfortunately, in the case of public cloud it can't be deployable due to the high cost of implementation. Most smart card-based approaches describe the use of the Deffie Hellman or elliptic curves principles that describes mathematical limitations at the generating function level [6,7,17].

### 3.5 Security Issues

The security of authentication protocols, which in particular manage trust relationships, is based on the security of cryptographic primitives and the design of the protocol schema to ensure confidentiality, integrity and authentication, and on the study of their vulnerability especially on the absence of faults in their constructions. In this section, we present the different approaches based on the principle of TTP and TPA with their limits and the different types of topical attacks defined by the international security organizations (ISO, OWASP, CSA ...) in the cloud environment (Table 1).

**Table 1.** Analysis of different protocol.

| | Analysis | | | |
|---|---|---|---|---|
| Techniques | Limits | Authentication type | Attack | Authentication requirement |
| Techniques | Limits | Authentication type | Attacks | Authentication requiremen |
| SSO | - Authority<br>- Trust | Strong Authentication | - Man in the middle<br>- DDOS | Mutual authentication |
| Oauth | - Authority<br>- Trust | Strong Authentication | - Man in the middle<br>- DDOS | Mutual authentication |
| Two factors authentication | - Storage/compute authentication of parameters | Strong Authentication | - Session Hijacking<br>- DDOS<br>- Dictionary | Mutual authentication Encryption Keys per user and per session |
| Smart card | - Session Recovery<br>-deploy ability | Strong Authentication | - DDOS<br>- Stolen/Lost smart card | Mutual authentication Encryption Keys per user and per session |

### 3.6 Conclusion

Relationships between organizations are characterized by their need for competition and cooperation without a common trusted agent [22]. This requires the invention of large scale systems, called distributed systems. These are autonomous systems whose entities of interest are mainly human beings (users, operators), machines (computers, peripherals) or processes (execution of the system programs, execution of the application programs). At present, information has become very easy to circulate thanks to the evolution of the Internet and emerging technologies whose management of trust becomes an essential priority preserving the confidentiality of data [24,26]. Although it is a combination of several distinct entities in Cloud as a specific case of distributed systems, the problem becomes more and more difficult to apply [3,9,13]. In this paper, we tried to describe and analysis from a computer security point of view most adopted authentication protocols for cloud computing, we deduce that authentication in cloud environment still remains debatable and must be completed at the proposed security policy level, while implementing and integrating all the communicating actors and services defined by the cloud to provide a policy that required a strong mutual authentication between these different actors.

## References


1. Ahmad, S., Ehsan, B.: The cloud computing security secure user authentication technique (multi level authentication). IJSER **4**(12), 2166–2171 (2013)
2. El Balmany, C., Asimi, A., Tbatou, Z.: IaaS cloud model security issues on behalf cloud provider and user security behaviors. Procedia Comput. Sci. **134**, 328–333 (2018). Elsevier
3. Bhadauria, R., Chaki, R., Chaki, N., Sanyal, S.: A survey on security issues in cloud computing, pp. 1–15 . arXiv preprint arXiv:1109.5388 (2011)
4. Celesti, A., Tusa, F., Villari, M., Puliafito, A.: Three-phase cross-cloud federation model: the cloud SSO authentication. In: 2010 Second International Conference on Advances in Future Internet, pp. 94–101. IEEE (2010)
5. Chari, S., Jutla, C.S., Roy, A.: Universally composable security analysis of OAuth v2.0. IACR Cryptology ePrint Archive **2011**, p. 526 (2011)
6. Chatterjee, S., Hankerson, D., Knapp, E., Menezes, A.: Comparing two pairing-based aggregate signature schemes. Designs Codes Crypt. **55**(2–3), 141–167 (2010)
7. Diffie, W., Van Oorschot, P.C., Wiener, M.J.: Authentication and authenticated key exchanges. Designs Codes Crypt. **2**(2), 107–125 (1992)
8. Fan, C.I., Lin, Y.H.: Provably secure remote truly three-factor authentication scheme with privacy protection on biometrics. IEEE Trans. Inf. Forensics Secur. **4**(4), 933–945 (2009)
9. Fong, P.W.: Relationship-based access control: protection model and policy language. In: Proceedings of the First ACM Conference on Data and Application Security and Privacy, pp. 191–202. ACM (2011)
10. Groß, T.: Security analysis of the SAML single sign-on browser/artifact profile. In: 19th Annual Computer Security Applications Conference, Proceedings, pp. 298–307. IEEE (2003)

11. He, D., Chan, S., Chen, C., Bu, J., Fan, R.: Design and validation of an efficient authentication scheme with anonymity for roaming service in global mobility networks. Wireless Pers. Commun. **61**(2), 465–476 (2011)
12. Joshi, M., Moudgil, Y.S., et al.: Secure cloud storage. Int. J. Comput. Sci. Commun. Netw. **1**(2), 171–175 (2011)
13. Kumar, R., Pandey, A.: A survey on security issues in cloud computing. Int. J. Sci. Res. Sci. Eng. Technol. (IJSRSET) **2**(3), 506–517 (2016)
14. Li, X.Y., Zhou, L.T., Shi, Y., Guo, Y.: A trusted computing environment model in cloud architecture. In: 2010 International Conference on Machine Learning and Cybernetics, vol. 6, pp. 2843–2848. IEEE (2010)
15. Li, X., Niu, J., Kumari, S., Liao, J., Liang, W.: An enhancement of a smart card authentication scheme for multi-server architecture. Wireless Pers. Commun. **80**(1), 175–192 (2015)
16. Naik, N., Jenkins, P.: An analysis of open standard identity protocols in cloud computing security paradigm, pp. 428–431. IEEE (2016)
17. Oswald, E.: Enhancing simple power-analysis attacks on elliptic curve cryptosystems. In: International Workshop on Cryptographic Hardware and Embedded Systems, pp. 82–97. Springer (2002)
18. Ranchal, R., Vijayachandra, J., Sagarika, P., Prathusha, B.: Protection of identity information in cloud computing without trusted third party. In: 2010 29th IEEE Symposium on Reliable Distributed Systems, pp. 368–372. IEEE (2010)
19. Ranjith, G., Vijayachandra, J., Sagarika, P., Prathusha, B.: Intelligence based authentication-authorization and auditing for secured data storage. Int. J. Adv. Eng. Technol. **8**(4), 628 (2015)
20. Rizvi, S., Razaque, A., Cover, K.: Third-party auditor (TPA): a potential solution for securing a cloud environment. In: 2015 IEEE 2nd International Conference on Cyber Security and Cloud Computing, pp. 31–36. IEEE (2015)
21. Sarr, A.: Authenticated key agreement protocols: security models, analyses, and designs. Ph.D. thesis, Université Joseph-Fourier-Grenoble I (2010)
22. Sharma, G.K., Hon, L.K.-M., Burjoski, J.D., Schneider, K.C.: Method and system for third party client authentication. Google Patents. US Patent 8,918,848, 23 December 2014
23. Sun, S.T., Beznosov, K.: The devil is in the (implementation) details: an empirical analysis of OAuth SSO systems. In: Proceedings of the 2012 ACM conference on Computer and Communications Security, pp. 378–390. ACM (2012)
24. Tbatou, Z., Asimi, A., Asimi, Y., Sadqi, Y.: Kerberos v5: Vulnerabilities and perspectives. In: 2015 Third World Conference on Complex Systems (WCCS), pp. 1–5. IEEE (2015)
25. Tbatou, Z., Asimi, A., Asimi, Y., Sadqi, Y., Guezzaz, A.: A new mutuel kerberos authentication protocol for distributed systems. IJ Netw. Secur. **19**(6), 889–898 (2017)
26. Tianfield, H.: Security issues in cloud computing. In: 2012 IEEE International Conference on Systems, Man, and Cybernetics (SMC), pp. 1082–1089. IEEE (2012)
27. Wang, B., Li, B., Li, H.: Oruta: privacy-preserving public auditing for shared data in the cloud. IEEE Trans. Cloud Comput. **2**(1), 43–56 (2014)
28. Wang, C., Ren, K., Lou, W., Li, J.: Toward publicly auditable secure cloud data storage services. IEEE Network **24**(4), 19–24 (2010)
29. Wang, D., He, D., Wang, P., Chu, C.H.: Anonymous two-factor authentication in distributed systems: certain goals are beyond attainment. IEEE Trans. Dependable Secure Comput. **12**(4), 428–442 (2015)

30. Wang, G., Yu, J., Xie, Q.: Security analysis of a single sign-on mechanism for distributed computer networks. IEEE Trans. Industr. Inf. **9**(1), 294–302 (2013)
31. Wang, Q., Wang, C., Ren, K., Lou, W., Li, J.: Enabling public auditability and data dynamics for storage security in cloud computing. IEEE Trans. Parallel Distrib. Syst. **22**(5), 847–859 (2011)
32. Wang, Z.: Security and privacy issues within the cloud computing. In: 2011 International Conference on Computational and Information Sciences, pp. 175–178. IEEE (2011)
33. Yang, F., Manoharan, S.: A security analysis of the OAuth protocol. In: 2013 IEEE Pacific Rim Conference on Communications, Computers and Signal Processing (PACRIM), pp. 271–276. IEEE (2013)
34. Yang, G., Wong, D.S., Wang, H., Deng, X.: Formal analysis and systematic construction of two-factor authentication scheme (short paper). In: International Conference on Information and Communications Security, pp. 82–91. Springer (2006)
35. Yang, G., Wong, D.S., Wang, H., Deng, X.: Two-factor mutual authentication based on smart cards and passwords. J. Comput. Syst. Sci. **74**(7), 1160–1172 (2008)
36. Yang, H., Zhang, Y., Zhou, Y., Fu, X., Liu, H., Vasilakos, A.V.: Provably secure three-party authenticated key agreement protocol using smart cards. Comput. Netw. **58**, 29–38 (2014)
37. Zissis, D., Lekkas, D.: Addressing cloud computing security issues. Future Gener. Comput. Syst. **28**(3), 583–592 (2012)

# Fetal Electrocardiogram Analysis Based on LMS Adaptive Filtering and Complex Continuous Wavelet 1-D

Said Ziani[1(✉)] and Youssef El Hassouani[1,2]

[1] ESIM, Département de Physique, Faculté des Sciences et Techniques, Université Moulay Ismail, Boutalamine, BP 509, 52000 Errachidia, Morocco
ziani9@yahoo.fr, hassouani@yahoo.fr

[2] LPMR, Département de Physique, Faculté des Sciences, Université Mohamed Premier, 60000 Oujda, Morocco

**Abstract.** Extracting clean fetal electrocardiogram (ECG) signals is very important in fetal monitoring. In this paper, we proposed a method for fetal ECG extraction and characterization based on wavelet analysis and the least mean square (LMS) adaptive filtering algorithm. First, abdominal signals and thoracic signals were processed by the LMS algorithm. The abdominal signal was taken as the original input of the LMS adaptive filtering system, and the thoracic signal as the reference input. Finally, the processed wavelet coefficients were. The results indicated that the proposed algorithm can be used for extracting automatically fetal ECG from abdominal signals.

**Keywords:** Fetal electrocardiogram · Weiner filters · Complex continuous wavelet

## 1 Introduction

Fetal ECG (FECG) Fig. 1 has an important significance for fetal monitoring. The commonly used method to detect FECG is using abdominal signal. However, the obtained FECG is easy to be affected by maternal ECG (MECG) and noise and pure FECG cannot be acquired. In order to eliminate the interference of MECG and noise and extract accurate FECG, many scholars have proposed methods to solve this problem, such as, JADE algorithm, segmentation of time scale image [1–5] (Ziani, Jbari Bellarbi), autocorrelation techniques, ICA, NMF, SVD. Recently, blind source separation (BSS) method has been introduced into FECG extraction domain. Especially, single channel blind source separation (SCBSS) has attracted attention because of its less require for observed signal. This paper focused on the extraction end characterization of fetal electrocardiogram FECG based on adaptive filtering and wavelet transform. This method is advantageous since it is based only on the analysis of two abdominal and thoracic signals, in contrast to other methods which need a large number of abdominal ECG recordings. And this algorithm could accurately apply for real life examples.


Y. Farhaoui (Ed.): BDNT 2019, LNNS 81, pp. 360–366, 2020.
https://doi.org/10.1007/978-3-030-23672-4_26

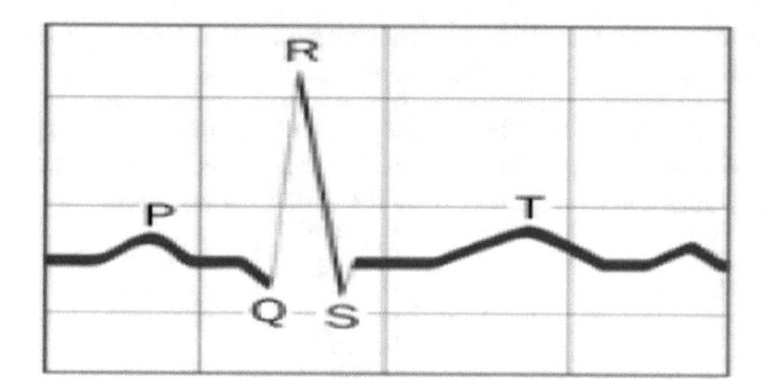


**Fig. 1.** Schematic representation of normal ECG

## 2 Theoretical Background

### 2.1 Least-Mean-Square Adaptive Filters LMS

Its advantages are the following: it is done straightforward, does not require matrix inversion, and it does not require correlation measurements [6–9]. The LMS algorithm uses instantaneous estimates of the vector based on sample values of the input [x (n)] and the error e(n). Therefore the following expression is used for calculating the instantaneous estimate of the gradient vector:

$$\nabla(\mathrm{n}) = -2\mathrm{e}(\mathrm{n})[\mathrm{x}(\mathrm{n})] \tag{1}$$

Along the LMS algorithm, the filter coefficients are updated along the direction of the gradient vector estimate according to following expression:

$$[h(n+1)] = [h(n)] + \frac{1}{2}\mu\left[-\widehat{\nabla}(n)\right] = [h(n)] + \mu e(n)[x(n)] \tag{2}$$

The gradient vector estimate of the LMS algorithm only requires knowledge of the data [x(n)] and no cross-correlation estimate. The error signal is defined by the following expression:

$$\mathrm{e}(\mathrm{n}) = \mathrm{f}(\mathrm{n}) - [\mathrm{x}(\mathrm{n})][\mathrm{h}(\mathrm{n})] \tag{3}$$

The adaptive filtering process ends up working very similar to the steepest-descent algorithm. One has to initialize the coefficient values as [h(0)] at n = 0, therefore a value of zero is given as an initial guess, but as the algorithm is started the coefficient values of the filter will begin to converge to the correct value. At a time n, the coefficients are updated in the following way:

- From [h(n)], the input vector (input signal) [x(n)], and the desired output f(n) we calculate the error signal by the expression given by Eq. (3)
- The new estimate [h(n + 1)] is obtained by the expression given in Eq. (2)

– The index n is incremented by one, and the process is iterated until one reaches a steady state, hence the system has converged.

It is important to notice that the LMS algorithm provides only an approximation to the optimum Wiener solution. But on the other hand, one is able to avoid estimating the mean-square gradient vector expressed by the following equation:

$$\nabla(n) = \begin{bmatrix} \frac{\partial \varepsilon(n)}{\partial h(0,n)} \\ \vdots \\ \frac{\partial \varepsilon(n)}{\partial h(M-1,n)} \end{bmatrix} \tag{4}$$

It can be done by replacing group averages by time averages. Despite the fact that one can manipulate the algorithm in a very simple form, the convergence behavior of the LMS algorithm is very complicated to analyze. This is because the gradient estimate is data dependent and stochastic in its character.

### 2.2 Complex Wavelet Transform

**(1) Continuous Wavelet Transform CWT**

The continuous wavelet transform CWT of a continuous signal is defined as [9–11].

$$\mathrm{T(a,b)} = \frac{1}{\sqrt{a}} \int_{-\infty}^{+\infty} x(t)\psi^*(\frac{t-b}{a})dt \tag{5}$$

This equation contains both dilated and translated wavelet $(\frac{t-b}{a})$ and the x(t) signal. The normalized wavelet is often written more compactly as

$$\psi_{\mathrm{a,b}} = \frac{1}{\sqrt{a}}\psi\left(\frac{t-b}{a}\right)$$

A wavelet is a function *(t)* which satisfies the following conditions [7–9].

- *Wavelet must have finite energy:*

$$\boldsymbol{\epsilon = \int_{-\infty}^{+\infty} |\psi(t)|^{2} dt < \infty} \tag{6}$$

- *The Fourier transform of (t)*

$$\widehat{\psi(\mathrm{f})} = \int_{-\infty}^{+\infty} \psi(t)e^{-i2\pi ft}dt \tag{7}$$

- *Admissibility constant*

$$Q = \int_{0}^{+\infty} \frac{\left|\widehat{\psi(\mathrm{f})}\right|^2}{f} df \tag{8}$$

**(2) The Time-Scale Image TSI:**

At a specific scale a and location b the relative contribution of the signal energy is given by the two-dimensional wavelet energy density function:

$$\boldsymbol{\epsilon}(\boldsymbol{a},\boldsymbol{b}) = |\boldsymbol{T}(\boldsymbol{a},\boldsymbol{b})|^2 \tag{9}$$

A plot of $\epsilon(a, b)$ is known as a ***scalogram*** or ***Times- scales image TSI.***

## 3 Methods

As an illustration of adaptive filtering, let us imagine the measurement of the cardiac activity of a fetus with an electrocardiogram (ECG) taken from the mother's abdomen and which, of course, is disturbed by the ECG of this one. This measurement requires the use of 2 sensors. With the first, we measure the reference signal x (n) representing, if possible, only the ECG of the mother. With the second, we measure the signal y (n) which is the ECG of the fetus disturbed by the cardiac activity of the mother. The signals of the filter scheme (Fig. 2) are then as follows:

1. x (n) = the maternal ECG measured near the heart, thoracic signal,
2. y (n) = the fetal ECG disturbed by that of the mother,
3. $y_p$ (n) = the maternal ECG near the fetus, abdominal signal,
4. $y_w$ (n) = the estimate of the maternal ECG near the fetus,
5. e (n) = the fetal ECG,
6. $\epsilon$ (n) = the estimate of the fetal ECG.

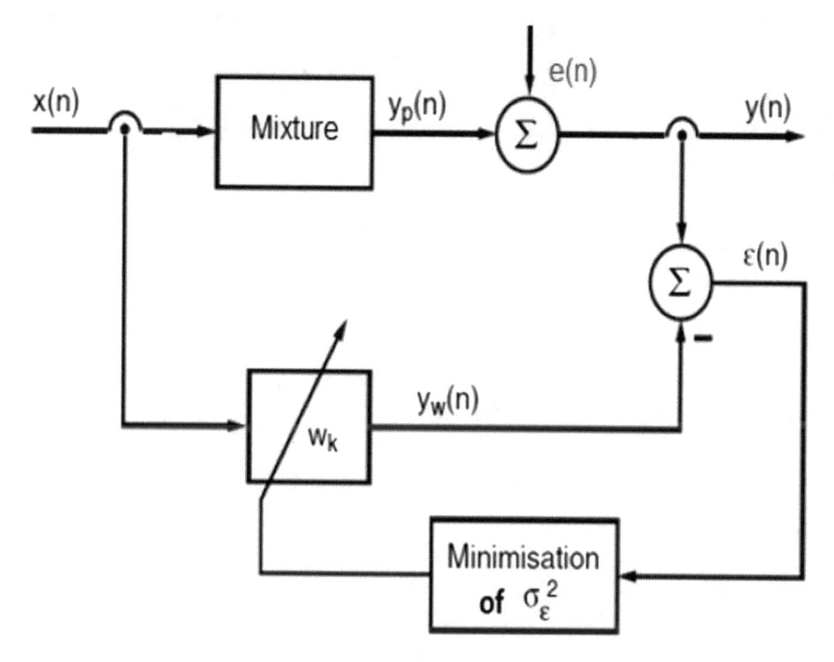


**Fig. 2.** Adaptive filter

# 4 Results and Discussion

## 4.1 ECG Recording and FECG Estimate

The data is extracted from the DaISy database [12]. (Database for the Identification of Systems). The sampling frequency is 250 Hz. We used the MATLAB 2015 a on Windows 7. The mixed signals are parameterized with T = 2500 s as in Fig. 3.

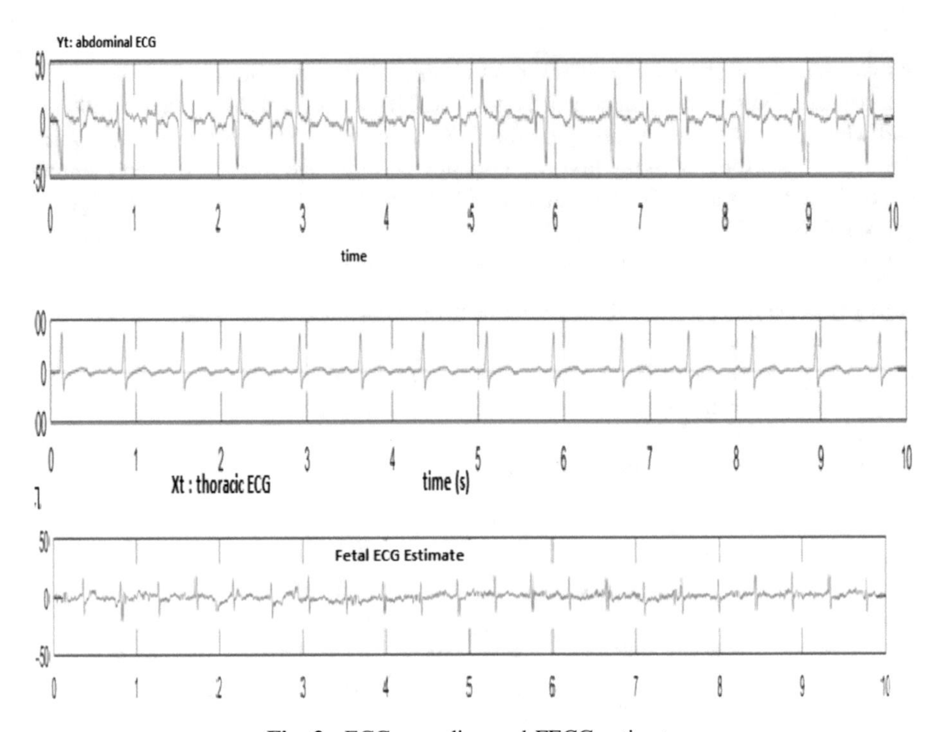


**Fig. 3.** ECG recording and FECG estimate

## 4.2 FECG Characterization

By using Cgau2 wavelet, the time scale image of FECG signal gives a good characterization of the complex QRS as shown in Fig. 3.

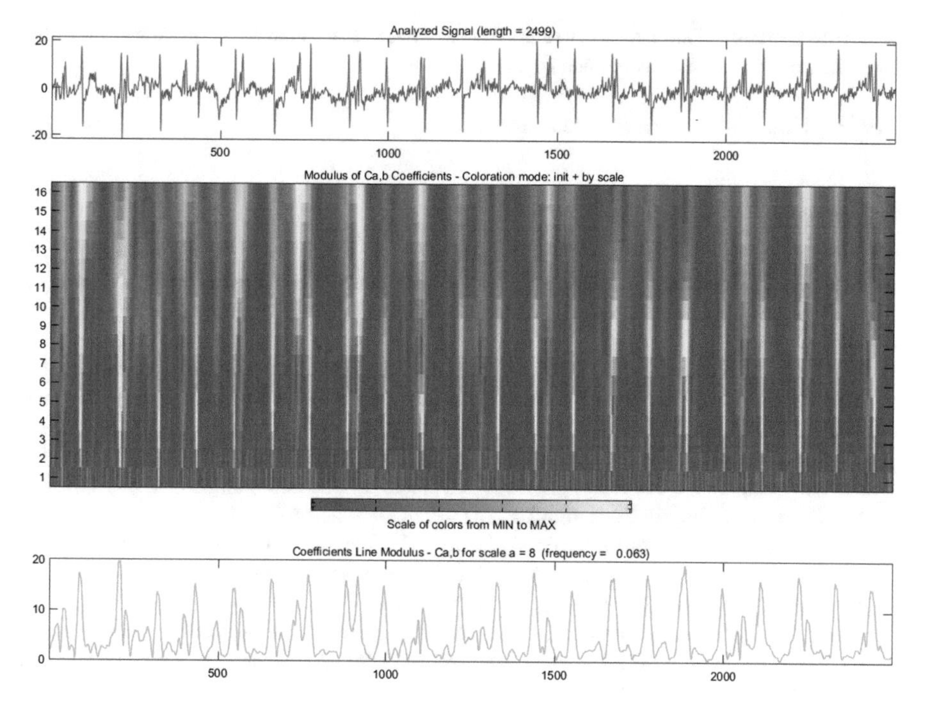


**Fig. 4.** QRS complex characterization of FECG estimate

# 5 Discussion and Conclusion

The LMS algorithm has established itself as an important functional block of adaptive signal processing. Compared to the other methods of analysis, this approach offers some highly desirable features:

- Simplicity of implementation is that in software or hardware form.
- Ability to operate satisfactorily in an unknown environment.
- Ability to track time variations of input statistics.

Indeed, the simplicity of the LMS algorithm has made it the standard against which other linear adaptive filtering algorithms are benchmarked. The formulation of the LMS algorithm presented in this paper has been from the viewpoint of spatial filtering. It may equally well be applied to solve temporal filtering problems. In the latter case, the filter takes on the form of a tupped-delay-line Filter.

The application of the continuous wavelet transform demonstrated the quality of fECG extraction. Figure 4 shows the detection the R peaks relative only to the fetal electrocardiogram and this constitutes an automatic criterion for measuring the reliability of the algorithm and especially if it is to be implemented in an embedded system such as FPGA or DSP.

**Acknowledgments.** The authors would like to thank the anonymous reviewers for their insightful comments and recommendations.

**Conflict of Interest Statement.** None declared.

## References

1. Ziani, S., El Hassouani, Y., Farhaoui, Y.: An NMF based method for detecting RR interval. In: International Conference on Big Data and Smart Digital Environment. Accessed 22 Feb 2019
2. Ziani, S., El Hassouani, Y.: Fetal-maternal electrocardiograms mixtures characterization based on time analysis. In: 4th International Conference on Optimization and Applications (IEEE), 25–26 April 2019
3. Ziani, S., Jbari, A., Bellarbi, L., Farhaoui, Y.: Blind maternal-fetal ECG separation based on the time-scale image TSI and SVD – ICA methods. In: The 2nd International Workshop on Big Data and Networks Technologies (BDNT2018). Science Direct
4. Ziani, S., Jbari, A., Bellarbi, L.: Fetal electrocardiogram characterization by using only the continuous wavelet transform CWT. In: International Conference on Electrical and Information Technologies (ICEIT), 15–18 November 2017
5. Ziani, S., Jbari, A., Bellarbi, L.: QRS complex characterization based on non-negative matrixfactorization NMF. In: 2018 4th International Conference on Optimization and Applications (IEEE) (2018)
6. Widrow, B.S.D.: Stearns 1985: Adaptive Signal Processing Algorithms. Prentice Hall, Upper Saddle River (1985)
7. Stearns, S.D., David, R.A.: Signal Processing Algorithms in Matlab. Prentice Hall, Upper Saddle River (1996)
8. Sameni, R., Clifford, G.D.: A review of fetal ECG signal processing; issues and promising directions. Open Pacing Electrophysiol. Ther. J. **3**, 4–20 (2010)
9. Mallat, S.: Wavelet Tour of Signal Processing, 2nd edn., pp. 162–219, 249–254. Academic Press, San Diego (1998)
10. Daubechies, I.: Ten Lectures on Wavelets. Society for Industrial and Applied Mathematics, Philadelphia (1992)
11. Abdelliche, F., Charef, A.: Application d'une nouvelle famille d'ondelettesfractionnaires basées sur la fonction de Cole-Cole. 1ère Conférence Nationale sur lesSystèmes d'Ordre Fractionnaire et leurs Applications, 18–19 Mai 2010, Skikda, Algérie (2010)
12. De Moor, B., De Gersem, P., De Schutter, B., Favoreel, W.: DAISY: a database for fication of systems. J. A Spec. Issue CACSD (Comput. Aided Control Syst. Des.) **38**(3), 4–5 (1997)

# Recommender System for Orientation Student

Ahajjam Tarik(✉) and Yousef Farhaoui

Faculty of Sciences and Techniques, Department of Computer Science,
Moulay Ismail University, IDMS Team, Errachidia, Morocco
ta.ahajjam@gmail.com, y.farhaoui@fste.umi.ac.ma

**Abstract.** This paper addresses the problem of orientation of high school students using a recommendation system that works through Learning machine algorithms.

**Keywords:** Data science · Machine Learning · Orientation · Recommendation · High school · Data analysis

## 1 Introduction

Today the school break has taken a prominent place in the education system (case of Morocco), it is for this purpose that the Ministry of Education concerned by high intention guiding students to branches and sectors appropriate to the student's skills. For this purpose a system was conducted that addresses improving the orientation by an intelligent model using Machine Learning algorithms.

During these years the recommendation systems plays a role very matter in the web field end of filtered information to users.

At the end of the first year of high school, students are asked to choose a die based on their preferences and abilities, also taking into account the studies considered after the baccalaureate. The choice of the sector in the first year of the Bachelor concerned the minds of students.

Indeed, this decision is important because it will guide the student later in its draft guidance to select a die in the second year of the Bachelor and therefore his studies after the baccalaureate.

Choosing the right path is not easy. This implies to reflect on itself.

The student must join the industry that corresponds him best and which is consistent with his desires, interests, aptitudes and skills.

No bachelor dies easier than another. It all depends on his future project and consideration of effort required to achieve it. It is imperative to learn about trades and training available after each sector, to make the right choice.

In this document we will carry out an intelligent system that will help students to choose the proper channels to their skills using Learning Machine algorithms.


Y. Farhaoui (Ed.): BDNT 2019, LNNS 81, pp. 367–370, 2020.
https://doi.org/10.1007/978-3-030-23672-4_27

## 2 Proposed Approach

We offer an intelligent system that meets the needs of students. To do this will require a system that will predict the baccalaureate average student through his notes of the core curriculum using Machine Learning techniques.

So the project is the design and implementation of a system of orientation of students of the core curriculum towards technical or scientific branches of the first degree.

The purpose of this system is to have a good orientation which will allow the student to have a good note as an existing template that contains all students who have passed their baccalaureate degrees in the region.

## 3 Approach Used

### 3.1 The Research Data

The research data is the most important phase in the field of artificial intelligence to build a model that is the basis of our work.

The data are taken from the test center located at the Regional Academy of Education and Training of Guelmimoued Noun region.

### 3.2 Data Preprocessing

Data is at the center of Machine Learning algorithms. Therefore, to better prepare these data, will have better performance.

Most of the time, Machine Learning, the DataSet come with different orders of magnitudes. This difference in scale can lead to lower performance. To overcome this, preparatory treatment of the data exists.

For DATASET was undergone several treatments including:

- Clear the missing feature.
- Reduce the size of variables that correlates strongly in them or duplicated.

Finally format the data for which is in the form of a matrix with explanatory variables and a dependent variable.

$$\begin{pmatrix} y_1 \\ \vdots \\ y_m \end{pmatrix} = \begin{pmatrix} x_{11} & \cdots & x_{1n} \\ \vdots & \ddots & \vdots \\ x_{m1} & \cdots & x_{mn} \end{pmatrix}$$

### 3.3 Model Building

After having created the data matrix, it remains only to design a model that will allow making decisions according to a precise prediction is using linear regression, decision tree and Random Forest.

### 3.4 Comparison

Linear regression is a linear model, which means that it works very well when the data has a linear shape. However, when the data has a non-linear, a linear model can not capture nonlinear entities.

In this case, we used decision trees, which is a better way to capture the non-linearity in the data space division and with an average error rate.

However if multiple decision tree of good results can be better than using a single tree that is why a new model was created with the Random Forest algorithm.

The following table measures the performance of each algorithm used.

| | Data | | | Algorithm | | |
|---|---|---|---|---|---|---|
| | Total | Training 67% | Testing 33% | Random Forest accuracy % | Decision tree accuracy % | Linear regression accuracy % |
| Branch 1 | 310 | 208 | 102 | 44.25 | 29.87 | 26.80 |
| Branch 2 | 7392 | 4953 | 2439 | 33.66 | 21.64 | 28.15 |
| Branch 3 | 726 | 486 | 240 | 32.29 | 13.70 | 22.98 |
| Branch 4 | 449 | 301 | 148 | 28.38 | 16.30 | 22.78 |
| Branch 5 | 13 | 9 | 4 | 13.74 | 14.20 | 9.13 |

### 3.5 Conclusion

This document aimed to design and implement a smart solution to a policy recommendation system students the core curriculum, building an intelligent model.

The study implemented three regression methods that each algorithm gave a result, we finally decided to chose the best that is regression by random forests algorithm.

### 3.6 Perspective

We can distinguish the short-term outlook and long-term. In the short term, we will implement the approach proposed in the region. In the long term to spread the system to overwrite all paid which generates a large amount of data that requires the use of Big Data technology.

## References

1. Isinkaye, F.O., Folajimi, Y.O., Ojokoh, B.A.: Recommendation systems: principles, methods and evaluation. Egypt. Inform. J. **16**, 261–273 (2015)
2. Thai-Nghe, N., Drumond, L., Krohn-Grimberghe, A., Schmidt-Thieme, L.: Recommender system for predicting student achievement. Procedia Comput. Sci. **1**, 2811–2819 (2010)
3. Hamsak, H., Indiradevi, S., Kizhakkethottam, J.J.: Student academic performance prediction model using decision tree and fuzzy genetic algorithm. Procedia Technol. **25**, 326–332 (2016)
4. Duan, L., Street, W.N., Xu, E.: healthcare information systems: data mining methods in the establishment of a clinical recommender system. Enterp. Inf. Syst. **5**(2), 169–181 (2011)

# Author Index




Y. Farhaoui (Ed.): BDNT 2019, LNNS 81, pp. 371–372, 2020.
https://doi.org/10.1007/978-3-030-23672-4

Printed in the United States
By Bookmasters